The Cuticles of Plants

The Cuticles of Plants

J. T. MARTIN, O.B.E.
B.Sc. (Birm.), D.Sc. (Lond.), F.R.I.C.
Reader and Assistant Director, Department of Agriculture and Horticulture, Long Ashton Research Station, University of Bristol

and B. E. JUNIPER
M.A. (Oxon.), D.Phil. (Oxon.)
Senior Research Officer in the Botany School, Oxford University and Fellow of St. Catherine's College, Oxford

New York · St. Martin's Press

First published 1970
First published in the United States of America in 1970
by St Martin's Press Inc.,
175 Fifth Avenue, New York, New York

First published in Great Britain by
Edward Arnold (Publishers) Ltd.

Library of Congress Catalog Card Number
77-111413

Printed in Great Britain by
R. & R. Clark, Ltd., Edinburgh

Preface

The leaves of plants are adapted to expose the maximum surface area of green tissue to sunlight. For maximum productivity a plant needs the maximum absorption of light and a high rate of exchange of gases, but these are in conflict with a third desirable feature, minimum loss of water. The plant achieves in the leaf a compromise of structure to meet these requirements. It retains a large surface area, but covers it with a cuticle which is translucent to light, relatively impervious to water and yet permits gas exchange through openings in the cuticle, the stomata. Thus, unlike animals, many important physiological processes are controlled at the surface of plants. The cuticle holds a position of importance as the bounding layer between the body of the plant and its environment. The evolution of the cuticle may well have been a crucial factor in the successful colonisation of land by plants.

Besides having a major influence on the processes named above, the cuticle contributes materially to the well-being of the plant. It provides the aerial parts with an envelope which holds the cells compact and firm and protects them from injuries by wind, physical abrasion or excessive insolation. It prevents the cells from becoming waterlogged by rain and shields their water-soluble components from being leached. The cuticle is implicated in the defence of plants against attack by pathogens and recent evidence suggests that it may play a part in warding off attack by insects. It may protect the plant from frost. Crucial actions concerning fertilisation take place at the cuticle. Among its more exotic roles it provides part of the trap mechanism of insectivorous plants, a lens for concentrating light in shade plants and a water-catching system.

In agriculture, plants are sprayed with chemicals to supply nutrients, to control growth, to protect the wanted or to kill the unwanted. The behaviour and efficiency of applied chemicals are greatly influenced by the nature of the cuticles of the plants on which they are used. In the past, the cuticle has been relatively a neglected structure, but in recent years interest in it has been stimulated by investigations on surface phenomena in which it exercises a governing role. Examples of such phenomena are the water repellency of surfaces (as depicted in the cover illustration) and the penetration of chemicals into plants. Research has been greatly helped by the application of modern methods of microscopy and analysis. The electron microscope reveals a fascinating array of varying ultra-structures and analysis has shown a great complexity of cuticular components. Much is now known of the changes which cuticles undergo during the growth of plants, the biosynthesis of cuticles and the value of cuticular components as chemo-taxonomic criteria. The study of fossil cuticles has linked the present with the remote past.

An outstanding feature that is not always fully appreciated is the great diversity in form and composition shown by the cuticles of plants of different species. For this reason we have purposely entitled the book *The cuticles of plants* rather than *The plant cuticle*. Some cuticles are thin and fragile and others thick and tough; each contributes to fitting the plant to its environment. Conflicting views have arisen from many investigations of surface phenomena in which cuticles are involved; in few of these have the physical and chemical characteristics of the cuticles concerned been defined. Records even of cuticle thickness are limited in number. Methods are now available for the detailed examination of cuticles; their application would help to dispel misconceptions and resolve conflicting reports. It is also hoped that by bringing together information on the cuticle, it will receive in botanical textbooks somewhat more detailed treatment than it normally gets.

The book deals with all aspects of the biology and chemistry of plant cuticles. Early work on the cuticle is included. Its chief purpose is to provide background information which will be of value to botanists, chemists, agriculturists and others concerned with crop protection, surface physiology and related studies. No attempt is made to deal in detail with the principles and practice of crop protection; these are adequately covered by other publications. Selection from the mass of information now available on cuticles is inevitable; the aim is to present an over-all picture of the present state of knowledge. The cuticle is an integral part of the epidermis of the plant, but not the only protective device; other surface structures and components therefore are also considered. Brief descriptions of other protective coverings such as bark are included.

The trivial names of the chemical constituents of cuticles have been retained and used when appropriate; they are generally more familiar to botanists than the systematic names and chemicals isolated earlier from cuticles and designated by trivial names were probably mixtures for which modern systematic names would be inappropriate. The systematic and trivial names of some cuticular components and the basis of chemical nomenclature are given on p. xix and the systematic names, synonyms and vernacular equivalents of plants on p. xiii. Terms which may not be generally familiar, on the first occasion on which they are used are printed in ***bold italic*** and are explained in a glossary (p. 296).

We express our indebtedness to our colleagues Mr E. A. Baker, Mr. R. F. Batt, Dr P. J. Holloway, Dr M. F. Roberts, Dr A. M. S. Silva Fernandes and Dr. N. D. Hallam for valuable discussion and assistance and to Dr G. S. Hartley and Dr R. J. Hamilton for helpful comment and criticism.

We are indebted to the following for their permission to reproduce illustrations: Professor H. N. Barber, Dr D. E. Bradley, Dr R. C. Brian and Mr N. D. Cattlin, Dr F. A. L. Clowes, Dr P. Echlin and the Cambridge Instrument Company, Dr W. Franke, Dr D. M. Hall, Dr N. D. Hallam,

Dr P. J. Holloway, Dr R. N. Konar, Dr B. Landau-Schachar, Professor von O. L. Lange, Dr M. C. Ledbetter, Dr L. Leyton, Dr T. P. O'Brien, Professor E. Schnepf, Dr D. S. Skene and Mr P. R. Williams.

1969

J. T. M. and B. E. J.

ABBREVIATIONS

SI units (Système Internationale d'Unités) are used (British Standards Institution Booklet No. 3763: 1964).

The following abbreviations are used in the illustrations and in the legends of the illustrations.

μm	= micrometre or micron: 1/1000 of a millimetre
nm	= nanometre or millimicron: 1/1000 of a micrometre
Glut	= the tissue was fixed in glutaraldehyde
Osm	= the tissue was post-fixed in osmium tetroxide
Pb	= the section was stained with lead citrate
Ur	= the section was stained with uranyl acetate
Cr	= a carbon replica was made of the original surface
Au	= the carbon replica was shadowed with evaporated gold
Pd	= the carbon replica was shadowed with evaporated palladium
S Cr/Pt	= a simultaneous carbon and platinum replica was made of the original surface
Ms/Sc	= the original surface was coated with a heavy metal and the whole object was photographed under the scanning electron microscope
F/E/Cr	= the original material was frozen, etched under vacuum and a carbon replica made of the etched surface

Where illustrations are not acknowledged they are by Dr B. E. Juniper. Figs. 2.14–2.18 and 10.1 are by Mr E. A. Baker.

Contents

Binomials and Vernaculars

Any plant mentioned by a vernacular name has, on the first occasion it is used, also been identified by its Latin binomial. Often, however, vernacular names cannot be accurately translated into Latin binomials. Often, too, binomials have been changed since the original work was published. We have, therefore, in the following list given first the Latin binomial now widely used, any commonly used synonym in lighter type, and the vernacular name by which the plant is known.

Latin	Vernacular
Abies concolor	Colorado White Fir
magnifica	Red Fir or Californian Red Fir
Acacia spp.	Wattle or Mimosa
Acer pseudoplatanus	Sycamore
Achillea millefolium	Yarrow
Aegopodium podagraria	Ground Elder
Aesculus hippocastanum	Horse-chestnut
turbinata	Japanese Horse-chestnut
Agave sisalana	Sisal Hemp
Agrostis stolonifera	Creeping Bent or Fiorin
Allium ampeloprasum (A. porrum)	Leek
cepa	Onion
Antirrhinum majus	Snapdragon or Antirrhinum
Arbutus unedo	Strawberry Tree
Arctostaphylos uva-ursi	Bearberry
Avena sativa	Oat
Bellis perennis	Daisy
Benincasa hispida	Wax Gourd
Berberis spp.	Barberry
Beta vulgaris	Beet or Sugar Beet
Betula lenta	Cherry, Sweet or Black Birch
pendula (*B. verrucosa*)	Silver Birch
Borago officinalis	Borage
Brassica campestris (*B. rapa*)	Turnip
Brassica hirta (Sinapis alba)	White Mustard
napus	Rape, Cole or Swede

Brassica oleracea var. ***acephala***	Kale
var. ***botrytis***	Cauliflower
oleracea var. ***capitata***	Cabbage
var. ***gemmifera***	Brussels Sprout
var. ***italica***	Sprouting Broccoli
rapa (***B. campestris***)	Turnip
Buxus spp.	Box
Calluna vulgaris	Ling or Heather
Cannabis sativa	Hemp
Cardamine pratensis	Cuckoo Flower or Ladies Smock
Carthamus tinctorius	Safflower or False Saffron
Castanea sativa	Sweet Chestnut
Catalpa bignonioides	Indian Bean
Cedrus atlantica	Atlantic Cedar
Ceroxylon (*Klopstockia*) ***andicolum***	Andean Wax Palm
Chaenomeles (*Cydonia*) ***lagenaria***	Quince
Chamaecyparis obtusa	Hinoki Cypress
Chamaenerion ***(Epilobium) angustifolium***	Rose-bay Willowherb
Chelidonium majus	Greater Celandine
Chenopodium album	Fat Hen
Chrysanthemum coccineum	Pyrethrum
segetum	Corn Marigold
Cicer arietinum	Chick Pea
Cinchona spp.	Peruvian Bark or Quinine Bark Tree
Cinnamomum zeylanicum	Cinnamon Tree
Citrus aurantifolia	Lime
aurantium	Orange
decumana (*C. paradisi*)	Grapefruit or Pomelo
grandis	Pummelo
nobilis	Mandarin
Cladium mariscus	Common or Prickly Sedge
Clematis vitalba	Old Man's Beard or Traveller's Joy
Cocos (*Syagrus*) ***coronata***	Ouricury or Ouricuri
Coffea arabica	Coffee
Coix lacryma (*C. lacryma-jobi*)	Job's Tears
Colchicum autumnale	Autumn Crocus
Convolvulus arvensis	Bindweed
Copernicia cerifera	Carnauba or Indian Wax Palm
Copernicia hospita	Cuban Palm
Cortaderia spp.	Pampas Grass
Corylus avellana	Hazel Nut
Crataegus oxyacantha	Hawthorn
Cucumis spp.	Cucumber, Melon
Cucurbita spp.	Marrow, Squash, or Pumpkin
Cydonia ***(Chaenomeles) lagenaria***	Quince
Cymbopogon citratus	Lemon Grass
Dactylis glomerata	Cock's-Foot Grass
Datura stramonium	Thorn Apple
Dianthus caryophyllus	Carnation

Digitalis purpurea	Foxglove
Dipsacus fullonum	Teasel
Dracaena draco	Dragon Tree
Dryopteris filix-mas	Male Fern
Endymion non-scriptus (*Scilla non-scripta*)	Bluebell or Wild Hyacinth
Epilobium angustifolium	Rose-bay Willowherb
Eragrostis curvula	Love Grass
Eriophorum spp.	Cotton Grass
Euphorbia antisyphilitica	Candelilla
cerifera	Candelilla
Fagus sylvatica	Beech
Festuca glauca (*F. ovina* var. *glauca*)	Blue Fescue
Ficus elastica	Indian Rubber Tree or Rubber Plant
spp.	Fig
Fragaria spp.	Strawberry
Fraxinus spp.	Ash
Fuchsia spp.	Fuchsia
Galanthus nivalis	Snowdrop
Ginkgo biloba	Maidenhair Tree
Gossypium spp.	Cotton
Hamamelis spp.	Witch Hazel
Hedera helix	Ivy
Helianthus annuus	Sunflower
Helxine soleirolii	Mother-of-thousands
Hevea brasiliensis	Para-rubber or Caoutchouc Tree
Hordeum vulgare	Barley
Humulus lupulus	Hop
Hyacinthus orientalis	Hyacinth
Ilex aquifolium	Holly
Irvingia oliveri	Cay-cay Tree
Kleinia ***(Senecio) articulata***	Candle Plant
Laburnum spp.	Laburnum or Golden Chain
Lagerstroemia spp.	Crape Myrtle
Lamium album	White Dead-nettle
amplexicaule	Henbit
Lapsana communis	Nipplewort
Ledum groenlandicum	Labrador Tea
Lemna spp.	Duckweed
Ligustrum vulgare	Privet
Limonium (*Statice*) spp.	Sea Lavender
Linum usitatissimum	Cultivated Flax
Lolium *multiflorum*	Italian Rye-grass
perenne subsp. ***multiflorum***	Italian Rye-grass
subsp. ***perenne***	Perennial Rye-grass

Lonicera spp.	Honeysuckle
Lupinus albus	White lupin
Lycopersicum esculentum	Tomato
Lycopodium clavatum	Common Club Moss
Malus sylvestris (*Pyrus malus*)	Crab Apple
Marrubium vulgare	White Horehound
Medicago arabica	Spotted Bur Clover or Spotted Medick
sativa	Lucerne or Alfalfa
Morus rubra	Red or American Mulberry
Musa paradisiaca	Pisang Wax Tree
subsp. ***sapientum***	Banana
sapientum	Banana
Myrica cerifera	Wax-myrtle
cordifolia	Cape Berry
pensylvanica	Bayberry or Candleberry
Narcissus pseudonarcissus	Wild Daffodil
Nepenthes spp.	Pitcher Plant
Nerium oleander	Oleander
Nicotiana tabacum	Tobacco
Olea spp.	Olive
Pennisetum glaucum (*P. typhoideum*)	Pearl, Indian or African Millet
Phaseolus aureus	Green or Golden Gram or Mung
coccineus (*P. multiflorus*)	Runner Bean
vulgaris	Kidney, Pole, Haricot, String or French Bean
Phragmites communis	Reed
Phyllitis scolopendrium	Hart's-tongue Fern
Picea abies	Norway, Common or White Spruce
pungens	Colorado Spruce
Pinguicula spp.	Butterwort
Pinus lambertiana	Sugar Pine
ponderosa	Western Yellow Pine
sylvestris	Scots Pine
Pistia stratiotes	Nile Cabbage or Water Lettuce
Pisum sativum	Pea
Plantago spp.	Plantain
Platanus × hybrida (*P. × acerifolia*)	London Plane
Poa colensoi	New Zealand Blue Tussock Grass
Populus gileadensis (*P. candicans*)	Balm-of-Gilead
Populus nigra	Black Poplar
Populus tremula	Aspen
spp.	Poplar
Primula auricula	Auricula
Prosopis spp.	Mesquite
Prunus andersonii	Desert Peach
armeniaca	Apricot
avium	Gean or Wild Cherry
cerasus	Sour Cherry

Prunus domestica	Plum
laurocerasus	Cherry Laurel
serotina	Black Cherry
Pseudotsuga menziesii (*P. taxifolia*)	Douglas Fir
Pteridium aquilinum	Bracken
Punica granatum	Pomegranate
Pyrus communis	Pear
Quercus suber	Cork Oak
Raphanus spp.	Radish
Raphia pedunculata	Raphia or Raffia
vinifera	Raphia, Raffia or Bamboo Palm
Rhamnus purshiana	Californian Bearberry
Rheum rhaponticum	Rhubarb
Rhizophora spp.	Mangrove
Rhus succedanea	Wax-tree of Japan
typhina	Staghorn Sumach
Ribes nigrum	Blackcurrant
uva-crispa (*R. grossularia*)	Gooseberry
Ricinus communis	Castor-Oil Plant or Castor Bean
Rosa damascena	Damask Rose
Rubus idaeus	Raspberry
phoenicolasius	Japanese Wineberry
Rumex spp.	Dock
Saccharum officinarum	Sugar Cane
Salix spp.	Willow
Sapium sebiferum	China Tallow-tree
Sambucus nigra	Elder
Schinopsis spp.	Quebracho
Secale cereale	Rye
Sempervivum tectorum	Houseleek
Senecio articulata (*Kleinia articulata*)	Candle Plant
Silene pendula	Drooping Catchfly
Simmondsia californica	Jojoba Wax Tree
Sinapis alba	White Mustard
Sinapis arvensis	Charlock
Solanum tuberosum	Potato
Sorghum vulgare	Sorghum or Millet
Spinacia oleracea	Spinach
Statice ***(Limonium)*** spp.	Sea Lavender
Stellaria spp.	Chickweed or Stitchwort
Stipa tenacissima	Esparto grass
Strychnos spp.	Natal Orange
Syagrus ***(Cocos) coronata***	Ouricury or Ouricuri
Syringa vulgaris	Lilac
Tamarix pentandra	Tamarisk
Taraxacum spp.	Dandelion
Thea sinensis (*Camellia thea or C. sinensis*)	Tea
Thlaspi arvense	Pennycress

Trifolium spp.	Clover
Triticum durum	Durum or Hard Wheat
vulgare (*T. aestivum*)	Wheat
Tropaeolum majus	Nasturtium
Tulipa sylvestris	Wild Tulip
Ulex europaeus	Gorse
Urtica dioica	Nettle
Vaccinium macrocarpon	American Cranberry
oxycoccos	Cranberry
Vicia faba	Broad Bean
Vinca major	Greater Periwinkle
Vitis vinifera	Grape
var. ***sultana***	Sultana Grape
Zea mays	Maize or Corn (amer.)

Table of systematic and trivial names of *n-aliphatic* constituents of plant waxes C_{12}–C_{35}

***n*-alkanes and *n*-alkane derivatives**

Carbon no.	Alkane	Alkanoic acid	Fatty acid	Fatty Alcohol
12	Dodecane	Dodecanoic	Lauric	Lauryl
13	Tridecane	Tridecanoic	—	—
14	Tetradecane	Tetradecanoic	Myristic	Myristyl
15	Pentadecane	Pentadecanoic	—	—
16	Hexadecane	Hexadecanoic	Palmitic	Cetyl
17	Heptadecane	Heptadecanoic	Margaric	Margaryl
18	Octadecane	Octadecanoic	Stearic	Stearyl
19	Nonadecane	Nonadecanoic	—	—
20	Eicosane	Eicosanoic	Arachidic	Arachidyl
21	Heneicosane	Heneicosanoic	Medullic	Medullyl
22	Docosane	Docosanoic	Behenic	Behenyl
23	Tricosane	Tricosanoic	—	—
24	Tetracosane	Tetracosanoic	Lignoceric	Lignoceryl
25	Pentacosane	Pentacosanoic	—	—
26	Hexacosane	Hexacosanoic	Cerotic	Ceryl
27	Heptacosane	Heptacosanoic	—	—
28	Octacosane	Octacosanoic	Montanic	Montanyl
29	Nonacosane	Nonacosanoic	—	—
30	Triacontane	Triacontanoic	Melissic (Myricic)	Melissyl (Myricyl)
31	Hentriacontane	Hentriacontanoic	—	—
32	Dotriacontane	Dotriacontanoic	Lacceroic	Lacceryl
33	Tritriacontane	Tritriacontanoic	—	—
34	Tetratriacontane	Tetratriacontanoic	Geddic	Geddyl
35	Pentatriacontane	Pentatriacontanoic		

The number of carbon atoms in a chain is indicated by the systematic name, and the position of a substituent group by the carbon atom involved numbered from one end of the chain. The prefix ***n*** indicates an unbranched chain. The carbon atoms of an acid are numbered from the head (carboxylic) end of the chain. The position of a double bond is given by the number of the first of the pair of carbon atoms involved; **ane** or **an** in a name indicates saturation, **ene** or **en** unsaturation, **ol** an alcohol, **one** a ketone and **oic** an acid.

The following examples demonstrate nomenclature:

Octacosan-1-ol

$CH_3(CH_2)_{26}CH_2OH$

Nonacosan-10-ol-15-one

$CH_3(CH_2)_{13}CO(CH_2)_4CHOH(CH_2)_8CH_3$

Eicosane-1,20-dioic acid

$COOH(CH_2)_{18}COOH$

***n*-Alkene derivatives (*n*-alkenoic acids)**

Octadec-9-enoic acid	oleic
$CH_3(CH_2)_7CH{=}CH(CH_2)_7COOH$	
Octadec-9,12-dienoic acid	linoleic
$CH_3(CH_2)_4CH{=}CH\ CH_2CH{=}CH\ (CH_2)_7COOH$	
Octadec-9,12,15-trienoic acid	linolenic
$CH_3CH_2CH{=}CH\ CH_2CH{=}CH\ CH_2CH{=}CH\ (CH_2)_7COOH$	

1

Introduction

> The structure of leaves is very simple. It consists of an outer skin or cuticle which is full of pores, the upper surface being varnished as it were. Cellular tissue is seen when the cuticle is removed.
>
> From *The Library of Agricultural and Horticultural Knowledge*
> J. Baxter, Lewes, 1834

This quotation aptly defines the cuticle as an outer skin. All aerial parts of plants are bounded by protective skins. These include the cuticles of fleshy tissues such as petals, leaves and stems, the peel or rinds of fruits and the barks of trees. The skins vary in thickness and complexity and are built up from materials which are formed in the plant cells and migrate to the surface. The cuticles, which are usually the most delicate of the skins, consist of non-cellular membranes. The peels or rinds, as separated mechanically from the flesh of fruits, are usually composed of thickened membranes carrying attached layers of cells. The barks are the most elaborate and complex form of surface incrustation, and culminate in the cork deposits of the cork-oak tree (*Quercus suber*).

The structural components which characterise the skins are the chemically related polymers cutin and suberin. Wax and tannin are other important constituents. Cuticles contain cutin; barks and cork suberin. The walls of pollen grains and moss and fern spores contain another substance, sporopollenin. In some fruit skins and in barks and cork, a cellular framework is heavily cutinised or suberised and the non-cellular membranous structure which characterises the simpler cuticles is lost. The algae and fungi are believed to be the only groups of plants which may lack cutin or suberin in their outer tissues; at what stage in the evolutionary development of plants the polymers first become used as protective devices is not known. Lee and Priestley (1924) observed that a cuticle is normally present on the surfaces of the leaves and shoots of Bryophyta and all vascular plants. Only the higher plants are known with certainty to produce cutin and suberin; Crisp (1963) suggests that their invention may well have been one of the crucial factors in the successful colonisation of the land by plants.

The cuticle lies over the leaf epidermal cells as a continuous membrane (Fig. 1.1C). There is evidence that the membrane in thinner form extends within and lines substomatal cavities (Fig. 1.2). A cuticle also occurs between cells as in salt glands, certain seed coats and some of the glands of insectivorous plants. Little is known of the extent to which other apertures, such as hydathodes, are lined internally with cuticle-like structures. The cuticle thus covers the epidermal cells, internally seals off apertures and extends within the cellular tissues. Fleshy aerial organs are therefore provided with virtually complete envelopes. Exceptions to this rule will be discussed in Chapter 4.

There has been much discussion as to whether cuticles occur on the roots of plants. The position is confused and is not helped by difficulty in identification and in defining what constitutes a cuticle. The outer wall of a root may be impregnated with lipid material; if suberin, the root is regarded as suberised, but if cutin the root may be said to have a cuticle. Whether the root has a cuticle or not may rest upon the definition, due to Lee and Priestley, that a cuticle consists of oxidised fats. The outer walls of the epidermal cells of some young roots are sufficiently impregnated with lipid material to permit the isolation of continuous membranes. However, it is generally thought that these fatty layers represent the precursors of suberin and not those of cutin. The arguments for and against cutinised roots are discussed more fully in Chapter 4.

Definition and nomenclature

The cuticle lies over and merges into the outer wall of the epidermal cell. It has long been known that the cuticle is multilayered. Early workers found

---

FIG. 1.1 Mosaic of electron micrographs of a section through leaf epidermal cells of *Eucalyptus papuana*. The epicuticular wax (*W*) and the cuticle (*C*) can be seen covering the epidermal cells. Large intercellular spaces (*IC*) occur between the mesophyll cells (*MC*) in which prominent chloroplasts can be seen. *M* is the nuclear membrane of one of the epidermal cells. Glut/Osm/Pb/Ur. (Courtesy N. D. Hallam)

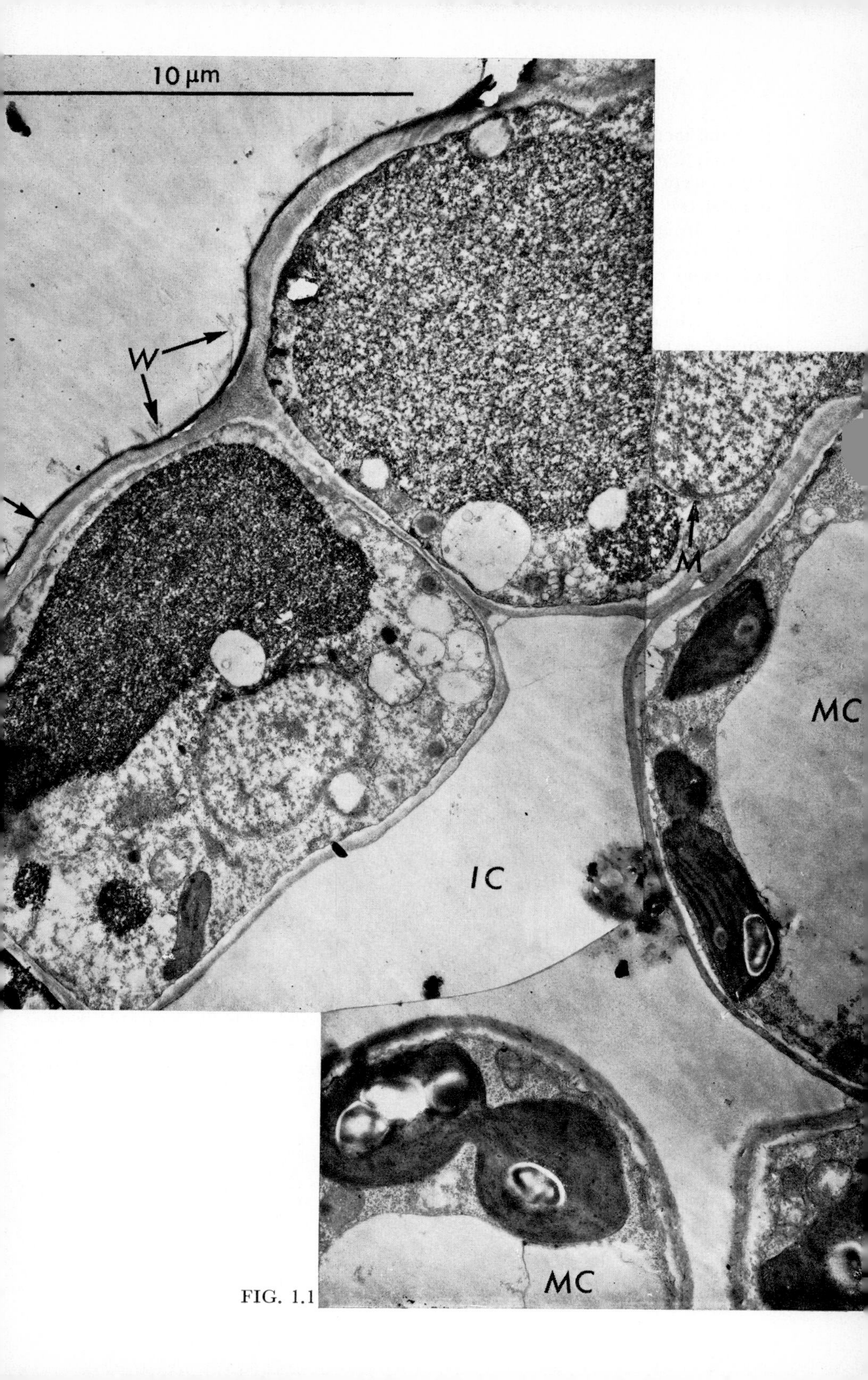

FIG. 1.1

that the membrane consists of an outer layer of cutin and an inner cutinised lamella. The work of Frey (1926), Anderson (1928, 1935) and others led to the concept of a two-layered cuticle attached to the outer wall of the epidermal cell through an intermediate zone of material identified as pectinaceous by positive staining with ruthenium red (see, however, p. 79). The pectic layer came to be regarded as a kind of cement fixing the cuticle to the epidermal cell wall. Such a layer is, as we shall see, not always present. Roelofsen (1952) defined the two-layered membrane lying above the pectic layer as the cuticular membrane. This is composed of an inner cuticular layer consisting essentially of a cellulosic framework incrusted with cutin between its ***microfibrils*** and an outer part, usually thinner, named the cuticle proper, made up chiefly of cutin adcrusted on the cuticular layer.

Esau (1953) referred to the process of incrustation of cellulose as cutinisation, and the adcrustation of the cuticle proper on the cuticular layer as cuticularisation. Sitte and Rennier (1963) examined by light and electron microscopy the cuticles of leaves of many species of plant. They demonstrated four surface layers with changes in gradient densities from wax on the surface through cutin and pectin to the cellulose of the periclinal epidermal cell wall. From their results with *Ficus elastica* they proposed that the definitions of 'cuticle' and 'cuticular layer' should be modified. They suggested that the term 'cuticle' should refer only to the oldest, outermost layer which never contains cellulose, and that 'cuticular layer' should denote the inner cutin-containing layer, regardless of whether or not it contains cellulose. Sitte and Rennier showed that often the cellulose content of the cuticular layer is small; the cellulose is difficult to detect and is present only in the inner lamellae of the cuticular layer. Goodman (1962a) described the morphology of the apple leaf cuticle as revealed by the electron microscope and designated as the true cuticle the cutin-wax layer which lies over the less dense cell wall-cutin complex. Crisp (1965) defined the outer epidermal wall as composed of stratified layers of cellulosic, pectinaceous and lipoidal materials, the outermost lipoidal layer being the cuticle. Following Esau, he defined the deposition of the lipoidal material as 'cuticularisation' and the deposition and accumulation of cutin as 'cutinisation'. The terms cuticular membrane (denoting the tissue lying above the pectinaceous layer), cuticle proper (denoting the outer layer of the membrane composed chiefly of cutin), and cuticular layer (denoting the inner layer of the membrane composed of cutin incrusted on cellulose) have been accepted and used by most workers.

In this book the words cutinisation and cuticularisation will be used as defined by Esau. The multi-layered structure that is usually separated from the epidermal wall by a layer of pectin will be described as the *cuticular membrane*. This comprises an inner *cuticular layer* and an outer layer, the *cuticle* proper, made up mainly or entirely of cutin.

Early work on the morphology and chemistry of the cuticle is reviewed in Chapter 3 and the anatomy and morphology of the cuticle and barks are described in Chapter 4.

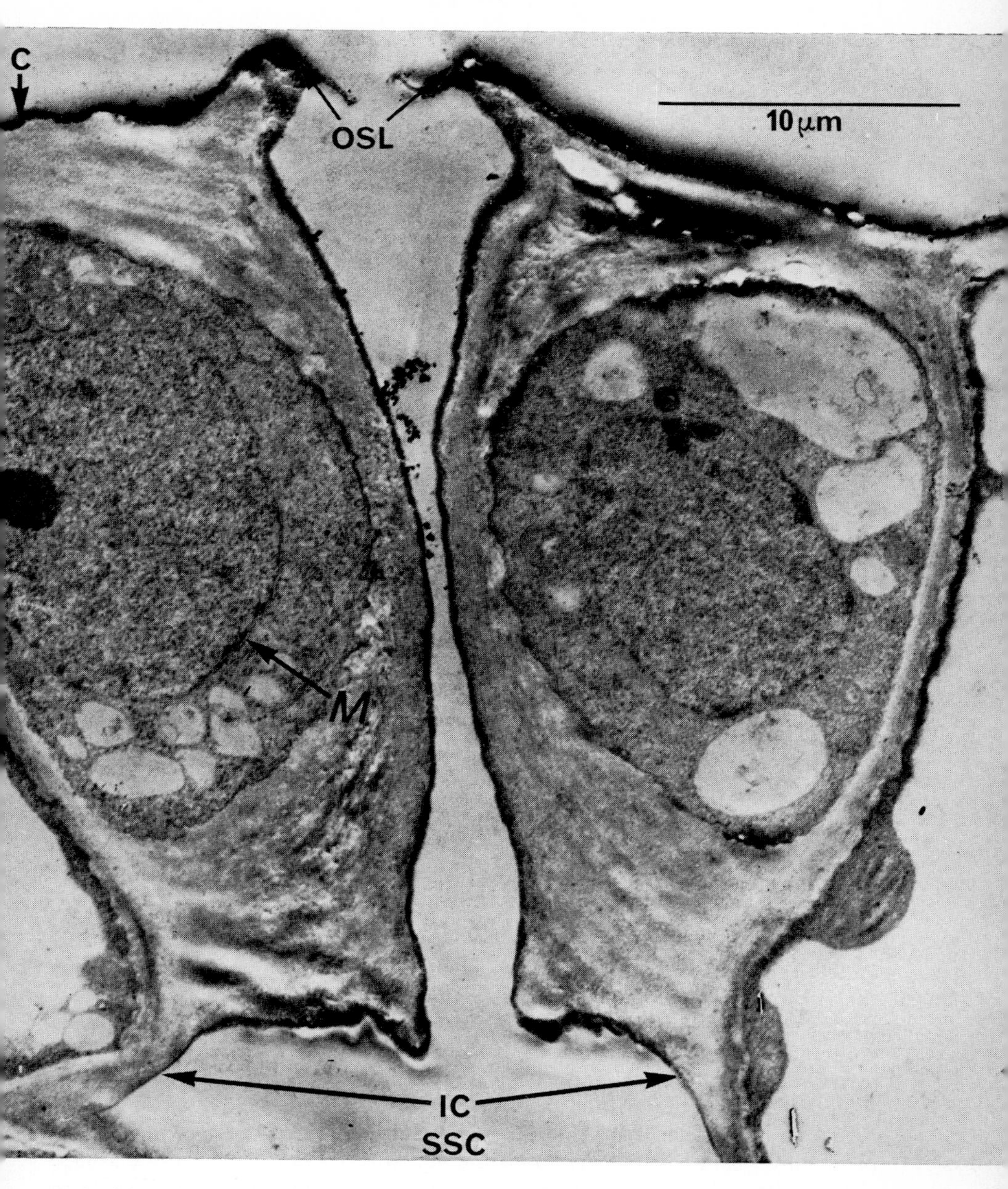

FIG. 1.2 Electron micrograph of a section through a stoma of *Phaseolus vulgaris*. The cuticle (C) forms prominent lips (OSL) over the outer stomatal cavity and extends inside the leaf (IC) coating the walls of the substomatal chamber (SSC). The membrane of the nucleus of the guard cell on the left hand side has (M) been cut exactly at right angles to the plane of the section thus showing its double nature clearly. Glut/Osm/Pb/Ur

The wax bloom

Wax is an important component of the cuticle. The term wax is used to denote a class of substances which qualitatively have certain physical properties such as plasticity in common (Hatt and Lamberton, 1956) rather than to define a precise chemical entity. The wax is embedded within and sometimes exuded over the surface of the cuticle. Schieferstein and Loomis (1959) found that epicuticular wax was present on the mature leaves of about half of many species tested.

Some plants exhibit a prominent waxy bloom. This is due to the reflection and scattering of light on the surface by waxy deposits whose dimensions are close to or only slightly above the wave-length of light. The bloom of the cabbage leaf (*Brassica oleracea* var. *capitata*) (see Figs. 1.1 and 2.5) is due to ubiquitous wax rodlets 1–2 μm long and 0·5–1·0 μm in diameter. The bluish bloom of the leaves of certain varieties of love grass (*Eragrostis curvula*) is attributed to the effect of wax exudates in the form of flat branched ribbons; the more densely the wax branches are packed the more intense is the blue colour (Leigh and Matthews, 1963). A bloom, however, does not necessarily indicate excessive waxiness; moreover, surfaces lacking a bloom may be appreciably waxy. The non-glaucous mutant of cauliflower (*Brassica oleracea* var. *botrytis*) carries about 60% of the amount of wax present on the normal glaucous plant. The leaves of the glossy mutants of pea (*Pisum sativum*), cauliflower, *Eucalyptus urnigera* and blue tussock grass (*Poa colensoi*) have wax deposits which are either smooth films on the cuticle or platelets which lie flat on the surface, whereas the waxes of leaves of the normal plants have the shape of rods or filaments growing outwards and presenting many light-scattering surfaces (Hall, Matus, Lamberton and Barber, 1965). Plum (*Prunus domestica*) fruits are no more waxy, but show a more definite bloom than apple (*Malus sylvestris*) (see Fig. 4. 25) or pear (*Pyrus communis*); the fine wax protuberances on a plum scatter light more effectively than the platelets on the apple or pear.

The effect of environment on cuticle development

Many attempts have been made to correlate the degree of waxiness and cuticle thickness with environmental conditions. Reports have conflicted and the subject is still controversial. Lee and Priestley (1924) concluded that species growing under high light intensities or with reduced moisture contents develop thick cuticles, but later Priestley (1943) decided that little or no relationship exists between the degree of ***xeromorphism*** and cuticle thickness. Plants indigenous to arid and hot regions have been believed to have waxy cuticles; McNair (1931) for example pointed out that more wax-producing plant families are found in the tropics than in temperate zones. Kurtz (1958), however, believes that this view is untenable; the majority of the ***xerophytes*** and succulents of S. Arizona (N. America) contain only small amounts of wax. Other workers also found no definite correlation between xeromorphic adaptation and the amount of surface wax present on plants. A quantitative comparison of surface and sub-surface wax in the mesophyte *Nicotiana glauca* and the xerophyte *Agave americana* indicated that the sub-surface wax may be of much greater significance to the plant in withstanding desiccating conditions than the wax on the surface (Schieferstein and Loomis, 1959).

Plants grown under glasshouse or controlled environment conditions may differ in cuticle characteristics from others of the same variety grown in the open. Differences have been noted in the response of plants under glass and in the open to treatment with herbicides. Glass may filter out light rays necessary for the proper development of the cuticle. Tribe, Gaunt and Wynn Parry (1968) in work on oat (*Avena sativa*) and barley (*Hordeum vulgare*) grown under controlled conditions found that the deposition of cuticle was directly proportional to light intensity and inversely proportional to relative humidity. The pattern of the lipid constituents of the cuticle, however, remained constant under different environmental conditions. The effect of controlled environment and also the influence of the nutritional status of the plant upon the development of the cuticle have received little or no attention and merit detailed examination.

The effects of environmental conditions are intimately bound up with the physiological functions which the cuticle performs. These are discussed in Chapter 7.

The chemistry of the cuticle

The waxes

The application of modern analytical techniques has done much to elucidate the nature of the components of the cuticle. The waxes are complex mixtures, of which common constituents are long-chain hydrocarbons, alcohols, ketones, fatty and hydroxy-fatty acids and esters. Some waxes contain appreciable proportions of ***aldehydes.*** Chain lengths usually extend from 21 to 35 carbon atoms; among the fatty acids, those containing 16 to 18 carbon atoms predominate. The hydrocarbons are mostly ***alkanes,*** but branched-chain compounds (the monomethyl ***iso-*** and ***anteiso-alkanes*** and dimethyl-alkanes) and ***alkenes*** are also found. The alcohols are ***primary alcohols, secondary alcohols*** and ***α,ω-diols***; the ketones, ***monoketones*** and ***β-diketones. Hydroxy-β-diketones*** occur in some waxes. The fatty acids include ***monocarboxylic acids*** (saturated and unsaturated) and ***dicarboxylic acids.*** The alcohols and fatty acids occur in the free state and also combined in the form of long-chain esters which may contain something of the order of 50 carbon atoms. Hydroxy-fatty acids are present and ***estolides***, composed of comparatively short chain (C_{12}–C_{16}) hydroxy-fatty acids have long been recognised as characteristic components of some waxes. The waxes of some plants are said to contain glycerides.

Cyclic compounds of different kinds occur in waxes. ***Terpenoid*** compounds which include ***diterpene*** hydrocarbons, diterpene diols and ***triterpene*** acids and methyl ethers are prominent in many. It has been suggested that the estolide, formed from about four molecules of hydroxy-fatty acids, that occurs in the wax of Colorado spruce (*Picea pungens*) has a cyclic form.

Flavones and ***sterols*** have also been identified in waxes. Some of the compounds recently found have been isolated from natural sources for the first time.

The composition of the cuticular wax is thus very complex. More than fifty compounds occur in apple wax. Two chief groups of waxes, however, can be defined; those which consist essentially of long-chain compounds, among which esters and hydrocarbons are prominent, and those which contain, in addition to long-chain compounds, appreciable proportions of cyclic compounds of terpenoid type. Within each group compositions differ greatly; the alkane content of the total leaf wax, for example, can vary from 0·2%, in that of *Eucalyptus largiflorens*, to 92% in that of *Solandra grandiflora*. Moreover, the compositions of cuticular waxes may change as leaves or fruits grow, or may differ on the upper and lower surfaces of leaves. Evidence has also been obtained that the composition of the wax may differ in different zones of the cuticle. It is certainly true that upper and lower surfaces and different parts of the same leaf may differ in the fine structure of their wax (see p. 99). Cuticular wax deposits differ greatly in amount; the leaves of some vegetables are virtually non-waxy and some fruits are very waxy.

Various classes of components of wax blooms have been used as taxonomic criteria, a purpose for which they are eminently suitable. The waxes can be easily isolated in a form uncontaminated by cellular components, are capable of rapid and accurate analysis and have been regarded as end-products of metabolism. Special attention has been given to the distribution in waxes of hydrocarbons, especially alkanes, and α,ω-diols in relation to botanical classification and the work has met with considerable success. It seems that a particular class of constituents provides a valuable taxonomic criterion if it constitutes a major fraction of the wax. The methods used for the analysis of waxes are described in Chapter 2 and the chemotaxonomic studies in Chapter 5.

The cuticular membrane

Cutin is regarded as the chief structural component of the membrane. On hydrolysis, cutin gives a mixture of saturated aliphatic monocarboxylic, hydroxy-monocarboxylic, dicarboxylic and hydroxy-dicarboxylic acids mostly of C_{16} and C_{18} chain lengths. It is a complex polymer, with the constituent acids interconnected by ester, peroxide and ether linkages. Hydroxy-fatty acids with ester linkages predominate in the polymer; cutin may thus be regarded as a much more elaborate form of the estolides found in some waxes. It is assumed that the acids occur in the polymer in the form in which they are isolated. This, however, is by no means certain. Crisp (1965), from a consideration of the numbers of hydroxyl groups present in the isolated acids and in the polymer, suggests that the acids found may be artefacts due to the isolation procedure.

A ***tannin***-like material occurs in cuticular membranes, sometimes as a major component. In such cases, cutin may not exceed 50% of the membranes. The chemical nature of the tannin is not known; it appears to be of the non-hydrolysable or ***condensed tannin*** type. Small amounts of free phenolic compounds occur in the membranes. The presence of a protein constituent in apple fruit cuticle has been reported, but has not been substantiated. Cellulose is embedded within the membranes in the zone adjacent to the epidermal cells; other material within this zone is regarded, from staining tests, as pectin, but its chemical identity has not been firmly established.

Other protective coverings

Wax and tannin occur with suberin in barks and cork deposits. A wide variety of cyclic compounds, including terpenoids and ***lignans***, have been isolated from barks. Suberin is chemically related to cutin. Similar hydroxy-fatty and fatty acids are obtained from suberin by hydrolysis, but some acids characteristic of suberin are of greater chain length than those of cutin. Suberin and cutin have long been confused botanically; differentiation between them on a chemical basis now seems possible. Little is known of the chemical nature of sporopollenin, the structural component of the walls of moss and fern spores and pollen grains. It is one of the most resistant materials in nature; the persistence of spores and pollen in fossilised plant deposits has been ascribed to its presence.

The chemistry of waxes and other secretions and of cutin, suberin, barks and sporopollenin is discussed in Chapter 5. Much work has been done on the biosynthesis and development of cuticular components and this is reviewed in Chapter 6.

Functions of the cuticle

The cuticle has an important role as a structural element, holding the cellular tissues compact and firm. Above all, it holds an important position as the bounding layer between the body of the plant and its environment. Functions ascribed to the cuticle include the conservation of water in the plant, the prevention of loss of plant components by leaching and the protection of the plant from injuries due to wind and physical abrasion, frost and radiation. The nature of the cuticle also greatly influences the deposition and subsequent behaviour of pesticides, growth regulators, foliar nutrients or other chemicals used on plants. The cuticle provides the first potential barrier to attack by fungi, insects or other pathogens. The various ways in which the cuticle is involved in phenomena at the plant surface can, for convenience of discussion, be classified in three broad categories: physiological functions, interactions with chemicals and interactions with pathogens.

Physiological functions

The most important function of the cuticle is probably to supplement the action of the stomata in regulating the passage of water from within the plant to its environment. Movement in the reverse direction also is involved; some plants may benefit from the possession of a poorly developed leaf cuticle, either overall or in localised areas, by absorbing water from rain or dew. It is generally believed that transpiration *via* the cuticle is relatively unimportant compared with loss through the stomata. Ketellapper (1963) echoes this belief by stating that at least 90% of the water loss from a leaf occurs by diffusion through the stomata. Other workers, on the other hand, believe that cuticular transpiration is generally larger than it is thought to be, and raise doubts on the efficiency of the cuticle as a protection against transpiration. Much evidence is now available to show that the degree of impregnation of the cuticle with wax, and not the thickness of the cuticle, is the important factor in controlling cuticular transpiration. Furthermore, the chemical composition of the wax markedly affects its efficiency as a water barrier. No generalisation can be made on the role of the cuticle in preventing the loss of water from plants; each case needs to be examined on its merits.

The leaching of nutrients and other substances from leaves is an important matter that, although recognised, has received comparatively little attention. As with transpiration, the wax component of the cuticle is undoubtedly the agent which reduces losses. Some leaves freely exude droplets when the surface wax is removed by washing with solvent. Carbohydrates including free sugars and polysaccharides, amino and other acids and inorganic nutrients are leached by rain, dew and mist. Mecklenburg and Tukey (1963) found leaching to occur from the leaves of more than 100 species. The losses may be surprisingly large. Tukey and Morgan (1964) emphasise the importance of the subject and assign to the leaf 'an even broader role beyond the classical concepts of transpiration and photosynthesis; a dynamic role of uptake and loss of water and metabolites from plants, helping to adjust a plant to its environment and influencing its ecological distribution'.

Comparatively little is known about gaseous exchange through cuticles, in contrast to the wealth of information available on exchange through stomata. The extent and mechanism of cuticular penetration of carbon dioxide is of special interest in relation to photosynthesis. For many years the stomata were regarded as the sole route of entry of carbon dioxide, but recent work indicates that in some plants cuticular penetration may be considerable. Dorokhov (1963), for example, reports that at the normal concentration of carbon dioxide in the air the astomatous ***adaxial*** cuticle of the apple leaf transmits 20–30% of the total quantity absorbed. This uptake he describes as responsible for 'extra-stomatal cuticular photosynthesis'. The subject merits more detailed examination using a range of species of differing cuticle characteristics.

The cuticle protects the plant from damage by wind or abrasion. The wax deposits on the surface, however, may be eroded. The impacts of wind, particles of blown soil or heavy drops of rain damage the wax structure of pea leaves and lead to increased retention of sprays (see p. 250). A weed-killing spray may then be damaging, and a standardised test has been devised for farmers by which retention may be assessed before spraying; if necessary, application of the weedkiller is delayed for a few days to enable the protective wax deposit to regenerate (Amsden and Lewins, 1966). Wax is removed from clover leaves (*Trifolium*) by weathering, e.g. by the leaves brushing together in windy conditions and the leaves become more easily wetted. Half of the wax may be lost from the leaves by the action of winds and the transpiration rate is then increased; when deposits are abraded, a new flow of wax usually occurs within a few days but some leaves fail to regenerate the wax lost (see p. 250).

Suggestions have been made that wax deposits on leaves play a part in the prevention of damage by frost. Hall and Jones (1961) found some evidence that the presence and the form of the wax on leaf surfaces are related to hardiness to frost. A rodlet type of wax predominates on plants of *Eucalyptus urnigera* at the higher altitudes, and becomes progressively less conspicuous at the lower (Hall, Matus, Lamberton and Barber, 1965); the more glaucous populations of *Eucalyptus* spp. occur in the more frosty localities (Barber, 1955).

The cuticular wax may assist in controlling the temperature of leaves, and has long been thought to protect the cellular tissues from excessive ultra-violet radiation. The reflection of incident radiation by the leaves of *Eucalyptus* spp. controls their temperature; the reflectivity of the rodlet form of the wax of *E. urnigera* at the higher altitudes has probably been of selective significance in the origin and maintenance of the ***cline*** (Thomas, 1961). The cuticle and epidermal cells play a major part in the reflectance of infra-red radiation from leaves. From the beginning of this century, suggestions were made that the thickening of the cuticles of alpine plants was a response to the excessive ultra-violet radiation encountered at high elevations. Some workers have suggested that ultra-violet rays are absorbed during passage through the cuticle. Crisp (1965) makes the interesting observation that the formation of peroxide linkages in the final stages of polymerisation of cutin is enhanced by ultra-violet irradiation and that the production of lipid and other materials is stimulated.

The physiological functions of the cuticle are discussed in Chapter 7.

Interactions with chemicals

An important function of the cuticle is to water-proof the plant, i.e. to prevent its tissues becoming saturated. Some plants are strongly water repellent; droplets impacting upon them retain a more or less spherical shape and run off. Others are easily wetted by rain or dew. Two mechanisms are involved in water-proofing; the shedding of droplets by the bloom and the resistance to the penetration of water created by wax embedded within the cuticular membrane.

The water repellency of many plant surfaces raises problems in the deposition of chemicals used as pesticides or for other purposes. Surface-active agents (surfactants) are used in chemical sprays to assist wetting. The concentration of a surface-active agent required to ensure the maximum deposition of a chemical on different plants differs appreciably according to the water-repellent properties of their surfaces. Many factors are involved in water repellency. Macro- or microcorrugations of the surface may be responsible. Some leaves, although visibly smooth, have microcorrugations which make the surfaces extremely difficult to wet. When the superficial wax is the important factor in repellency, its effect depends upon its physical structure which, in turn, is dependent to some extent upon its chemical composition; water repellency is greatest when the wax has a rough surface in the form of projecting rods, plates or crystals. Of the many classes of compounds present in waxes, the alkanes are the least wettable; ease of wetting is enhanced by an increase in polarity due to substitution in the aliphatic molecule. Waxes with high contents of alkanes or ketones show a pronounced tendency to form crystalline deposits; the water repellency of these waxes is due not only to their hydrophobic constituents but also to their physical form.

Many chemicals freely enter leaves, but little is known of the extent to which they penetrate other aerial parts such as stems. The processes of uptake by leaves are complex. Much depends upon the properties of the chemical, e.g. whether water- or lipoid-soluble, the medium in which it is applied, the environmental conditions and the physical and chemical nature of the cuticle. The relative importance of the stomata and the cuticle as a route of entry has been much debated; the consensus of opinion, substantiated by many reports, now is that the main route for both water- and lipoid-soluble materials is provided by the cuticle. If penetration of the stomata does occur, negotiation of the thin, cuticular lining of the substomatal cavity is necessary for access to the mesophyll. Atmospheric conditions influence penetration; high humidity, for example, favours the entry of water-soluble chemicals into some leaves. The incorporation of surface-active agents in spray fluids leads to greater penetration which is ascribed to the more efficient wetting of the cuticular membrane. Although somewhat fanciful, the cuticle has been likened to a sponge, its cavities impregnated with wax, which expands under humid conditions and contracts under dry. Penetration has been associated with various pathways through the cuticle, and conduction from the cuticle to the cellular tissues with the ***ectodesmata***. These have achieved prominence from the work of Franke (1960). The absorption of chemicals through the cuticle is not merely a passive process of diffusion, but is activated by metabolic processes within the leaf. The nature of the cuticle has an important influence upon penetration; a waxy hydrophobic cuticle is likely to favour the entry of lipoid-soluble materials, but to provide a serious barrier to the penetration of water-soluble materials.

There has been a tendency in the past in questions of pest or disease control to regard the plant surface merely as a passive site of attack and to think only in terms of the interaction between an applied chemical and a parasite. Recent evidence indicates that the influence of the cuticle or surface components upon the chemical and, conversely, the effect of the chemical upon the cuticle cannot be ignored.

The interactions of the cuticle with chemicals are discussed in Chapter 8.

Interactions with pathogens

The cuticle has long been supposed to protect the plant from attack by fungi. Brown (1936) stated that 'the importance of the outer layers of the plant's body, whether cuticularised epidermis or cork, in preventing the entrance of parasitic fungi is well recognised'. Some fungi penetrate directly through the cuticular membrane; others invade through natural openings in the cuticle such as stomata and nectaries or through punctures or wounds. The cuticle can afford protection only against those fungi that invade by the seemingly more difficult route.

Many factors influence the efficiency of the cuticle as a barrier against attack. Some characteristics of cuticles favour an invader; others militate against it. Spores may be carried to leaf surfaces by currents of air or may be deposited in impacting droplets of water. A film of moisture on the surface may assist the adherence of spores and be necessary for germination. A leaf surface that is easily wetted is favourable to the deposition of a water-borne inoculum and the establishment of the infection; on the other hand, a water-repellent surface which prevents deposition may be an important factor in defence.

Since the classic work of Brown (1922) on the physiology of fungal parasitism it has been known that nutrients exuded or leached from leaves stimulate some fungi before they penetrate. Evidence has also been obtained that leaf exudates may contain substances that can inhibit the growth of fungi. The ability of a leaf to secrete materials depends upon its possession of a hydrophilic wax, and a leaf with this type of wax is easily wetted and retains water. On this type of surface a fungus (with the possible exception of a powdery mildew) finds itself in a favourable environment but under opposing influences, stimulatory and inhibitory. Substances present in the waxes of some leaves are known to suppress the growth of fungi; it has been suggested that the waxy layer of *Ginkgo biloba* (Darwin's 'living fossil') holds the secret of why the leaves are so resistant to penetration by common pathogenic organisms (see p. 266). When the nature of the cuticle is such that the leaching of nutrients to the surface is restricted (as in *G. biloba*), the cuticular wax may play a dominant part in protection.

The role of the cuticle as a mechanical barrier to invasion has been much discussed and opinions are conflicting. From a knowledge of its chemical nature, there is no reason to believe that the cuticle is structurally tough and a serious impediment; this is borne out by the ready penetration of some substantial cuticles. The fungal infection thread may make its way between cuticular lamellae or traverse less dense 'pectinaceous' or other pathways. Whether a pathogenic fungus facilitates its entry into the living leaf by dissolving the cutin has also been much debated. Wood (1960) summarises the evidence on cuticle penetration and concludes that entry is a mechanical process and does not depend upon substances secreted by the hyphae of the invading organism. He suggests that, so long as a fungus is able to penetrate the plant by other means, the breakdown of the cuticle would be disadvantageous; the impermeability of the surface layer to water would be destroyed, the young hyphae would be exposed to desiccation and other organisms would gain entry. The view that penetration is a mechanical process is still generally held despite reports that cuticle breakdown may be involved. The role of the cuticle in defence against fungi is reviewed by J. T. Martin (1964).

The cuticle may also protect plants from attack by insects. These feed on leaves in two ways: by sucking juices through a proboscis inserted into the phloem (e.g. aphids) or by chewing the leaf material (e.g. caterpillars). The cuticle in no way can afford protection from the latter form of attack; a well-developed cuticle may, however, deter sucking insects, but on this matter no definite information is available. Components of the outer layer of a leaf may contribute to making the surface unattractive to the alighting insect, and recent limited evidence suggests that the superficial waxes may play a part in the repellency of some plants to insects. Some plants have developed specialised mechanisms for trapping insects. An example is the well-known pitcher plant (*Nepenthes*) (see p. 272).

The cuticle may also afford protection from attack by bacteria and may limit the entry of viruses, but on these matters information is rather more fragmentary. The surface of the leaf as an environment for micro-organisms (the 'phyllosphere') is a fascinating subject for study which until recently has been somewhat neglected. The interactions of the cuticle with pathogens and the interplay between organisms on the plant surface are discussed in Chapter 9.

The cuticle in decay

Cutin has long been thought to be a persistent material. This view undoubtedly is derived from observations that the cuticles of plants may persist for long periods. Intact leaf cuticles containing cutin have been obtained from fossilised deposits and used extensively in taxonomic studies (see p. 279). Bandulska (1926, 1931), for example, examined the cuticles of leaves of recent and fossilised Lauraceae and Myrtaceae and showed that fossilised plants could be identified by comparing their cuticles with those of living genera. Other workers have confirmed the value of the cuticle in taxonomic studies. Sen (1954) found that the cutin of fossilised and living plant material responds similarly to optical and chemical tests, indicating that no significant physical or chemical change occurs in cutin during and after fossilisation. The so-called 'paper coal'—a coal with a papery texture—of the Moscow basin and Indiana consists largely of a mat of resistant cuticles and normal coal on oxidation may leave a residue of cutin.

Peats consist of the residues of vegetation which have persisted for long periods under boggy conditions. Lebedev (1959 a, b) has shown that the non-hydrolysable fraction of root-sedge peats contains about one-third of substances of the cutin-suberin group. Because of this, he points out that the term '***lignin***' usually applied to the non-hydrolysable fraction is incorrect. Some peats also contain appreciable proportions of waxes, which yield hydrocarbons and other components similar to those found in present-day plant waxes. *Iso*- and *anteiso*-***paraffins*** have been identified in the hydrocarbon fractions of Precambrian sediments; these compounds occur in contemporary plant waxes. This suggests that the indigenous hydrocarbons of Precambrian sediments one billion years old are of biological origin (see p. 281). The value of hydrocarbons and other classes of organic compounds as biological markers in a systematic search for chemical evidence of early life in ancient sediments has been discussed by Eglinton, Scott, Belsky, Burlingame and Calvin (1964).

The persistence of cuticular components in fossilised deposits can only be explained by the existence of conditions unfavourable to biological agencies at the time of formation of the deposits. A hot, desiccating environment may have been a factor. Cuticles are likely to have survived unchanged throughout long periods of time only under strictly anaerobic conditions; with sufficient oxygen available, their components would gradually have been decomposed by moulds and other organisms.

Considerable amounts of cuticular material are deposited each year in the leaf litter of deciduous plants. Waxes are present but do not build up in productive soil (Stevenson, 1966). Under forest conditions the decay of leaf litter is slow and the preserved membranes of leaves can be seen, but the cutin content of the litter is not known. Cutin, however, fails to accumulate in orchard or cultivated soils and attention has been given recently to the organisms responsible for its breakdown. Soil animals play an important part in comminuting plant material and making it more accessible to further attack, but whether they degrade cutin in the process is largely unknown. Some fungi and other micro-organisms are now known to be able to break down cutin. The degradation is ascribed to the action of a cutinase which, it is suggested, may in some instances also play a part in the process of pollination.

Fossil cuticles and the various aspects of the breakdown of cutin and other cuticular components are discussed in Chapter 10.

Commercial uses of cuticular components

For many years plant waxes have been used for a variety of purposes. Examples of commercially important waxes are carnauba wax obtained from *Copernicia cerifera*, candelilla wax from *Euphorbia* spp., ouricury wax from *Syagrus* (*Cocos*) *coronata*, esparto wax from *Stipa tenacissima* and sugar cane wax from *Saccharum officinarum*. Some barks yield tanning materials or medicinal preparations. The commercial exploitation of components of cuticles and barks is considered briefly in Chapter 11.

2

Methods of Research on Cuticles

The remarkable advances in recent years in microscopy and analysis have greatly assisted the elucidation of the structure and composition of the cuticles of plants. The earlier workers were handicapped by the inadequacy of the methods then available; the histologists were confined to observations on the effects of staining reagents on components of the cuticle that were ill-defined and chemists strove to unravel the nature of the components by methods that were clumsy and time-consuming. The electron microscope has revealed, for the first time, the ultrastructure of the cuticle and the delicate and diversified form of its surface wax, while modern analytical methods such as chromatography and spectroscopy permit the rapid separation and identification of cuticular constituents. This chapter describes the principles and applications of the physical and physico-chemical methods now used.

MICROSCOPICAL TECHNIQUES

Light microscopy

Microscopes, both simple and compound, were used early in the seventeenth century. Using a simple microscope Leeuwenhoek (1632–1723) discovered a range of micro-organisms never before suspected. Our knowledge of cells, however, dates from the publication just over 300 years ago of the *Micrographia* of Robert Hooke who used the word *cell* to describe the compartments that he saw in cork and other plant tissues. Hooke was also the first person to study in detail the surface of a leaf, and described the hairs on the leaves of nettle (*Urtica dioica*).

The modern light microscope is a direct development of the compound microscope used by Hooke. Simple microscopes, after Leeuwenhoek, have played very little part in microscopy. The compound microscope, with resolving power improved to about 0·3 μm, reached the peak of its development almost 100 years ago and cannot be improved further because of the nature of light (Table 2. 1).

Microscopes were then developed which use light of shorter wavelengths to improve the resolution. The ultra-violet microscope enjoyed a period of popularity for very fine work; its resolution is about 0·15 μm, but it was expensive to make and difficult to operate. X-ray microscopes are available, and although their resolution is equivalent only to that of a light microscope, they benefit from the high penetrating power of X-radiation.

Polarised light microscopy

Three modifications of the light microscope have contributed to the study of plant cell structure; these have given rise to phase, interference and ***polarised light*** microscopy. Of these only polarised light microscopy has been used to any extent in the study of cuticles and much of our early knowledge, and indeed much that is still relevant about cell walls and cuticles, is due to it.

Plane-polarised incident light shows how the molecules are orientated. If the molecules have a non-random orientation, so that the material has a different pattern of components looked at along one axis compared with another then the plane-polarised light will distinguish between the axes of the substance. Its refractive index will then differ according to which axis is measured. Such a substance is called ***anisotropic***, to distinguish it from ***isotropic*** substances in which the molecules are randomly oriented, or evenly spaced in all directions. Anisotropic substances are ***birefringent***, which means that when ordinary unpolarised light passes through them it emerges in such a form that it may be resolved into two component rays of different refractive indices and whose planes of vibration are at right angles to each other. Birefringence is termed positive if the refractive index is greatest parallel to the axis of the ***organelle*** and negative if it is at right angles to the axis.

Transmission electron microscopy

Beams of electrons have wave-like properties and, as De Broglie predicted in 1924, can be used to produce a high resolution microscope. The wavelength of an electron beam at normal accelerating voltages is about 10^{-5} smaller than that of visible light; the resolution of an electron microscope is not, however, 10^5 better than that of a light microscope, but only about 10^3. The lenses for electron beams are not glass, but are electromagnetic or electrostatic fields; almost all electron microscopes use electromagnetic lenses. These unfortunately have to be made with small apertures because the method of image formation involves the deliberate exclusion of electrons scattered by heavy atoms and so the optimal resolution of about 0·5 nm is far worse than might be expected on theoretical grounds. Nevertheless the resolution attainable by any electron microscope is still better than can be exploited by the present methods of preparing specimens of biological material.

The science (or art) of electron microscopy is a recent development and at least so far as plant material is concerned almost all the work has been done in the last ten years. In an electron microscope the specimen must endure a high vacuum because an electron beam is stopped by appreciable numbers of atoms. Specimens must be dehydrated and are therefore dead; they must be prepared so as to withstand the impact of high speed electrons and the consequent heating. The possible effects of preparative techniques on specimens must therefore be borne in mind when interpreting the images obtained. Electrons cannot be seen, but photographic emulsions fortunately react to them in almost the same way as they do to photons. An image is formed on a photographic plate—the electron micrograph. The plate is developed in the ordinary way to give a 'negative'. As yet, there is no standard way of printing micrographs.

The formation of an image in the electron microscope is quite different from that in the light microscope. Contrast, brought about by the differential scattering of electrons, is augmented by the deposition of an alien metal in or onto the specimen. Generally heavy metals with atomic weights greater than that of molybdenum (96) are employed, although manganese (55) is a widely used exception. Metal can be used to introduce contrast in three different ways; shadowing, positive staining and negative staining. A discussion of

---

FIG. 2.1 Section through the cuticle of a holly (*Ilex aquifolium*) leaf under the polarising light microscope showing ***E*** the epidermal cell layer, ***H*** the hypodermis and ***P*** the palisade layer. I.C is the isotropic (dark in this micrograph) layer of the cuticle; B.CT the birefringent cutinised wall layer; I.P the isotropic 'pectin' layer and B.E and B.H the birefringent walls of the epidermal and hypodermal cells respectively. (Courtesy F. A. L. Clowes)

FIG. 2.2 Diagram of surface replication. The surface to be examined (***A***) is coated with a thin layer of electron translucent thermostable material (***B***). Carbon evaporated under high vacuum is commonly used. The parent material is then etched, washed or peeled away from the replica (***C***) and this may then be shadowed as in Fig. 2.3. Note that the evaporated carbon replicates re-entrant angles

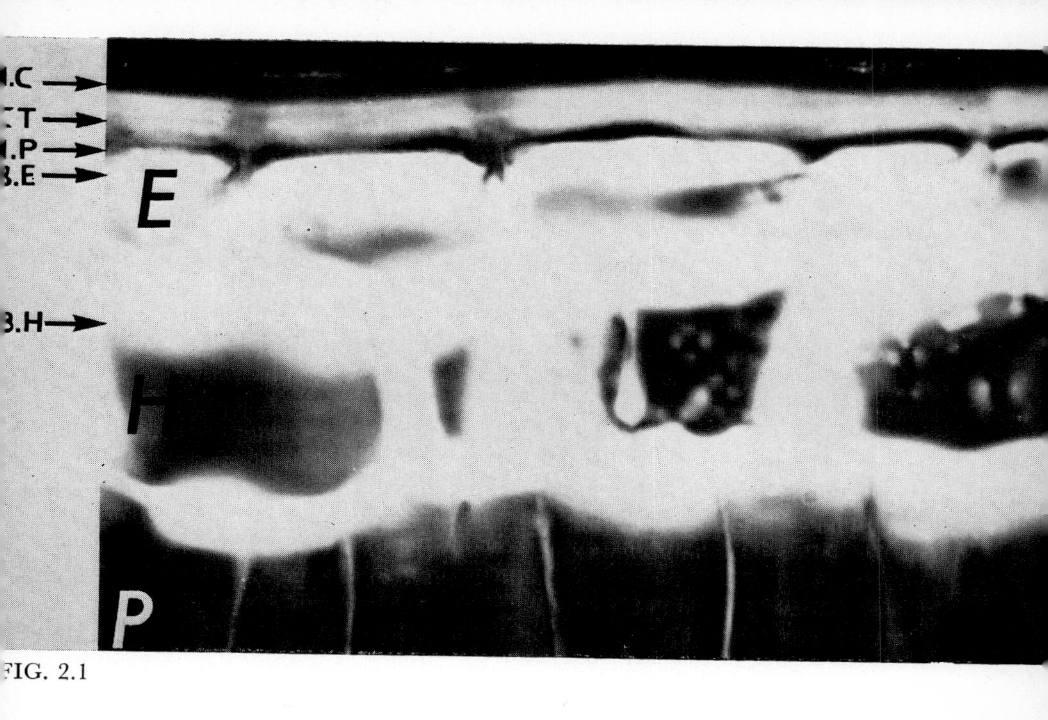

FIG. 2.1

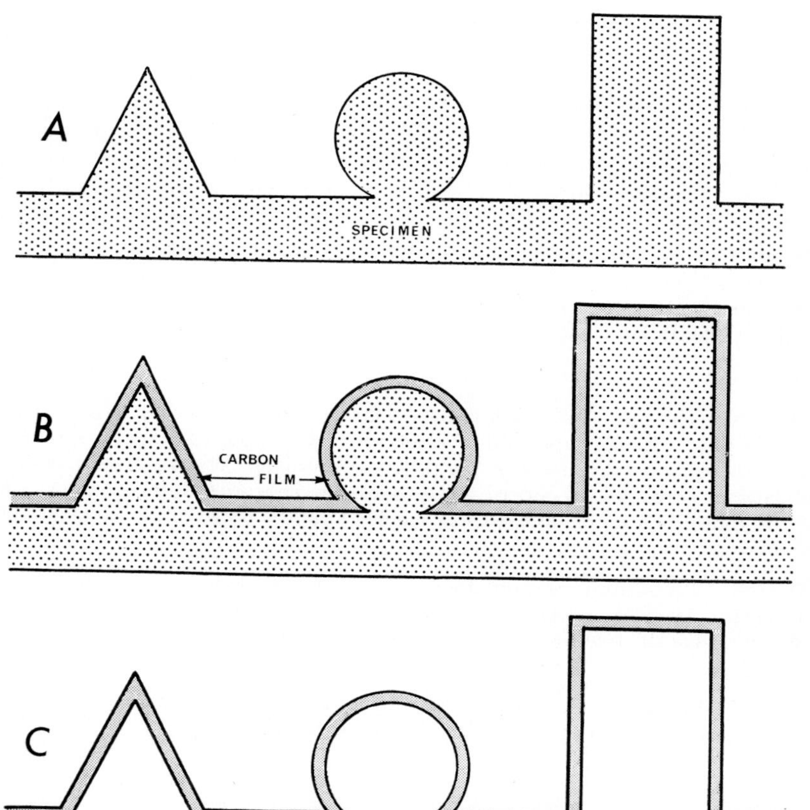

FIG 2.2

these techniques is given in Clowes and Juniper (1968). Only shadowing and positive staining will be considered here.

Metal shadowing

The first shadowing technique is almost as old as electron microscopy itself, and was devised by Williams and Wyckoff (1946). It is applicable only to specimens with relief such as surface replicas (Fig. 2.2)' ***freeze-etched*** surfaces, or to whole objects such as viruses, ***ribosomes*** or isolated microfibrils. A heavy metal, such as palladium, platinum or gold, is evaporated from a tungsten filament which has been heated in a high vacuum, onto the specimen to be examined (Fig. 2.3). The metal, evaporated from more or less a point source, is deposited on the 'windward' side of each target object, and since the metallic atoms travel only in straight lines, in the 'lee' of the object a region free of metal is formed. The heavy metal atoms in the 'fall-out' area impede the free passage of electrons, hence more electrons flow through the region lacking the metal coating than pass through the coated region, and as a result produce a darker image on the photographic emulsion (Fig. 2.4). The angle of deposition is measured and the height of object is then determined from the length of the cast shadow.

The technique has limitations, which, though unimportant at low magnifications, become progressively important and intrusive as higher resolution is demanded. In the first place, as Fig. 2.3 shows, the deposition from an angle of a coat of metal onto an object tends to distort the shape of that object. This is unimportant where, as in most projections from a leaf surface, the target object is relatively large in relation to the thickness of the metal coating. The thickness of the metal coating varies according to the electron scattering power required and the atomic weight of the metal concerned, but is generally 1–3 nm (Bradley, 1965). This is insignificant when compared with a leaf surface projection of 1 μm, but becomes serious when an object (e.g. tobacco mosaic virus) is only 17 nm in diameter. The second disadvantage of shadow-casting is that all metals deposited in this way tend to form crystals whose sizes vary both according to the metal used and the manner in which it is deposited. Metals such as gold and palladium used alone may produce crystals as

---

FIG. 2.3 Diagram of 'shadowing' with a heavy metal. A grid (G) is coated with a support film (SF) and upon this the specimen (S) is placed. Carbon replicas can usually be mounted directly upon the grid. A heavy metal (M) such as gold, palladium or platinum is evaporated under a high vacuum from a point source onto the specimen. If α is known it is simple to work out the height of the object from the length of the shadow cast.

FIG. 2.4 Electron micrograph of a shadowed carbon replica of the adaxial surface of a barley (*Hordeum sativum*) leaf. A 'shadow' is thrown by one of the wax projections (S). The ridge (R) free from wax projections is a feature of many grass leaf surfaces. Cr/Au/Pd

FIG. 2.5 Electron micrograph of a replica of a cabbage (*Brassica oleracea* var. *capitata*) adaxial leaf surface. S Cr/Pt. (Courtesy D. E. Bradley)

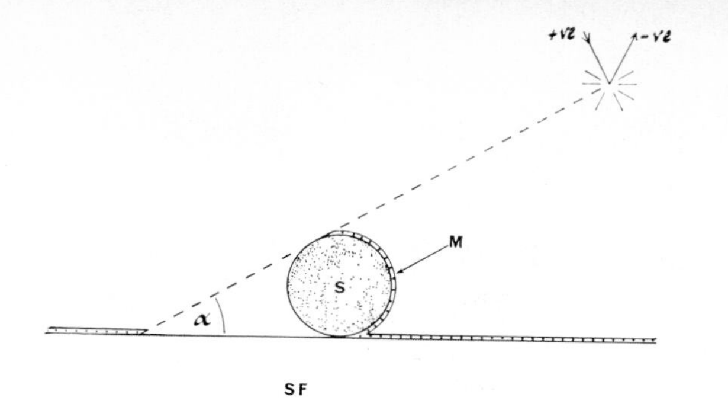

FIG. 2.3

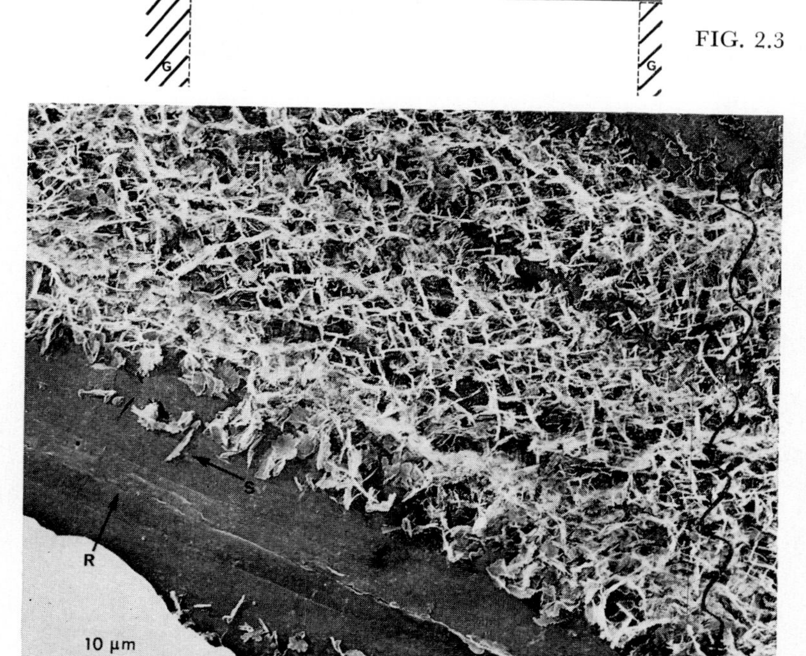

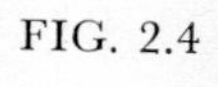

FIG. 2.4

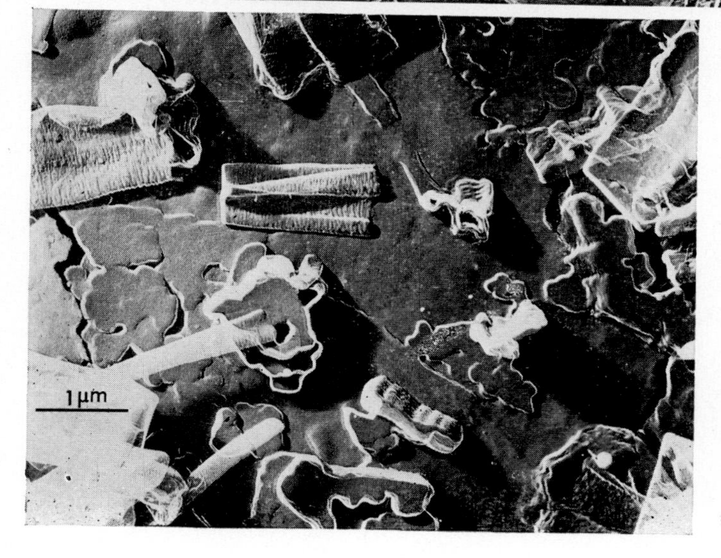

FIG. 2.5

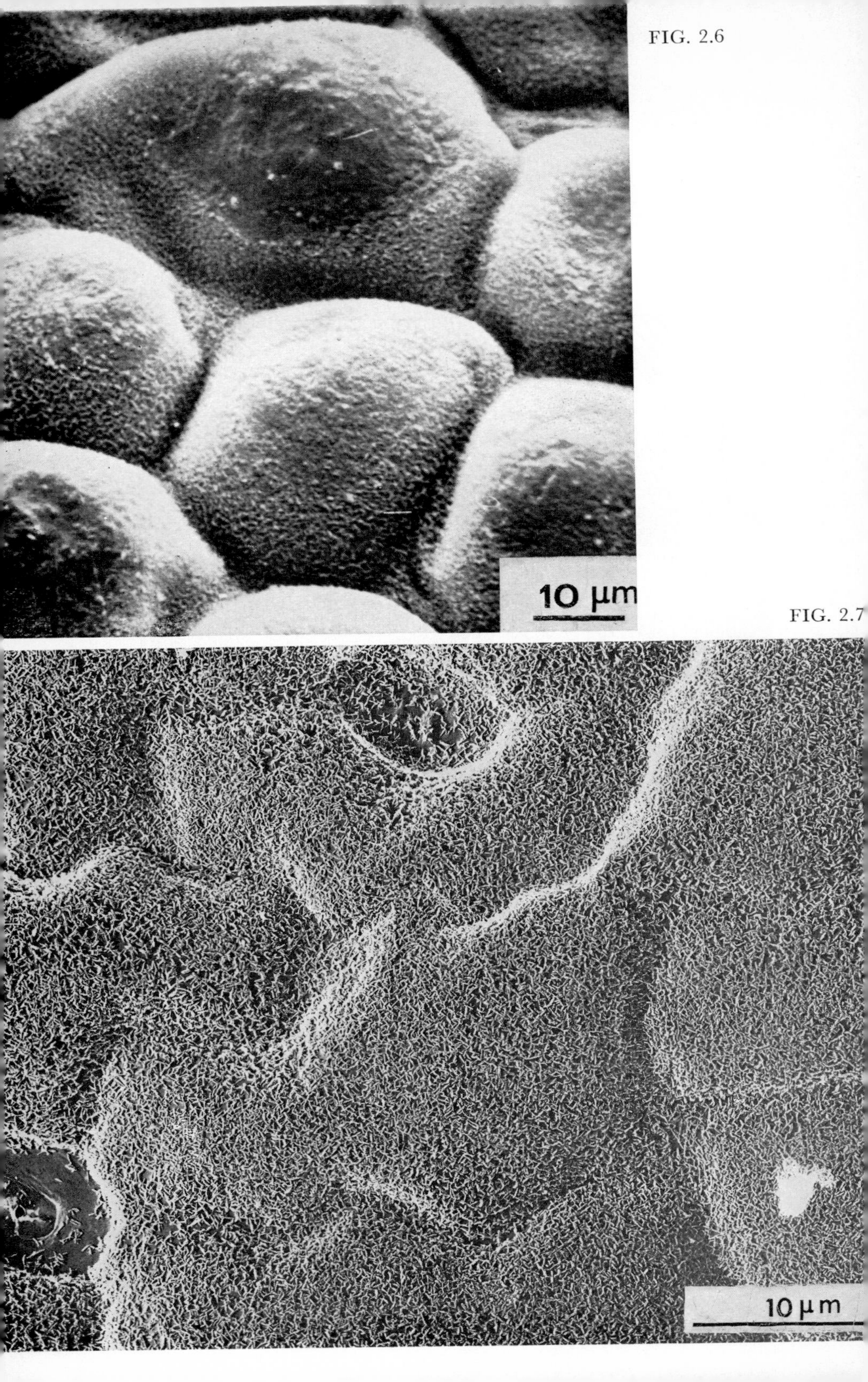

FIG. 2.6

FIG. 2.7

large as 4 nm. Alloys give better results and form crystals of only about 2·5 nm. The simultaneous platinum/carbon shadowing method developed by Bradley (1965) inhibits crystallisation so well that the particle size is probably below 1 nm and thus not resolved in the electron microscope (Fig. 2.5).

Staining

Tissue or positive staining for electron microscopy is in some ways identical to tissue staining for the light microscope in that the stains incorporated into the tissue confer contrast by virtue of the extent to which they combine with specific chemical groups. In spite of intense research into staining techniques for electron microscopy over the last twenty years, there is as yet no series of stains comparable with that available to the light microscopist. With few exceptions, the heavy metal stains available for the electron microscope, e.g. potassium permanganate, osmium tetroxide and lead and uranium salts are relatively unspecific. Progress however is being made in staining techniques. Clowes and Juniper (1968) give a list of the fixatives and stains in common use for electron microscopy of plant material.

Replicas

Each of the devices to improve contrast described above may be used in conjunction with different methods of preparing specimens. The most commonly used of these are the formation of replicas, the sectioning of impregnated or unimpregnated plant tissue and freeze-etching.

Many materials are used for producing replicas; carbon is now probably the most widely applied. The first application of the carbon replica technique to plant surfaces was made by Juniper and Bradley (1958). Replica techniques in general are discussed by Bradley (1965).

Mueller, Carr and Loomis (1954) were the first workers to use the electron microscope for looking at leaf surfaces. They used a double-stage plastic method, and their method involved the wetting of the plant surface with a neutral wetting agent. The technique of Juniper and Bradley avoided the difficulty of having to wet the surface by employing evaporated carbon as the material for a single-stage replica. The accuracy of this technique in faithfully preserving details of biological surfaces has since been confirmed by the scanning electron microscope (Figs. 2.6 and 4.31) and the freeze-etching technique (see Fig. 2.12). However, the replica technique of Juniper and Bradley is only successful on relatively waxy surfaces and is difficult to use on smooth wax-deficient or hairy surfaces. The modification by Williams and

---

FIG. 2.6 Scanning electron micrograph of the adaxial surface of a leaf of white clover (*Trifolium repens*). (Courtesy P. J. Holloway) Note the damage to the waxy layer on the 'crowns' of the epidermal cells. Ms/Sc.

FIG. 2.7 Low magnification electron micrograph of the adaxial surface of a pea leaf (*Pisum sativum* var. *Alaska*). Note the absence of wax projections around the stoma. The fine structure of the surface is very similar to that of *Trifolium repens* (Fig. 2.6). Cr/Au/Pd

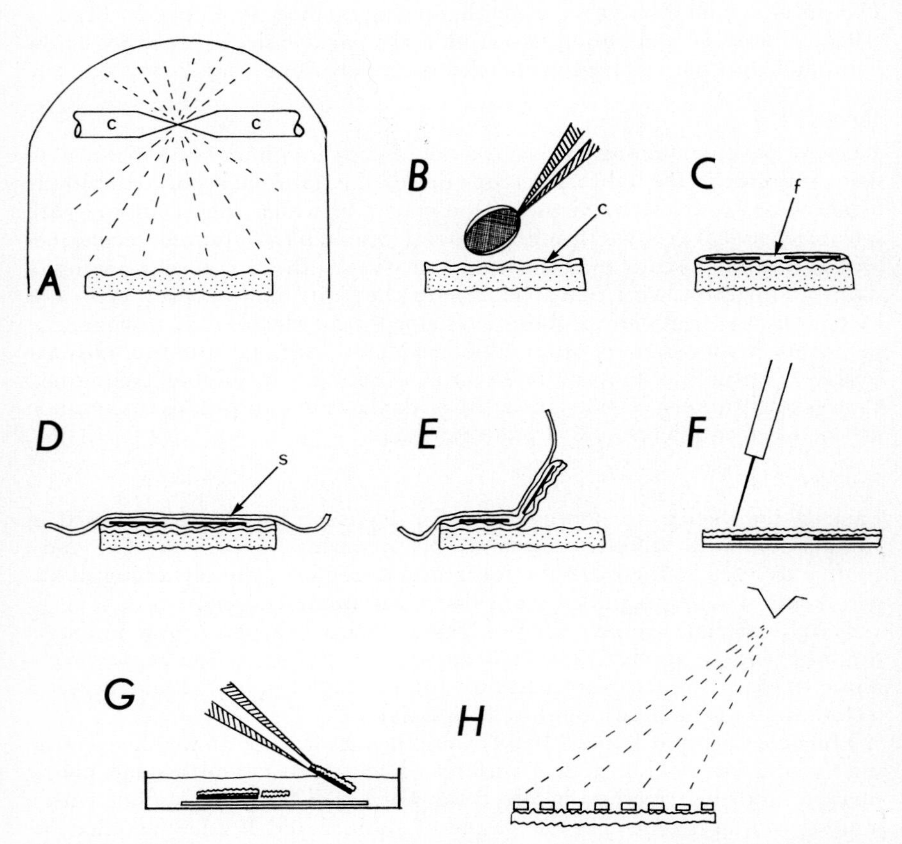

FIG. 2.8 Diagram of the carbon replica technique. In a vacuum chamber (***A***) about 15 nm of carbon is deposited on the fresh plant surface from a carbon arc (c). The distance from the arc to the leaf is about 15 cm. The vacuum is better than 10^{-3} mm Hg. The leaf is then removed from the vacuum. Specimen support grids (***B***) coated with 2% Formvar in chloroform are placed matt-side down on the carbon-coated surface. This surface is allowed to dry for a few seconds and then (***C***) the whole leaf is flooded with Formvar again (f) and allowed to drain and dry. A strip of cellulose adhesive tape (s) is pressed onto the Formvar layer (***D***). The tape is then pulled away from the leaf with the composite film and grids adhering to it (***E***). The film is cut with a sharp needle around the perimeter of the grids (***F***). The tape and grids are immersed in chloroform (***G***) and the grids lifted from the film. The remaining Formvar is washed from the grids with fresh chloroform. The replica can then be shadowed (***H***) as in Fig. 2.3. (Courtesy P. R. Williams)

Juniper (1968) is quicker, easier and effective on all plant surfaces so far examined. An outline of the technique is given in Fig. 2.8. Sometimes it is necessary to clean the carbon film in a bath of chromic acid to remove any dirt picked up from the leaf or any material which is not soluble in acetone or chloroform. The replica is finally shadowed with metal at a fairly high angle (45° is suitable for most plant surfaces) and viewed directly in the microscope.

The leaf or other plant material in the vacuum does not, surprisingly as it may seem, liberate gas vigorously enough to disturb the proper deposition of the carbon. This is probably due to the waxy cuticle effectively sealing the leaf except for the stomatal and petiolar apertures and cut edges. But for this the method would probably fail. The backed carbon layer does not stick strongly to the leaf and can be stripped away quite easily. The wax layer, which is almost always present to some extent on a leaf surface may be softened by the solvents used, and the wax/cell wall interface seems to form a natural fracture layer. The leaf or other plant surface does not appear to shrink to any measurable extent in the short time it is in the vacuum. The only possible source of artefact would seem to be the loss, under vacuum, of any volatile substances which might be present on the leaf surface.

The method is simple and reliable and, so far as our knowledge goes, there are no obvious inconsistencies between the results obtained by the replica method and such evidence as is given by the light microscope, the scanning microscope, or the very different preparative technique of freeze-etching (see below). The technique, however, does produce two different types of specimen, either a true replica or a pseudo-replica. The latter consists of a carbon replica with some or all of the original surface protrusions embedded in it. Treatment with strong organic solvents or chromic acid may remove much of the undissolved wax layer. A true replica and a pseudo-replica are shown in Figs. 2.9 and 2.10. However, the distinction is academic and in fact a pseudo-replica may be as informative as a true replica except, as can be seen, the undissolved portions of such a replica will appear as dense white regions in a negative print. In both cases white outlines to features are caused by large vertical thicknesses in the carbon film and must be interpreted accordingly. Despite the attractiveness of a replica and the apparent wealth of detail, it may not always be what it seems to be (Fig. 2.11, image P and diagram P_1). Here the replica has cast the surface of a solid rod of wax and by chance this rod lies with its long axis parallel to the electron beam. The vertical sides of the replica of the rod reinforce each other through the depth of the replica and hence show up as regions of high density (white in the micrograph), but the cap over the top is reinforced by nothing underneath and therefore gives in contrast the impression of a hollow rod, which it almost certainly is not. Were the depth of field of an electron micrograph not so great this difficulty would not arise. Unlike the light microscope in which the depth of field is similar to the resolution, the electron microscope has a depth of field very much greater than the resolution and about ten times greater than the maximum specimen thickness. This necessitates a completely different approach to the interpretation of electron and light micrographs.

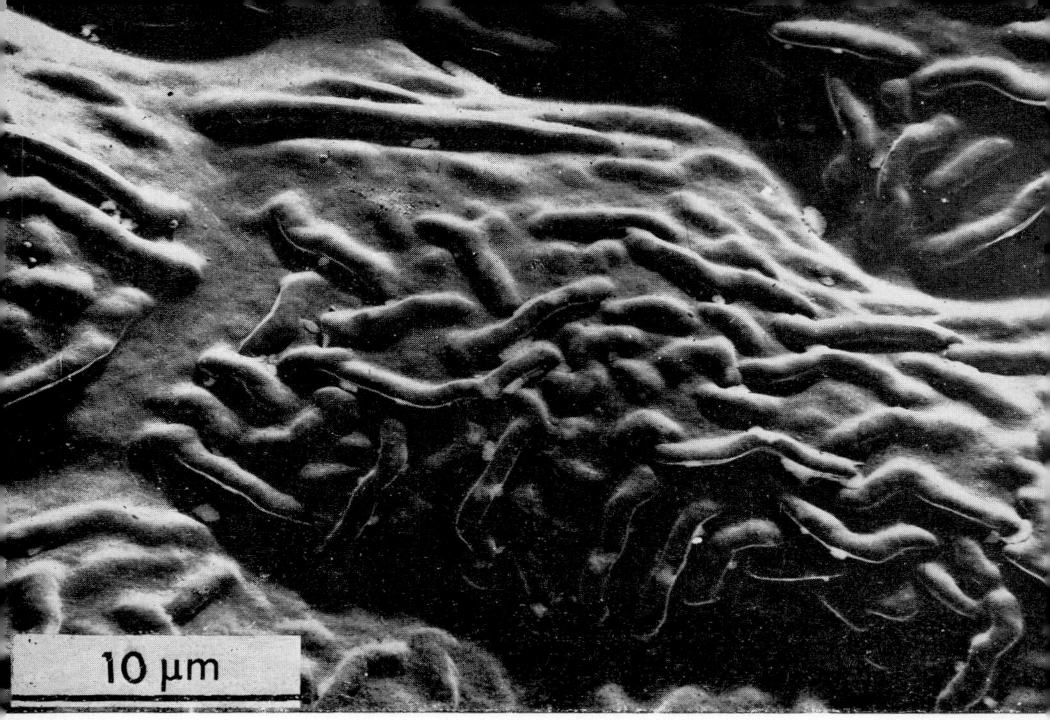

FIG. 2.9

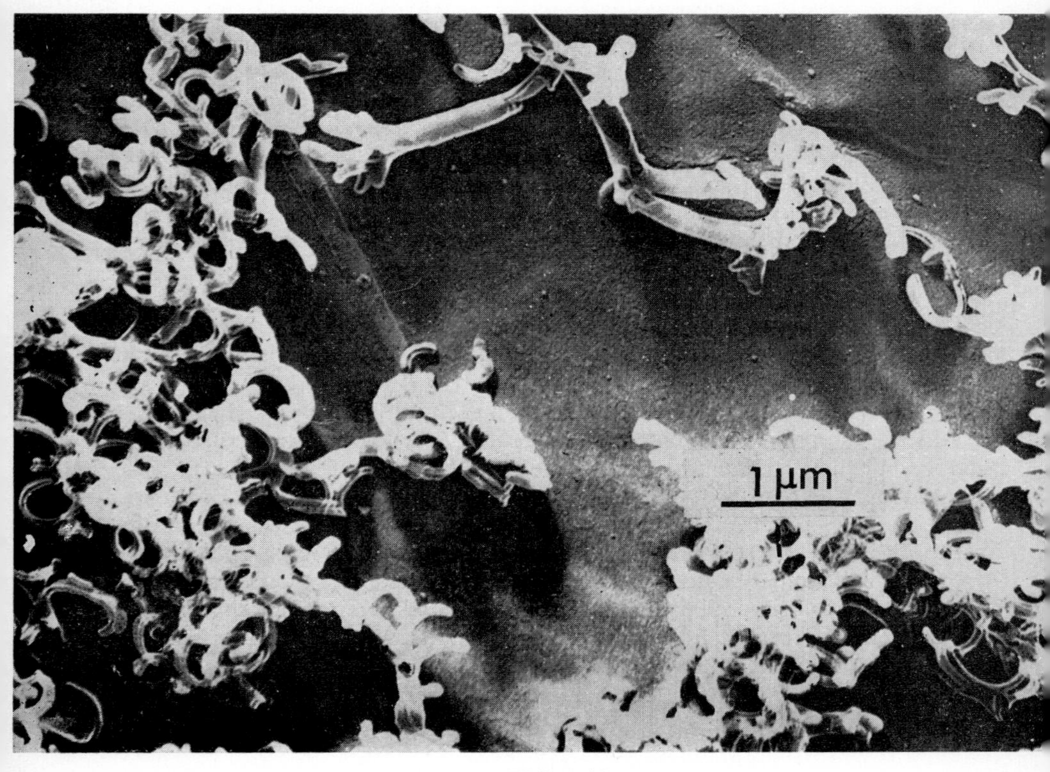

FIG. 2.10

The thickness of a replica or section in the electron microscope reinforces or may confuse the image presented to the eye, whereas in the light microscope only one tenth of the specimen may be in focus at any one time. Hence in an electron microscope section a good image will be obtained only if a membrane is cut exactly at right angles by the knife and is viewed exactly parallel to the electron beam (see Figs. 1.1M and 1.2M).

Sectioning

Tissue for sectioning is usually embedded in resins either of the methyl/butyl methacrylate type or aliphatic-based epikote resins such as 'Araldite' or 'Epon'. These are polymerised by the addition of accelerators, by heat, by ultra-violet light or combinations of these. The cross-linked polymers withstand cutting down to 50 nm, which is the thickness necessary for the penetration of the electron beam, whereas wax-embedded tissue can only be cut down to 1 μm. Sections for the electron microscope are cut on ultramicrotomes using glass or diamond knives; techniques of ultramicrotomy are discussed in detail by Glauert (1965). Some tissues, for example secondary wall material, are exceedingly difficult to section for the electron microscope.

The thickness of a section for the electron microscope (50 nm) means that few objects will be small enough to lie within a single section; the principal exceptions are, if correctly orientated, cellulose microfibrils which are 8–30 nm in diameter, ribosomes at 15 nm and most viruses. This means that most cell components may be cut by the knife in any plane and considerable difficulty may arise in the interpretation of images. Any membranes not cut exactly at right angles will show progressively more confused and less dense images as the angle away from that parallel to the beam is increased.

Electron microscopy of whole specimens is more or less self-explanatory and the only details of particular interest are those which concern the isolation of the individual objects, or the enhancement of contrast on or within the specimen, which have already been considered in the section on the achievement of contrast.

Freeze-etching

Freeze-etching is again essentially a surface technique in which the surface can be selected at will (Moor, Mühlethaler, Waldner and Frey-Wyssling, 1961) (Fig. 2.12). It is the most recent technique and perhaps the most difficult as far as interpretation is concerned. The technique so far has not been applied to any extent to the study of leaves and surfaces, but the absence of tissue manipulation which it involves and the apparent veracity with which it preserves membranes are bound to bring it into use in this field. Freeze-

---

FIG. 2.9 Electron micrograph of the surface of an adaxial epidermal cell from a leaf of horse-chestnut (*Aesculus hippocastanum*). Cr/Au/Pd

FIG. 2.10 Electron micrograph of the adaxial surface of a leaf of corn marigold (*Chrysanthemum segetum*). Cr/Au/Pd

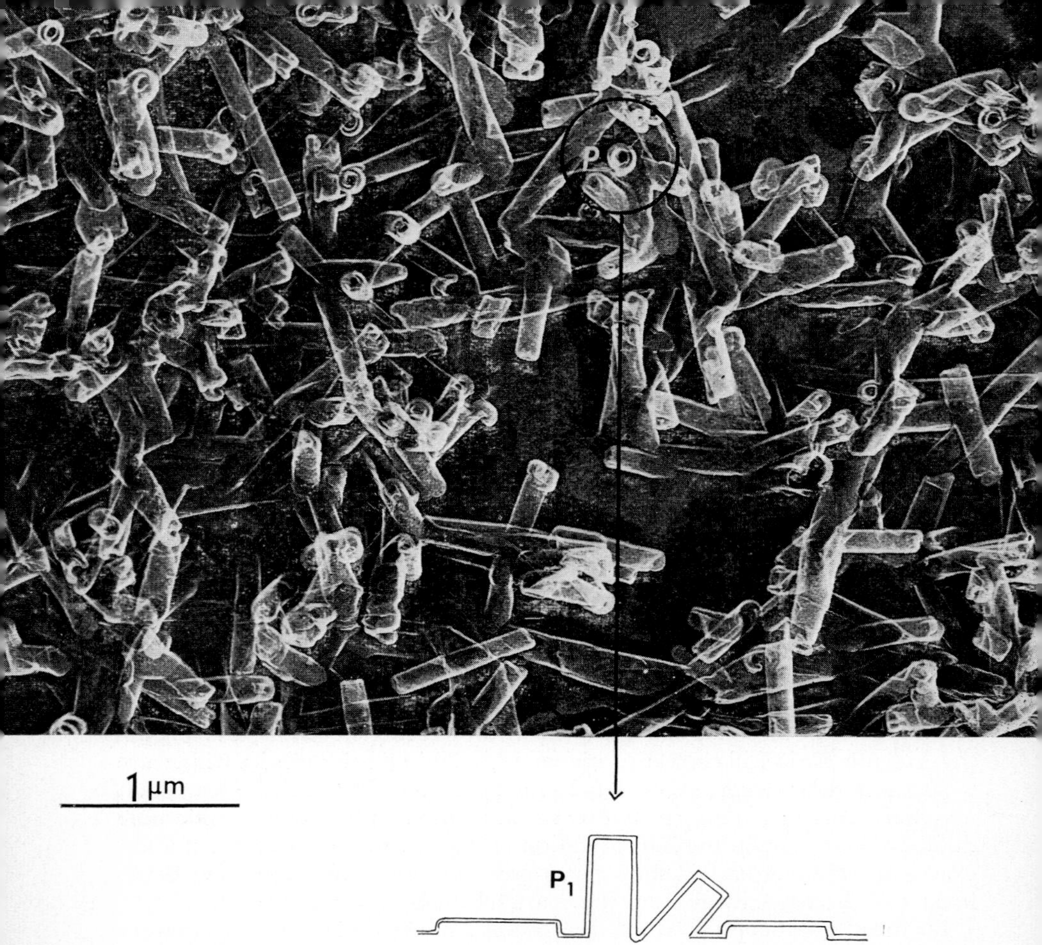

FIG. 2.11 Electron micrograph of the adaxial surface of a leaf of *Tulipa* sp. As the attached drawing indicates, the effect of replication is to suggest that what are probably solid rods (P in the micrograph) are hollow tubes (P_1 in the drawing). Cr/Au/Pd

---

FIG. 2.12 Electron micrograph of a frozen and etched adaxial cuticle of a leaf of white clover (*Trifolium repens*). *EW* shows the external wax projections. Note that, although the technique is entirely different, the structure of the wax projections is identical to that revealed by the replica and scanning techniques. *C*, the cuticle of the leaf; *MS*, the microfibrils of the epidermal cell wall; *WM*, the wax microchannels described on p. 112. (D.M. Hall, 1967b) F/E/Cr/Pt

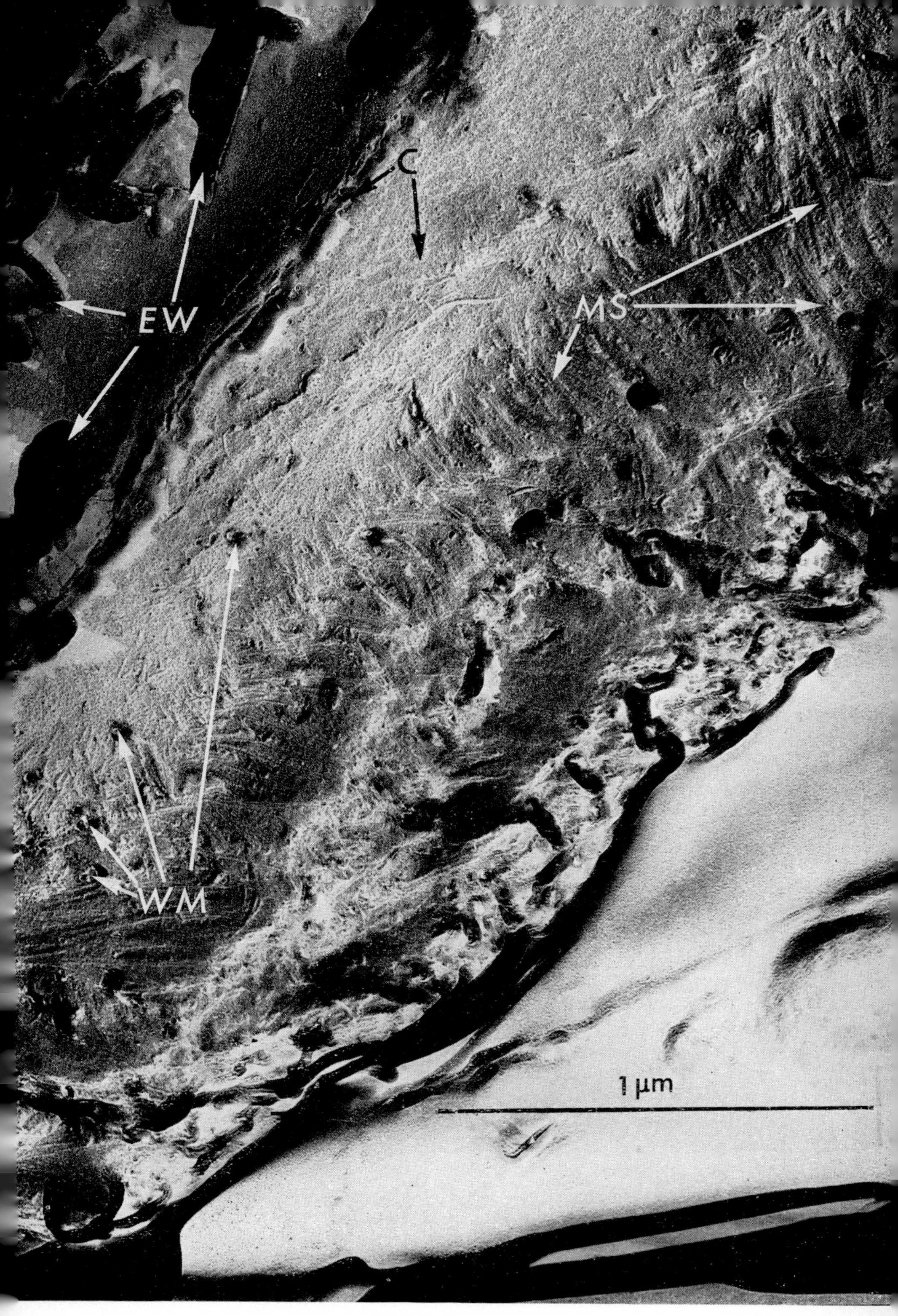

FIG. 2.12

etching overcomes the objections to using fixatives and embedding agents necessary for conventional sections. The information derived from the techniques has not contradicted, in any major feature, information from tissue replicated or fixed, stained and sectioned in the normal way. This supports the view that the basic structure of surfaces, walls and cytoplasm is faithfully preserved by such methods, and that many of the earlier fears of introducing artefacts are largely unfounded.

The tissue to be examined is either frozen directly or frozen after incubation for some time in glycerol and water to inhibit ice crystal formation and frozen too at a rate of fall of 100 °C per second so that the damage caused by ice crystals is reduced to a minimum. After freezing (Fig. 2.13 A) the tissue is placed still frozen in a vacuum and under vacuum is cut in a microtome (B). The surface is then etched by freeze-drying under high vacuum, to a depth into the tissue of a few tens of nm (C). The etched surface is then replicated in the usual way, still at about −100 °C and at a pressure of $2\text{–}3 \times 10^{-6}$ mm Hg (D). The vacuum is then broken, the replica floated away from the surface of the tissue on water and any organic material that remains on the surface of the replica is washed or etched away. The replica is picked up on a Formvar coated grid (E).

The tissue, frozen and under high vacuum, is splintered rather than cut by the knife. Thus the knife may never come in contact with much of the tissue surface. The knife may, in places, penetrate along lines of weakness into the tissue, thereby yielding cross-sections, and will commonly follow along membrane surfaces, thereby revealing images with a third dimension not possible with conventional sectioning techniques. A freeze-etched, replicated surface is therefore neither a section nor a replica of a complete surface, but an irregular marriage of them both (cf. Figs. 1.1, 2.6, 2.12 and 4.21). Since the plane of observation of a cuticle or a wall changes continually, interpretation is difficult and requires a great deal of experience.

Scanning electron microscopy

A new technique of using an electron beam to scan over a specimen, as opposed to transmission electron microscopy, has recently been developed (Smith and Oatley, 1955). This permits the direct examination of solid specimens, and in many cases is the easiest method to use. This is particularly so if the surface structure is very rough, has many re-entrant angles or has other characteristics that make impossible an examination in other types of microscope. The micrographs produced are similar to those produced by reflection light microscopy but with a better resolution and, like transmission electron microscopy, with a much greater depth of field (see Figs. 2.6 and 4.17). The resolution possible by the scanning technique lies between that of the transmission microscope and the light microscope (see Table 2.1, p. 33). For many specimens of which replicas cannot satisfactorily be made, the scanning electron microscope comes into its own (see Fig. 4.31).

In the apparatus, an electron beam is concentrated by condenser lenses to a diameter of about 10 nm. The beam is caused to move over a square section

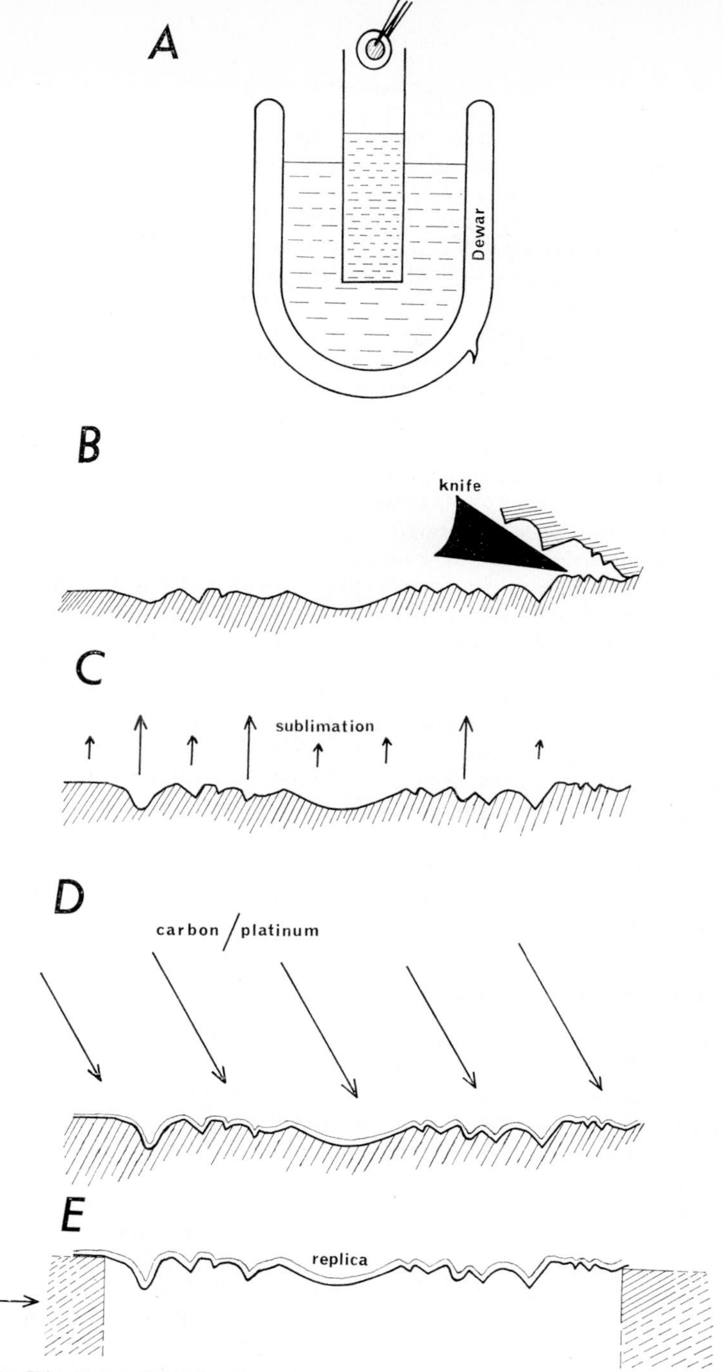

FIG. 2.13 Diagram of the freeze-etching technique. A, the specimen is frozen to −190 °C. B, the specimen at −180 °C and under vacuum is cut or fractured with a knife. C, the cut surface is allowed to sublime ('etch') at a slightly higher temperature (−95 to −100 °C), but still under vacuum. D, the specimen is recooled to −165 °C and the etched surface is then coated with a mixture of carbon and platinum. E, the carbon platinum replica is detached from the specimen, cleaned and mounted on a support grid for examination in the electron microscope

of the surface of the specimen in a zig-zag raster by the current from two pairs of reflecting coils. The same current also traverses the deflecting coils of the cathode ray tube to produce on its screen the zig-zag raster greatly magnified. The primary beam of electrons is partly absorbed and partly reflected. The image is produced by low-energy secondary electrons emitted from the surface as a result of the primary irradiation. The secondary electrons are picked up by an electron collector, fed to a scintillator-photo-multiplier and the collected signal produces the image on the screen. The brightness of each point on the screen is determined by the number of secondary electrons leaving the corresponding point on the surface of the specimen. Because of the point-to-point correspondence between the raster on the specimen and that on the screen, an image of the surface is built up on the screen. For biological specimens such as plant surfaces the useful maximum magnification is about ×20,000. An outstanding feature of the instrument is the great depth of field; although variable, for a similar magnification it is better than that of the light microscope by a factor of at least 300. To obtain the best resolution, non-conductive specimens need to be coated lightly and uniformly with a conductive substance such as gold/palladium.

The preparation of the specimen is normally very simple. A small piece is freeze dried and attached to a holder for coating, Amelunxen, Morgenroth and Picksak (1967) and Holloway (1967) show pictures of plant surfaces taken with the scanning electron microscope. Pictures of similar specimens obtained by the transmission and scanning techniques are shown in Figs. 2.6 and 2.7. With the exception of the better resolution obtained by the transmission technique there would not appear to be any significant difference between them.

Details of all these preparative techniques for transmission and scanning electron microscopy of plant tissue are given in Juniper, Cox, Gilchrist and Williams (1969).

Interference microscopy

Linskens (1966) and Linskens and Krner (1966) have devised a technique for examining the relief of plant surfaces under the light microscope. The method they use, interference microscopy, has normally been used for the study of metallic surfaces. A transparent replica of a plant surface is made with gelatin emulsion or a plastic sheet softened in acetone which is pressed on to the surface to take the impression. The replica is then placed in the so-called Zehender chamber, a mirrored surface with a coverslip, and the specimen is viewed under the interference microscope. Depending on the nature of the surface, interference pictures built up of lines of constant distance apart are obtained in the replica. In air the line distance is 0·54 μm, in water 1·5 μm and in certain oil mixtures 3 μm and 5 μm. The resolution of the method can thus be varied to suit the surface concerned. From the contour lines on the replica profile lines can be drawn which give the actual relief of the surface. The technique is non-destructive and is useful for recording the development and variation of the surface layer under different physiological conditions. It may also be used for the detection of pesticide residues. The use of the softened plastic sheet suffers from the defect that the solvent must destroy some, at least, of the fine structure of the surface.

Table 2.1 Resolving powers of microscopes and the human eye

Technique	Best resolution	Working resolution
The light microscope	0·25 μm	0·4– 1·0 μm
The scanning electron microscope	20·0 nm	30nm– 0·2 μm
The transmission electron microscope		
Gold/palladium shadowing	7·5 nm	8·0–10·0 nm
Platinum/carbon shadowing	1·8 nm	2·0– 2·5 nm
Sectioning	2·5 nm	3·0– 5·0 nm
Freeze-etching with platinum/carbon shadowing	2·5 nm	3·0– 5·0 nm
The human eye	80 μm	250 μm upwards

X-ray diffraction

Rays of light are bent at the edge of an opaque body and break up the edge of the beam, if it is monochromatic, into a series of light and dark bands. If it is not monochromatic the bands will be coloured. This is due to interference between the bent or diffracted rays. If the opaque body has a pattern as, for example, a grating consisting of opaque lines separated by transparent spaces, the diffraction of a monochromatic beam passing through it produces a diffraction pattern of black and white bands. These bands provide information about the dimensions of the pattern of the body if the wavelength of the light is known. This is true only if the wavelength of light (400–700 nm) is of the same order of size as the pattern in the body.

Diffraction phenomena occur with all kinds of electromagnetic waves from radio waves down to X-rays. The short wavelength of X-rays (less than 1 nm) has been exploited to give information about the three-dimensional lattice of atoms and molecules within crystals from the diffraction pattern produced on a photographic plate. The interatomic and intermolecular distances in biological material are of the same order of size (1·0–0·1 nm) as the wavelength of X-rays. For example the C—C distance in aliphatic compounds is 0·154 nm and the repeating distance of cellobiose units in cellulose is 1·03 nm. The wavelength of X-rays produced by accelerating electrons through a potential difference of 50 kV is 0·025 nm. Few biological solids are regular enough in their internal structure to give a diffraction pattern that can be interpreted, but DNA (deoxyribonucleic acid), a few proteins, cellulose and some plant waxes have had their structures revealed in this way. Even so, diffraction patterns may contain very many spots and the labour involved in interpreting these has restricted the use of this technique to a very few substances whose importance was thought to justify the labour involved.

The X-ray diffraction pattern is recorded on a photographic plate when the specimen that produces the diffraction is illuminated by a narrow beam of X-rays. The pattern is a series of concentric rings and spots or arcs around a centre point that marks the position of the direct beam. If the atoms or molecules are not regular in position the pattern appears as a series of concentric bands. The slightly imperfect arrangement of molecules in cellulose produces rings of arcs. Perfect orientation, as in some crystals, produces rings of spots. The distance between the spots and the central beam spot indicates the space between the repeating units of the pattern. The interpretation of the diffraction pattern is assisted by inserting heavy marker atoms chemically into the specimen. These may then be identified by their greater scattering power. Cellulose has been extensively examined (Preston, 1952) and leaf waxes and the suberin from cork have been examined by Kreger (1948, 1958). Kreger and Schamhart (1956) showed that the X-ray diffraction patterns of waxes scraped from plant surfaces give fairly accurate estimates of the nature and chain lengths of the chief constituents. The *n*-aliphatic long-chain compounds crystallise in monomolecular layers with the carbon chains vertical or in tilted positions. The chain lengths of the true wax esters determined by the method are C_{40}–C_{64}.

PHYSICAL AND CHEMICAL TECHNIQUES

Absorption spectroscopy

Absorption spectroscopy provides a rapid means of characterising the nature of individual compounds or classes of compounds obtained from cuticles by chromatographic or other processes. Its principles are as follows. If a beam of radiant energy impinges upon a substance or solution, the radiant energy may be absorbed entirely or in part. If the radiant energy is partially absorbed, the emergent beam when passed through a prism yields a spectrum which shows regions or bands of low transmission. Such a spectrum is called an *absorption spectrum*. Most organic and inorganic substances absorb and the absorption occurs at several characteristic wavelengths. The radiant energy commonly used in absorption spectroscopy consists of rays in the ultra-violet (0·185–0·38 μm), visible (0·38–0·78 μm) or infra-red regions (0·8–50 μm) regions. An infra-red range of 0·8–25 μm is the one normally exploited by commercial instruments.

Spectrometers for use with ultra-violet, visible or infra-red radiation are constructed of essentially the same basic units. Radiation from a suitable source passes through the sample to a monochromator which contains the prism and thence to a detector and recorder. In double beam spectrophotometers, one beam traverses the sample and the other a reference blank or solution. The sample may be presented as a dilute solution (in ultra-violet or visible light analysis) or as a concentrated solution, slurry or compressed admixture with a non-activated carrier such as potassium bromide (in infra-red analysis). The spectrum is obtained by plotting the percentage transmission against the wavelength λ (in microns) or the wavenumber v.

The wavenumber is the reciprocal of the wavelength and is expressed by $10^4 \div \lambda$ (in microns).

The absorption of radiant energy is a specific process related to characteristic molecular structures. The total energy of a molecule is made up of binding energy associated with the movement of electrons and kinetic energy associated with the vibrational and rotational activities of groups within the molecule. Absorption involves a transfer of energy to the absorbing medium; changes in the electronic energy give rise to absorption bands in the ultra-violet and visible regions, changes in the energy of vibrations of the atomic nuclei to bands in the near and middle infra-red (1–22 μm) region, and changes in energy associated with transition from one rotational state to another to bands in the infra-red region beyond 20 μm. The infra-red absorption is one of the most characteristic properties of a compound. Vibrating bonds of specific groups give rise to absorption bands at characteristic wavelengths. The infra-red absorption spectrum therefore provides a valuable means of identifying functional groups within a molecule of unknown chemical structure.

Infra-red analysis in the range 2·5–15·4 μm is of special value in identifying the components of waxes. Aliphatic compounds such as the paraffins give simple spectra and cyclic compounds such as the triterpenoid acids complex; characteristic spectra are shown in Fig. 2.14. The shape and

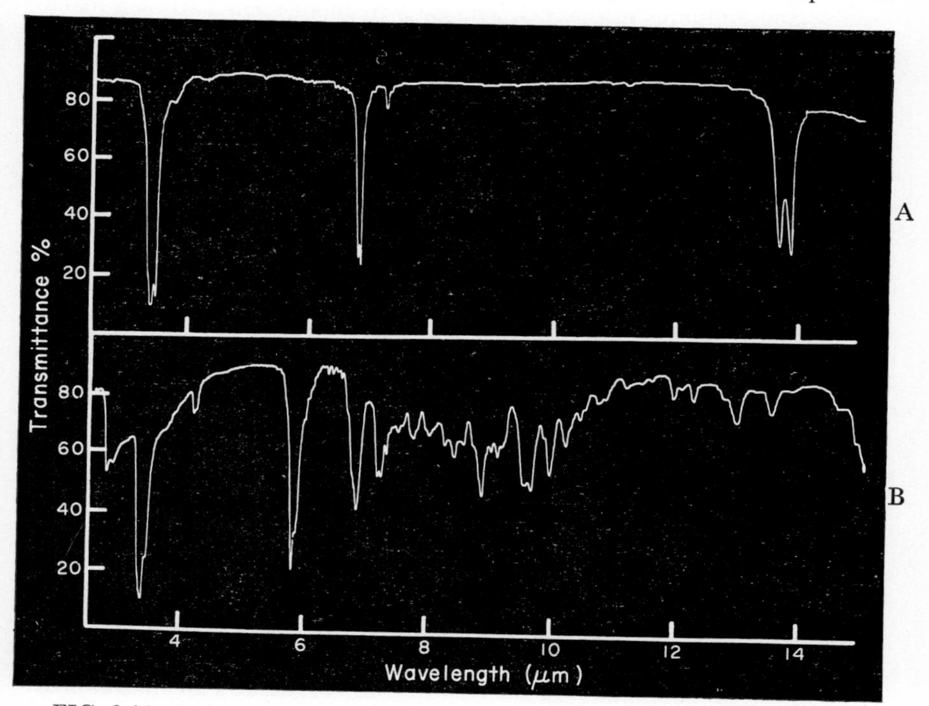

FIG. 2.14 Infra-red spectra of apple wax paraffins (*A*) and ursolic acid (*B*)

prominence of an absorption band in the range 3·3–3·6 μm indicates the type of C—H bond present. An O—H group gives a band at about 2·8–3·2 μm; if the group is free or unassociated the band is sharp. The C=O group in its various forms absorbs over the range 5·5–6·1 μm. If an O—H group is attached, as in a carboxylic acid, the band due to C=O occurs at 5·9–6·0 μm; peaks due to an ester C=O group are generally in the region of 5·7–5·8 μm. The spectra are most valuable when interpreted in the light of other physical and chemical properties of the material under examination.

Chromatography

Chromatography, now about 100 years old, in principle is a process by which a mixture of substances is resolved into its components by passage through a fixed sorbent bed. The method achieved prominence from the work of Tswett (1906, 1907) who observed the formation of sharply defined coloured bands when a mixture of plant or animal pigments in solution in an organic solvent was passed through a column of inulin. The term 'chromatography' was coined at this time and, although inconsistent with later practices involving the separation of colourless materials, has been retained. The pioneering work of Tswett and others with naturally occurring mixtures led to standard laboratory techniques for the examination of many kinds of organic and inorganic substances. The early methods were used on a macro scale; modern techniques employ a wide range of sorbent beds and eluting solvents and are capable of separating quantities of materials on the microgram scale.

The sorbent bed must be chemically unreactive with the mixture and the agent effecting the passage of the mixture through the bed may be a liquid or gas. The bed may take the form of a column of material held in a tube or a layer upon a plate. Two phases are involved; the stationary, sorptive phase and the mobile, eluting phase. Two kinds of chromatography are recognised. In the first, called *adsorption chromatography*, the separation of the components of a mixture is brought about by their differential adsorption upon a stationary phase which is a solid. In the second, known as *partition chromatography*, separation is effected by combining in one technique the principles of solvent partition and adsorption chromatography. When a substance in solution in a solvent is shaken with another solvent immiscible with the first, the solute distributes itself between the liquid phases according to its nature and the solvents used. In partition chromatography one solvent is held upon the sorbent bed to constitute the stationary phase and the other provides the mobile phase. The separation of the components of a mixture is then achieved by their differential partition between the stationary and mobile phases.

The phases used in adsorption chromatography are liquid-solid or gas-solid. The simplest form, as used by Tswett, consists of a column of sorbent such as alumina, cellulose powder, silica gel (silicic acid) or kieselguhr through which the mixture, in a solvent, is run. Each band is removed from the column and the adsorbed material extracted from it or, more commonly, the column is washed with a series of solvents of increasing power of elution and the fractions collected. Gas-solid chromatography is now little used; in this, the mixture is injected into a gas stream which ascends a sorbent bed.

Partition chromatography was developed by Martin and Synge (1941) who used silica gel containing water as the stationary phase and water-immiscible chloroform-butanol as the mobile to effect a separation of amino acids. A process involving water or a hydrophilic solvent held upon a sorbent bed and a mobile, water-immiscible or hydrophobic solvent is referred to as *normal partition* chromatography. The phases may be reversed; a hydrophobic material is supported on the bed and a hydrophilic solvent used as the mobile phase. This system is known as *reversed-phase partition chromatography*.

Paper chromatography

Paper chromatography was developed by Consden, Gordon and Martin (1944) who showed that sheets or strips of cellulose could be used alone or to support a stationary liquid phase. The method offered a great advantage over column chromatography in that a small amount of a mixture could be resolved. It thus became widely used. A drop of the mixture is applied at a point near one end of the impregnated paper and separation into components is effected by the mobile phase moving through the paper either upward by capillary attraction or downward under gravity—the ascending or descending techniques. The extent of movement of a component is expressed by an R_F value, which indicates the distance travelled by the component in relation to that travelled by the eluting solvent. The R_F value is dependent upon the nature of the component and the characteristics of the system used.

Thin-layer chromatography

Paper chromatography has now largely been superseded by *thin-layer* or *laminar chromatography*. The first recorded attempt to use an adsorbent in the form of a layer was by Izmailov and Shraiber (1938). Later, attempts were made to improve the separation of lipid mixtures by using silica gel supported on paper, but this acted as if it were a layer of silica gel alone. This led to the use of silica gel upon a glass plate; a binder, e.g. calcium sulphate, was needed to make the silica adhere. A micro technique similar to paper chromatography was then possible and showed advantages in quickness of separation and more confined locations of the separated components. Stahl, Schraeter, Kraft and Renz (1956) investigated the factors effecting resolution, and the introduction of new systems of absorbents and mobile phases led to dramatic developments in the analysis of lipid and other materials. If one solvent is unable to resolve a mixture of substances it is often possible to effect a separation by eluting, after drying, with a second solvent moving at right angles to the direction of the first.

The locations of the components on a test paper or thin-layer plate are detected by a variety of methods, e.g. by heating after spraying with chlorosulphonic-acetic acid or by applying a suitable staining reagent. Standard reference compounds may be run on the paper or plate alongside the mixture under test. After determining the positions of the components, quantitative measurements may be made by methods based, for example, on spot area and optical density, spectroscopy, fluorimetry (for fluorescent compounds) or, when using radioactive material, autoradiometry. The thin-layer technique has an important application in analysis by ***electrophoresis***.

Thin-layer chromatography may also be used for separating mixtures into components on a preparative scale. A band of the mixture is eluted through a thicker layer of sorbent concurrently with marker spots, and the separated zones are scraped off and extracted. A series of plates may be used to augment the yields of components.

Adsorption chromatography is usually used to separate the classes of compounds present in lipid mixtures and reversed-phase partition chromatography not only to separate the classes but also to identify individual compounds within a class. Both methods have been used with great success for the analysis of cuticular waxes which contain long-chain hydrocarbons, alcohols, ketones, acids and esters.

In a process of separation by adsorption, the hydrocarbons are little if at all adsorbed and thus migrate faster. Adsorption affinity is increased by the introduction of a functional group into an hydrocarbon chain in proportion to the polarity which it confers; thus esters migrate slower than hydrocarbons, ketones slower than esters, alcohols slower than ketones and acids least of all. In reversed-phase partition chromatography, the affinity of the hydrocarbon chain for the stationary lipophilic phase increases resistance to elution, whereas the presence of a polar group in the chain augments affinity for the mobile hydrophilic phase and so aids elution; thus the movement of the hydrocarbons is retarded and the acids migrate fastest.

The type of system used and the balance between the polar and non-polar characteristics of a compound or a class of compounds are thus the chief factors influencing its mobility. Other factors, such as the degree of ionisation of acids in the eluting solvent, play a part. The principles underlying resolution by chromatography are demonstrated in Figs. 2.15 and 2.16, which show the differences in mobility, under different systems, of hydroxy-fatty acids obtained from cutin. Figure 2.15 shows the separation of 9,10,18-trihydroxyoctadecanoic, 10,18-dihydroxyoctadecanoic, 18-hydroxy-octadecanoic and 10,16-dihydroxyhexadecanoic acids by adsorption chromatography on a silica gel layer with chloroform-acetic acid (9:1) as eluting solvent. Under these conditions ionisation of the acids is suppressed and the influence upon mobility of the polarity/non-polarity balance, as expressed by the number of hydroxyl groups present in relation to chain length, is seen. The order of mobility of the acids is hydroxyoctadecanoic (1 OH group, 18 C atoms) >dihydroxyoctadecanoic (2 OH groups, 18 C atoms) >dihydroxyhexadecanoic (2 OH groups, 16 C atoms) >trihydroxy-octadecanoic (3 OH groups, 18 C atoms). The affinity for adsorption from the hydrophobic solvent diminishes as the chain length dominates the polar characteristic. Figure 2.16 shows the resolution of the acids by reversed-phase partition chromatography, in this case on paper impregnated with castor oil and with 60% acetic acid as the mobile phase. An opposite effect is seen; tri-hydroxyoctadecanoic acid moves faster than dihydroxyhexadecanoic, and

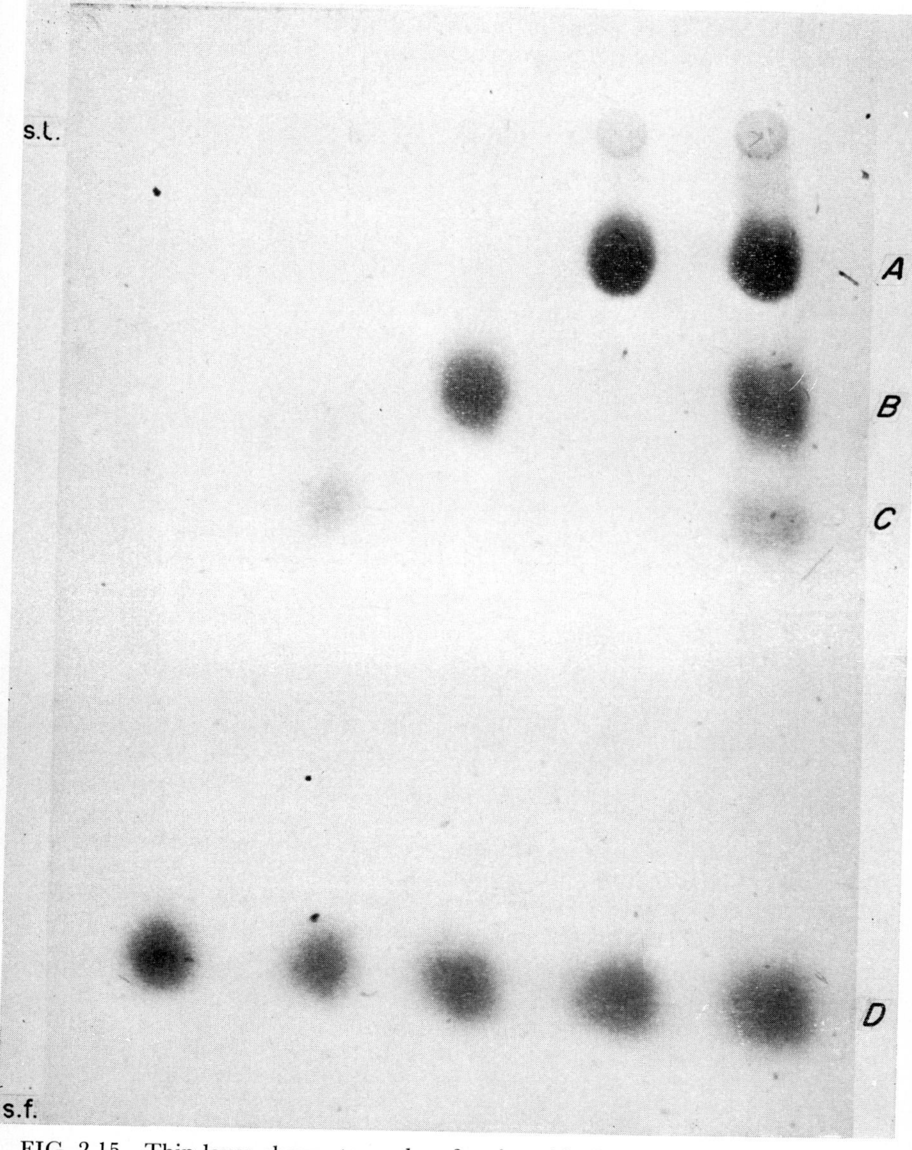

FIG. 2.15 Thin-layer chromatography of cutin acids. Kieselgel HR, mobile phase chloroform-acetic 9:1. s.l. starting line; s.f. solvent front. ***A***, 9,10,18-trihydroxyoctadecanoic acid, R_F 0.16; ***B***, 10,16-dihydroxyhexadecanoic acid, R_F 0.30; ***C***, 10, 18-dihydroxyoctadecanoic acid, R_F 0.38; ***D***, 18-hydroxyoctadecanoic acid, R_F 0.82

this faster than dihydroxyoctadecanoic. Here, the affinity for adsorption on the lipophilic stationary phase increases as the chain length dominates the polar characteristic. ***Vicinal*** dihydroxy acids occur in the form of ***threo*** and

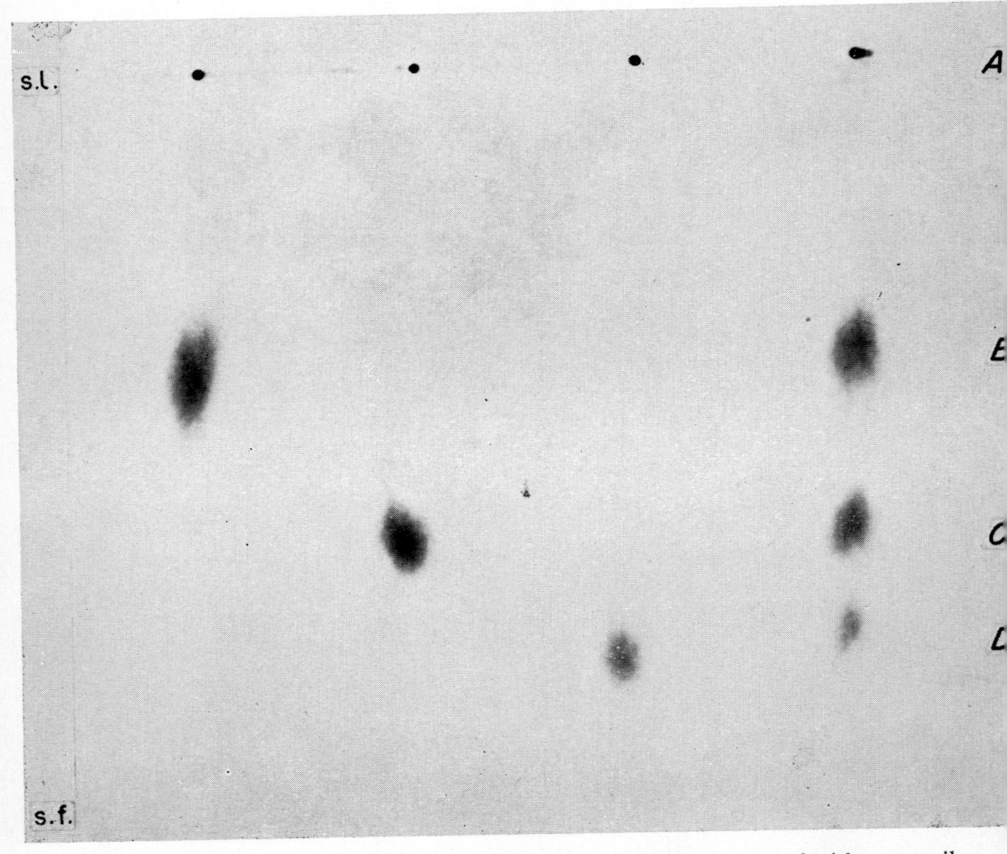

FIG. 2.16 Paper chromatography of cutin acids. Paper impregnated with castor oil, mobile phase 60% acetic acid. s.l. starting line; s.f. solvent front. ***A***, 18-hydroxy-octadecanoic acid, R_F 0·02; ***B***, 10,18-dihydroxyoctadecanoic acid, R_F 0·33; ***C***, 10,16-dihydroxyhexadecanoic acid, R_F 0·52; ***D***, 9,10,18-trihydroxyoctadecanoic acid, R_F 0·64

erythro stereoisomers; the *threo* and *erythro* isomers of 9,10,18-trihydroxy-octadecanoic acid have been found in the products of hydrolysis of apple cutin (Eglinton and Hunneman, 1968). The methyl esters of *threo* and *erythro* isomers of long-chain polyhydroxy acids can be separated on a layer of silica gel impregnated with boric acid, sodium borate or sodium arsenite; some form of chemical ***chelation*** with the impregnating agent is considered to be one factor, and probably the main factor, in producing the different patterns of migration (Morris, 1963).

Purdy and Truter (1963c) employed a combination of adsorption and reversed-phase partition chromatography for the isolation and identification

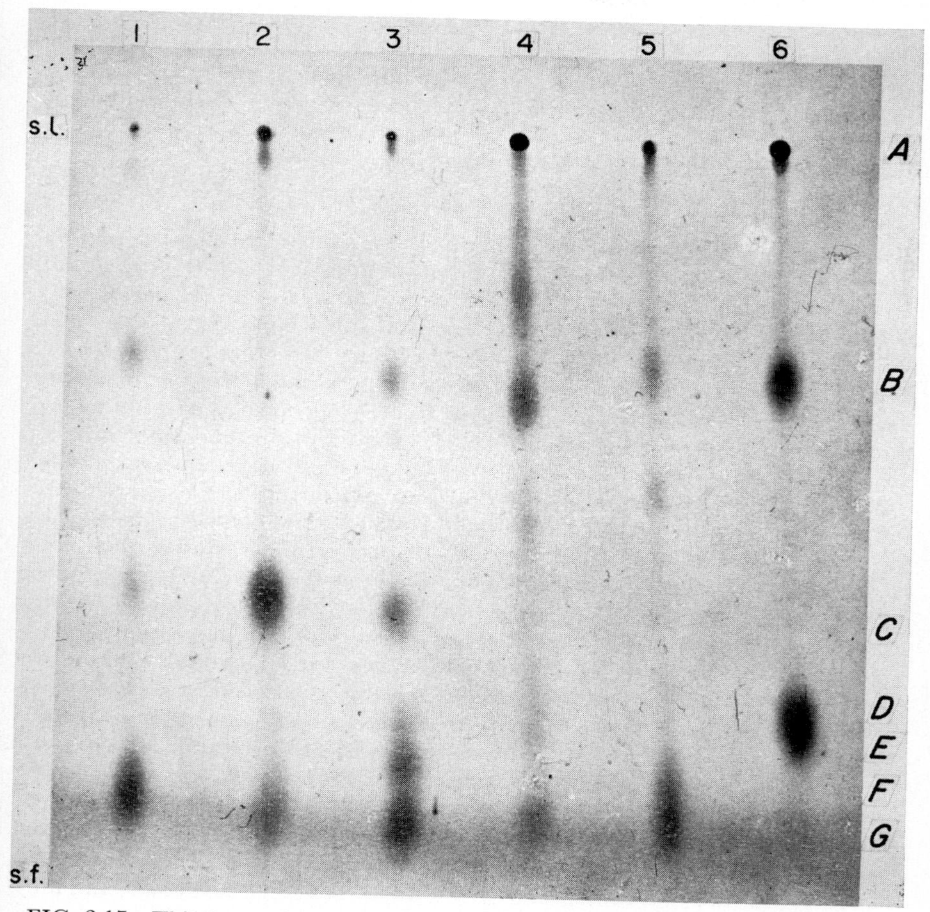

FIG. 2.17 Thin-layer chromatography of plant waxes. Kieselgel G, mobile phase methylene dichloride. s.l. starting line; s.f. solvent front. 1, pea; 2, nasturtium; 3, cabbage; 4, carnauba; 5, candelilla; 6, sugar cane. ***A***, fatty acids; ***B***, primary alcohols; ***C***, secondary alcohols; ***D***, aldehydes; ***E***, ketones; ***F***, esters; ***G***, hydrocarbons

of the components of the surface wax of the cabbage leaf. Holloway and Challen (1966) found adsorption thin-layer chromatography to be a valuable technique in the study of long-chain aliphatic and of cyclic constituents in natural waxes. Four systems employing two types of adsorbent and four eluting solvents were designed and, by using the systems in combination, over 60 waxes were successfully resolved into their constituent classes of compounds. The resolution of 6 surface waxes (pea, nasturtium *Tropaeolum majus*, cabbage, carnauba, candelilla and sugar cane) is shown in Fig. 2.17. The components were located by treatment with chlorosulphonic acid reagent and heating at 160 °C for 30 min. The pea wax consists chiefly of alkanes, with esters and primary and secondary alcohols. The nasturtium wax has a high content of secondary alcohols, with some alkanes and fatty acids. Alkanes, esters, ketones, primary and secondary alcohols and fatty acids are revealed in the cabbage wax. The candelilla wax has a high content of alkanes; the carnauba wax a low. The sugar cane wax contains aldehydes, primary alcohols, some fatty acids and a trace of alkanes.

Gas chromatography

Gas-liquid partition chromatography, now widely used, was developed by James and Martin (1952) who employed as sorbent a non-volatile liquid supported on an inert solid. When a mixture is injected into the gas stream, components move through the column of sorbent at rates dependent upon their respective volatilities and interaction with the non-volatile liquid phase. Components with the greatest solubilities in the liquid phase are retarded. A satisfactory separation depends upon a correct choice of the stationary liquid phase, a supporting solid of small, even mesh size and large surface area, satisfactory dimensions of the column, temperature and a suitable rate of gas flow. Properties such as the relative polarities of the components of the mixture and the stationary liquid, and interactions between them such as hydrogen bonding and other cohesion forces affect the resolution obtained.

In its basic units, the apparatus as normally used consists of a gas supply and flow control, a port for the injection of the mixture, a column of sorbent bed, a detector of the components as they emerge with the carrier gas, and a recorder. The components of the mixture and the liquid phase must be stable at the temperature used. The supporting solids consist of materials such as processed brick dust, diatomaceous earth, silicone-rubber or synthetic polymer; typical phases are hydrocarbon oils, silicones and polyglycols. Nitrogen and helium are the usual carrier gases. Various types of detector are used, according to the nature of the material under examination. Flame ionisation detectors are especially suitable for the analysis of wax components such as hydrocarbons. The effluent from the column mixed with hydrogen is burned in the detector unit, at the tip of a metal jet in an excess of air, and the effect is recorded by an electrometer system in a separate unit. The flame ionisation detector is essentially a carbon counting device which produces a current proportional to the number of ions or electrons formed in the flame. The detector is highly sensitive; the emergence of components is indicated by peaks on the chromatogram produced by the recorder, and quantitative assessments are made from the heights or areas of the peaks. The volatility of compounds can be improved by the formation of derivatives; fatty or hydroxy-fatty acids, for example, may be more satisfactorily resolved in the form of their methyl esters. This aspect of analysis by gas chromatography is discussed more fully on p. 50. Figure 2.18 shows the separation achieved of a mixture of the methyl esters of 16-hydroxyhexadecanoic acid, 10,16-dihydroxyhexadecanoic acid, 18-hydroxy-9,10-epoxy-octadecanoic acid and 9,10,18-trihydroxyoctadecanoic acid which are components of cutin.

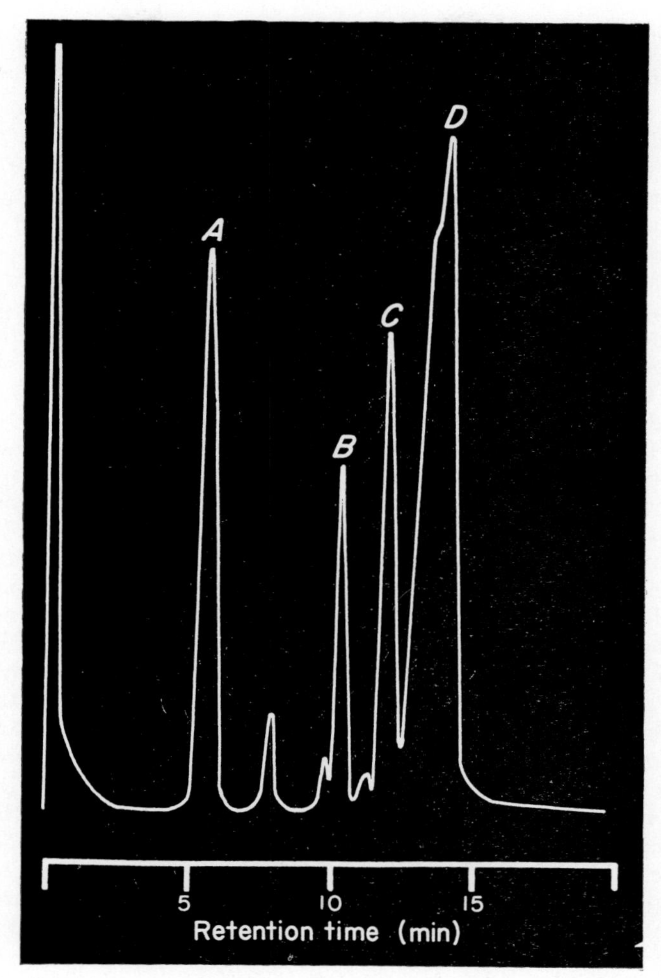

FIG. 2.18 Gas chromatography of cutin acid methyl esters. Column 6 ft × $\frac{1}{8}$ in., 1·5% OV 1 on 70/80 mesh Chromosorb G, temperature programmed 150–295 °C, carrier gas nitrogen, flame ionisation detector. ***A***, 16-hydroxyhexadecanoic; ***B***, 10,16-dihydroxyhexadecanoic; ***C***, 18-hydroxy-9,10-*epoxy*-octadecanoic acid; ***D***, 9,10,18-trihydroxyoctadecanoic acid

Gas-liquid chromatography may be used to obtain pure compounds on a preparative scale by increasing the capacity of the column, by automatically feeding the column continuously or by replicating the number of columns.

The principles and application of chromatography are fully discussed by Lederer and Lederer (1954), Truter (1963), Giddings and Keller (1965) and Stahl (1965); those of gas chromatography by Littlewood (1962) and Nogare and Juvet (1962); and those of absorption spectroscopy and gas chromatography by Willard, Merritt and Dean (1965).

Gas chromatography—mass spectrometry

Mass spectrometry has long been used to determine the structure of organic compounds and for the analysis of volatile chemical mixtures. In the mass spectrometer, the organic molecules are bombarded with electrons and transformed with characteristic series of positive ions due to the rupture of the chemical bonds in the molecules. Electrical signals given by ions of differing mass/charge (m/e) ratio are recorded to give the mass spectra. Data sheets of the mass spectra of known organic compounds have been published and are used in the identification of unknowns. The use of a computer has greatly assisted the identification of unknown compounds by permitting the ready comparison of their mass spectra with those of reference compounds stored on magnetic tape. Nagy, Modzeleski and Murphy (1965) used mass spectrometry in the examination of the hydrocarbon fraction of banana leaf wax. The mass spectra indicated approximately equal amounts of saturated aliphatic and of alkyl-substituted aromatic compounds.

The combination of gas chromatography with mass spectrometry has provided an analytical technique of great potential. Gohlke (1959) showed that it can rapidly and completely characterise organic chemical mixtures boiling below 350 °C. Lindeman and Annis (1960) used a gas chromatograph coupled directly with a conventional magnetic field mass spectrometer and showed that the system can identify and measure the components in multicomponent chromatographic peaks. The technique will undoubtedly become a major tool in research on cuticle components. The use of thin-layer chromatography in conjunction with combined gas chromatography and mass spectrometry permits the resolution, identification and determination of the components on a micro scale. Adsorption thin-layer chromatography resolves a complex mixture into its constituent classes of components, gas chromatography separates and measures the components of each class, and mass spectrometry identifies each component as it emerges from the gas chromatograph. Eglinton and Hunneman (1968) used combined gas chromatography–mass spectrometry in the analysis of hydroxy-fatty and fatty acids derived from apple cutin. For the determination of the acids present to the extent of 5% or more the procedure is potentially applicable to samples as small as 0·5 mg of cutin.

The isolation and analysis of cuticles

Considerable progress has been made during the past ten to fifteen years in the quantitative determination of the components of plant cuticles. Previously, quantitative data on cuticles were comparatively few and largely confined to those which could be easily isolated. Early chemical measurements on leaf cuticles include those of Skoss (1955) who extracted isolated cuticles with alcohol to obtain the wax and weighed the residue to assess the cutin, of Matic (1956) who resolved the constituent acids of *Agave* cutin, and of Schieferstein and Loomis (1959) who determined changes in the weights of cuticular membranes with increasing age of leaves. The growing awareness of the importance of cuticles in many surface phenomena and the development of rapid and sensitive analytical methods stimulated investigations, and the detailed compositions of many cuticles are now known. The methods in principle involve the isolation, breakdown and identification of the components of the surface and occluded fractions of the cuticular wax, the separation of the cuticular membrane from the cellular tissues, the removal of non-cutin components of the membrane and the degradation of the residual cutin and identification of its constituent acids. Among the methods used for the fractionation and identification of components chromatographic and spectroscopic procedures figure prominently.

The isolation of surface wax

Surface waxes may be removed, uncontaminated by cytoplasmic constituents, by a short-period immersion of the plant material in a suitable solvent. Surface waxes differ greatly in amount and in the relative proportions of aliphatic and aromatic and of polar and non-polar constituents. The suitability of a solvent depends upon the chemical composition of the wax; a non-polar solvent may efficiently remove a paraffinic type of wax, but not a wax containing a large proportion of polar compounds. Chloroform has the merit of dissolving, with few exceptions, the compounds known to occur in surface waxes and has been widely used in the immersion method.

Roberts, Martin and Peries (1961) showed that four successive immersions, each of 10 sec duration, of apple leaves in chloroform at room temperature removes the surface wax, uncontaminated by cellular constituents, more efficiently than similar treatment with ether, carbon tetrachloride, benzene, petroleum ether or *n*-hexane. The least satisfactory solvents for this type of wax were petroleum ether and *n*-hexane. The removal of wax from pea, kale (*Brassica oleracea* var. *acephala*) and nettle (*Urtica* sp.) leaves by immersion in solvent was examined by Dewey, Hartley and MacLauchlan (1962). Chloroform was more efficient than *cyclo*hexane, and the immersion procedure discriminated sharply between superficial substances which dissolved in up to 20 sec and others, in smaller quantities, which were extracted more slowly and with decreasing speed over a period of many hours. Eglinton, Gonzalez, Hamilton and Raphael (1962) isolated wax from leaves by three short-duration immersions in chloroform and demonstrated the efficiency of the method for the species examined. Tribe (1967) found that treatment of leaves of Gramineae with chloroform for 10–30 sec removed most of the surface wax, the short period being necessary to reduce contamination by cellular components; chloroform sprayed on to the leaves for 10–15 sec was equally efficient. Purdy and Truter (1961, 1963a) confirmed that four immersions, each of 10 sec, of leaves in ether remove surface waxes uncontaminated by cellular lipids. Materna and Ryšková (1966) isolated surface wax from spruce (*Picea* sp.) needles by an eight-fold washing with ether; additional washing was attended by the risk of extracting ether-soluble substances from within the tissues. Wax was obtained by washing the leaves of *Nicotiana glauca* and *Agave americana* three times with petroleum ether, and 'non-wax' constituents were removed from the extract with acidified alcohol (Schieferstein and Loomis, 1959). This procedure was adopted by Daly (1964) for the isolation of wax from the leaves of *Poa colensoi*. Wax may be obtained separately from the ***abaxial*** and adaxial surfaces by washing each in turn, with a suitable solvent (Baker, Batt, Silva Fernandes and Martin, 1964; Silva Fernandes, Batt and Martin, 1964).

In work on the waxiness of leaf surfaces in relation to water repellency Silva Fernandes (1965a) restricted the washing procedure to two immersions in chloroform, each of 2 sec, to ensure that only constituents located at the surface of the waxy coverings were obtained. Hallam (1967) found that the procedure inadequately removed the wax from *Eucalyptus* leaves and additional extraction by refluxing with petroleum ether was used. Lamberton (1964) and Horn, Kranz and Lamberton (1964) obtained the waxes from *Eucalyptus* and *Angophora* leaves by suspending them over boiling petroleum ether; the condensing vapour effectively washing the surfaces.

Surface waxes may be removed from fruits by similar procedures. Horrocks (1964) immersed apple fruit cuticle in ether; the wax obtained amounted to about 50% of the total cuticular material. The waxy deposits on apple fruits, or grapes (*Vitis vinifera*), are ten times heavier than those of the leaves and to ensure satisfactory recovery more drastic treatment is needed. The removal of wax by four successive immersions, each of 20 sec, of whole fruits in chloroform is not so clean-cut as from leaves (Batt and Martin, 1961) and the washing of cuticular discs, carrying minimum amounts of flesh, gives similar results (Batt and Martin, 1966). Four 10-sec immersions in chloroform were used by Dudman and Grncarevic (1962) to remove the surface wax structures from grapes, but Chambers and Possingham (1963) showed, by the electron microscope, that this treatment was inadequate. The wax was removed completely by hot chloroform (Radler and Horn, 1965).

The above procedures are designed to remove the waxes that are located on the surfaces of the cuticles of leaves or fruits. From the known surface area of the plant material used, an assessment is made of the amount of wax present on a unit area of surface. Waxes embedded within the cuticular membranes are more difficult to measure; they are obtained by extraction of the membranes after they have been separated from the cellular tissues. This approach, the only one possible, pre-supposes that no loss of wax occurs from the membranes during their isolation. In the separation of surface waxes from leaves, if the treatment is too drastic, contamination by cytoplasmic constituents is likely. In the recovery of the surface waxes of fruits, if the treatment is inadequate, wax is left on the surface; if too severe, wax is extracted from within the membrane. The isolation of the cuticular waxes as separate fractions has provided information of some interest; the fraction of apple wax removed by the first immersion in chloroform contains the bulk of the paraffins and the fractions obtained by the additional extractions are rich in the triterpenoid constituents (Silva Fernandes, Batt and Martin, 1964).

The isolation of the cuticular membrane

Strongly acidic substances were used from 1800 onwards to digest the cellular tissues of leaves and so release the acid-resistant membranes. The materials used included nitric and chromic acids, singly or as mixtures. These were gradually replaced by milder chemical reagents. Sando (1923) used dilute hydrochloric acid to remove the flesh from apple membranes, and Huelin and Gallop (1951a) a dilute solution containing ammonium oxalate and oxalic acid. This reagent causes the cellular tissues to disintegrate by dissolving pectin and has been extensively used to isolate cuticular membranes from leaves and fruits (Mazliak, 1958; Roberts, Batt and Martin, 1959; Schneider, 1960; Mazliak 1962, 1963d; Goodman and Addy, 1962b). The isolated membranes carry attached remnants of the epidermal cells which may be removed by treatment with zinc chloride-hydrochloric acid or cuprammonium solution. Holloway and Baker (1968b) effectively isolated cuticular membranes from leaves by direct treatment with zinc chloride-hydrochloric acid solution. Letham (1958) reported that ethylenediamine tetra-acetic acid (EDTA) has the ability to dissolve pectic substances of the ***middle lamella*** and this reagent was used by Schneider (1960) to obtain the membranes of apple fruits. Roelofsen (1952) used dilute hydrogen peroxide or ammonium oxalate solution to remove the cuticles from leaves of *Clivia nobilis*. The procedure using oxalate solution usually liberates the membranes within a few hours and is of particular value when the major interest lies in the cutin component of the membranes.

Alternatively, cuticular membranes may be isolated by enzymic treatment in a process akin to retting. This, although more time-consuming, is less drastic than chemical treatment and preferable in quantitative work. Skoss (1955) digested leaves of *Nicotiana glauca* anaerobically with *Clostridium roseum* obtained from the intestinal flora of a herbivore and Orgell (1955) obtained the membranes from *Convolvulus*, *Vinca* and *Philodendron* spp. by the use of pectic enzyme preparations. Scott, Hamner, Baker and Bowler (1958) used cellulose-degrading bacteria or pectinase, Schieferstein and Loomis (1959) and Schneider (1960) pectinase-cellulase mixtures and Huelin (1959) cellulase in the form of snail gut extract. Various pectinase preparations have since been used on leaves and fruits in what has become almost standard practice (see, for example, Johnston and Sproston, 1965). Discs are incubated at pH 4 and 37 °C with the enzyme until the cellular tissue disintegrates and can be removed from the cuticular membrane by gentle scraping; remnants of epidermal cells remain attached to the membrane as in the chemical treatment. The membranes of a leaf or fruit isolated by pectinase, cellulase or oxalate treatment are of similar weights (Baker, Batt and Martin, 1964). Preece (1962) removed the apple leaf cuticle with pectinase to reveal the mycelium of *Venturia inaequalis*.

Occasionally difficulty is met in removing leaf cuticular membranes by treatment with oxalate or pectinase solution. The membranes of *Salix scouleriana*, Old man's beard (*Clematis vitalba*) and *Vitis vinifera* strongly resist separation from the cellular tissues, but can be removed by direct treatment with zinc chloride-hydrochloric acid solution. Membranes often become detached from the adaxial and abaxial surfaces after differing periods of treatment; the adaxial leaf membrane of *Citrus aurantifolia* is much more readily detached than the abaxial, and the abaxial of *Lamium album* more readily than the adaxial. In some cases membranes are difficult to isolate because of thinness or a tendency to fragmentation during isolation; examples are those of strawberry (*Fragaria* sp.) and bean (*Phaseolus vulgaris*) leaves.

Before analysis, measurements are made of the surface areas of the plant materials to be examined. The wax extracts and cuticular membranes are dried and weighed to provide a measure of the weight of cuticle on a unit area of plant surface. The membranes are extracted with dilute aqueous alkali to remove tannin material and hydrolysed with alcoholic alkali to give a measure of the cutin (Martin, 1960).

The fractionation and identification of wax components

Column chromatography has been used extensively to sub-divide the wax into its constituent alkane, alcohol and other classes of components. Zetzsche and Lüscher (1937) used elution from alumina to separate and purify the neutral components of cork (*Quercus suber*) wax. Since then chromatography on alumina has been used by many workers, e.g. by Wanless, King and Ritter (1955) for the fractionation of pyrethrum (*Chrysanthemum* sp.) wax, Savidan (1956a) for carnauba (*Copernicia cerifera*) wax, Cole (1956) and Cole and Brown (1960) for ouricury (*Syagrus* sp.) wax, Carruthers and Johnstone (1959) for tobacco (*Nicotiana*) wax, Wiedenhof (1959) for carnauba, sugar cane and other waxes, Downing, Kranz and Murray (1961) for carnauba wax, Mazliak (1960a, 1961a and b, 1962, 1963d) for apple and carnauba waxes, Silva Fernandes, Baker and Martin (1964) for apple wax and Radler (1965a), for grape wax. Analysis of a complex wax is often aided by the removal of a major class of components before chromatography. As an example the procedure adopted by Horn, Kranz and Lamberton (1964) may be cited. Some *Eucalyptus* waxes contain large proportions (*c.* 50%) of long-chain β-diketones. These were first removed by extraction, as copper complexes, from a solution of the wax in petroleum ether (Horn and Lamberton, 1962), and free acids were separated by extraction with dilute sodium hydroxide in aqueous ethanol. The residual wax was then chromatographed on alumina; hydrocarbons were eluted with petroleum ether, unhydroxylated esters with benzene, a triterpene lactone acetate, alcohols and hydroxylated esters with benzene-ether mixtures, and triterpene acids with chloroform. Elution from alumina with a hydrocarbon solvent is especially useful for the isolation of the alkane fractions of waxes and has found many applications. Alumina, however, has disruptive effects on some wax constituents. Aldehydes in waxes are destroyed by alumina and must be chromatographed on silicic acid; this was used for the fractionation of grape wax by Radler and Horn (1965). Mecklenburg (1966) eluted the alkanes of the waxes of *Solanum* spp. from activated silicic acid with *n*-hexane. Tulloch and Weenink (1966) also used a column of silicic acid to fractionate wheat wax. Alkanes were eluted with petroleum ether, and esters and β-diketones, alcohols and hydroxy-β-diketones were obtained successively by elution with petroleum ether containing increasing amounts of ether. Downing, Kranz and Murray (1961) and Mazliak (1963a) separated hydroxy-fatty acids (as methyl esters) from normal acids by the use of a column of magnesium trisilicate at 50 °C. The classes of compounds obtained by column chromatography are identified by infra-red analysis and other methods.

Column chromatography permits the resolution of waxes on a macro scale, but has receded in importance in favour of micro methods using the thin-layer chromatographic technique. Thin layers of absorbent have now replaced paper for chromatographic purposes. Many systems have been described for the separation and identification of wax components. Adsorption thin-layer chromatography, usually carried out on layers of silicic acid (silica gel) or alumina, fractionates the wax into its constituent classes of components (Mangold and Malins, 1960; Purdy and Truter, 1961, 1963a; Kaufmann and Das, 1963; Hessler and Sammet, 1965; Kolattukudy, 1965; Radler and Horn, 1965). The value of the technique has been demonstrated by Holloway and Challen (1966) and Holloway (1967). Over sixty natural waxes were successfully resolved; silica gel and aluminium oxide were used as adsorbents and four solvents as eluants and the use of a minimum of two systems effectively separated all the classes of constituents. Reactions used in the identification of the classes on the chromatograms are summarised by Holloway and Challen (1966). Reagents used by Holloway (1967) included iodine in petroleum ether and fluorescein-bromine for the detection of unsaturated compounds, 2,4-dinitrophenylhydrazine for carbonyl compounds, an acidified ethanolic solution of vanillin for higher alcohols and ketones, and acetic anhydride-sulphuric acid (Liebermann-Burchard reagent) and chlorosulphonic acid in glacial acetic acid for triterpenoids and steroids. The identities of primary and secondary alcohols were confirmed by thin-layer chromatography of their acetates, fatty acids by the chromatographic examination of their methyl esters, and ketones by chromatography after reduction to alkanes. Esters were confirmed by the liberation, on saponification, of acids and alcohols which were identified by their chromatographic behaviour and by infra-red spectroscopy of derivatives. Adsorption thin-layer chromatography on a preparative scale, using several hundred mg of wax, permitted the isolation of the various classes for further analysis by gas chromatography. The results of Holloway's examination of natural waxes are summarised in Chapter 5.

Reversed-phase thin-layer chromatography was used for the identification of the individual components of classes by Kaufmann and Das (1963) and Purdy and Truter (1963b, c). Kartnig and Scholz (1965) separated wax acids from alcohols by means of an ion exchange resin, and identified the components of the fractions with the help of thin-layer chromatography. The practical details of thin-layer techniques and the R_F values obtained in the separation and identification of homologous series of paraffins, mono-basic acids, alcohols and esters are given by Hessler and Sammet (1965).

Gas chromatography is now widely used for the identification of the individual components of the classes of wax constituents. Attempts to analyse whole wax by direct injection into the gas chromatograph have met with limited success; the most useful procedure, in general, is the fractionation of the wax by preparative scale thin-layer chromatography followed by the gas chromatographic analysis of each fraction. Alkane, alkene, alcohol, ketone, aldehyde and ester fractions may be chromatographed directly; some of these and other fractions more effectively after conversion to derivatives. Ludwig (1966) identified the esters in carnauba and ouricury waxes by gas chromatography; compounds of chain length C_{44} to C_{64} were found, with C_{56} predominating. Tribe (1967) isolated fractions of Gramineae waxes on a preparative scale by thin-layer chromatography; the alkanes, alkenes and alcohols were analysed by gas chromatography without further treatment, but the fatty acids were methylated and the aldehydes reduced to hydrocarbons. Mazliak (1961b, 1962) analysed the methyl esters of the fatty acids derived from the esters of carnauba wax and the diacetates of the diols of the wax. Kranz, Lamberton, Murray and Redcliffe (1960) reduced the aldehyde, alcohol and acid fractions obtained from sugar-cane wax to the corresponding hydrocarbons before chromatographic analysis. In the analysis of *Eucalyptus* wax Horn, Kranz and Lamberton (1964) prepared other derivatives for chromatography; primary alcohols were oxidised to acids (analysed as the methyl esters), secondary alcohols to ketones, and fatty acids obtained from the saponification of esters were converted to the methyl esters. The structures of the β-diketones of the wax were determined by gas chromatographic analysis of the methyl ketones produced on alkaline hydrolysis. A wide choice of derivatives of alcohols is available; among these the trimethylsilyl ethers ($R—O—Si(CH_3)_3$) are especially valuable since they are easily chromatographed and display characteristic mass spectrometric fragmentation patterns. The trimethylsilyl group ($—Si(CH_3)_3$) is readily introduced into a wide variety of organic compounds containing active hydrogen. The derivatives of alcohols were prepared by reaction with hexamethyldisilazine in ethyl acetate solution with pyridine as catalyst and the reaction mixture was chromatographed directly (Vandenheuvel, Gardiner and Horning, 1965). The reagent now commonly used consists of hexamethyldisilazine and trimethylchlorosilane in anhydrous pyridine (see, for example, Brieskorn and Reinartz, 1967b; Eglinton and Hunneman, 1968). The trimethylsilyl derivatives were used by Duperon, Vetter and Barbier (1964) to examine the aliphatic alcohols of the unsaponifiable fraction obtained from the male fern (*Dryopteris* [*Polystichum*] *filix-mas*) spores. A combination of thin-layer and gas chromatographic procedures was used by Brieskorn and Reinartz (1967a) in the analysis of tomato (*Lycopersicum esculentum*) fruit wax; hydrocarbons were gas-chromatographed directly, fatty acids as the methyl esters and triterpenoid and steroid components as the trimethylsilyl ethers.

Methods have been developed for the separation and identification of the components of mixtures of hydrocarbons of different structural types. Alkenes may be separated from alkanes by the use of silica gel impregnated with silver nitrate; the alkenes are selectively held on the silver ions as weak ***adducts***. A combination of column and thin-layer procedures incorporating this separation was used by Šorm, Wollrab, Jarolimek and Streibl (1964) to isolate from waxes ***homologous*** series of alkenes whose structures were determined by hydrogenation and chromatographic analysis. Herbin and Robins (1968a) isolated the hydrocarbons of *Aloe* waxes by chromatography on alumina, and identified alkenes in the mixtures by the shorter retention times, in the gas chromatograph, than those of alkanes of the same chain lengths, by the reversal of the retention time relationship in another chromatographic system and by the disappearance of peaks after hydrogenation with concurrent increases in the corresponding peaks for alkanes. Earlier gas chromatographic techniques failed to resolve the peaks due to isomeric alkanes; Streibl and Konečný (1967) identified the C_{26} and C_{28} 2- and 3-methylalkanes in the waxes of poplar (*Populus*) and other plants by the use of a capillary column. Branched chain and cyclic hydrocarbons can be separated from straight-chain compounds by forming adducts of the latter with urea or by the use of a molecular sieve. Nagy, Modzeleski and Murphy (1965) utilised the formation of urea adducts in the fractionation of the hydrocarbons of banana (*Musa*) leaf wax and examined the compounds isolated by mass spectrometry, infra-red spectroscopy and fluorescence under ultra-violet light. Streibl, Jarolimek and Wollrab (1964) and Jarolimek, Wollrab and Streibl (1964) report the synthesis and gas partition chromatography of some higher saturated and unsaturated hydrocarbons.

Downing, Kranz and Murray (1960) and Downing, Kranz, Lamberton, Murray and Redcliffe (1961) used a Linde 5A molecular sieve attached to the outlet of the gas chromatograph to separate the straight chain and singly branched hydrocarbons derived from wool wax and bees' wax. Kaneda (1967) used the molecular sieve to separate the normal hydrocarbons of tobacco from the branched, and identified constituents by mass spectrometry, X-ray diffraction analysis, determinations of melting point and infra-red spectroscopy.

Non-chromatographic methods have been used in the examination of plant waxes and their constituents. Findley and Brown (1953) used molecular distillation supplemented by functional group analysis to characterise some plant waxes, and Lamberton and Redcliffe (1960) removed aldehydes, alcohols, acids and hydrocarbons from sugar cane wax by distillation under reduced pressure. Thermal micro methods have been used to characterise naturally occurring wax acids and alcohols (Kartnig, 1967). Alkyl esters may be determined spectrophotometrically by reaction with hydroxylamine and ferric perchlorate; a violet-purple colour is produced (Goddu, Leblanc and Wright, 1955). Hydroxy-fatty acids in the fruit skin of apple and other members of the Rosaceae give a red colour on heating with hydrochloric acid in acetone (Brieskorn and Schneider, 1961). Ursolic and oleanolic acids occur together in many plant waxes and are difficult to separate; Brieskorn and Hofmann (1962) have described a spectrophotometric method based on the Liebermann-Burchard reaction with acetic anhydride and sulphuric acid for the determination of the proportion of each compound in a mixture. Methods of isolation, identification and determination of hydroxy-fatty acids are reviewed by Radin (1965).

The determination of cutin

Amongst the earliest measurements of cutin were those of König (1906) and König and Rump (1914). The plant material was digested with 72% sulphuric acid or cuprammonium solution to remove cellulose and with ammoniacal hydrogen peroxide to remove lignin, and the residue was regarded as cutin. Zetzsche and Scherz (1932) used a similar method; they extracted lipid material from dried leaves with solvent and assessed cutin from the residue obtained after treatment with sulphuric acid and hydrogen peroxide. Lüdtke (1961) determined cutin in retted fibre by progressive digestion with hypochlorite solution, 72% sulphuric acid and sulphite solution. The method of Zetzsche and Scherz was used by Kausch and Haas (1965) to determine cutin in the sun and shade leaves of copper beech (*Fagus sylvatica* var. *atropurpurea*). Markley and Sando (1933) found that treatment with sulphuric acid led to a partial decomposition of the cutin; they extracted isolated apple fruit cuticle with dilute aqueous acid and alkali and measured cutin by the loss in weight of the membrane on saponification with alcoholic alkali. Skoss (1955) extracted isolated leaf cuticles with alcohol and weighed the residue to obtain a value for the cutin.

The ready saponification of cutin with alcoholic alkali provides the most dependable basis for an analytical method. Huelin and Gallop (1951a) determined cutin in the apple cuticle from the yield of ether-soluble acids liberated on hydrolysis of the membrane in preference to the loss in weight. Matic (1956) showed that hydroxy-fatty acids comprise more than 80% of the ether-soluble acidic fraction obtained, in 50% yield, from the leaf cutin of *Agave americana*, and Huelin (1959) obtained an 80% yield of hydroxy acids from purified apple fruit cutin. Roberts, Batt and Martin (1959) assessed cutin in the isolated cuticular membranes of brassica, beet (*Beta* sp.), banana, tomato, strawberry, blackcurrant (*Ribes nigrum*), laurel (*Prunus laurocerasus*), rhododendron and euonymus leaves by both the loss in weight of the membranes on hydrolysis and the recovery of the ether-soluble cutin acids. The values for cutin obtained from the yields of the acids were 76 to 94% of those from the losses in weight during saponification. Similar proportionate values were found in assessments of cutin in the membranes of apple leaves and fruits (Richmond and Martin, 1959). The assessment of cutin from the loss in weight of the membrane on hydrolysis is valid only if cutin alone is solubilised by the alcoholic alkali. Any residual tannin or other alkali-soluble material in the membrane before hydrolysis leads to over-evaluation; on the other hand, assessment from the yield of ether-soluble acids under-evaluates due to the loss of a water-soluble fraction. Baker, Batt and Martin (1964) formulated the analytical conditions necessary for the removal of tannin without loss of cutin; provided they are adhered to, the loss in weight of the membrane on hydrolysis provides a dependable assessment of cutin. In later work, the evaluation of cutin from the yield of ether-soluble acids is preferred; the details of the analytical method are given by Baker and Martin (1967).

Gas chromatography combined with thin-layer chromatography has proved invaluable for the determination of the amounts and identities of the acids liberated from cutin. Crisp (1965) fractionated, by gas chromatography, the hydroxy acids of *Agave* cutin as the methyl esters or methyl esters of the acetoxy derivatives. Thin-layer and column chromatography, and gas chromatography of the methyl esters of the trimethylsilyl ethers were used by Brieskorn and Reinartz (1967b) to separate and identify the hydroxy acids of tomato cutin. Eglinton and Hunneman (1968) separated the acids from apple fruit cutin on a preparative scale by thin-layer chromatography of their methyl esters and determined the amounts and identities of the acids by gas chromatography and combined gas chromatography-mass spectrometry of the methyl esters of the fatty acids and of the methyl ester-trimethylsilyl ether derivatives of the hydroxy-fatty acids. The location of double bonds in unsaturated compounds was determined by a novel method involving hydroxylation with osmium tetroxide followed by gas chromatography-mass spectrometry of the trimethylsilyl ethers of the resulting ***vic-dihydroxy*** compounds (see also Eglinton, Hunneman and McCormick (1968). Wilk, Gitlow and Clarke (1967) showed that trifluoroacetate derivatives of aliphatic and aromatic hydroxy compounds give large responses with the electron capture detector; Holloway and Baker (1968a) use electron capture gas chromatography (nickel 63 detector) of the trifluoroacetate derivatives ($-OH \rightarrow -OOCCF_3$) to identify and measure the hydroxy acids derived from cutin. The method is extremely sensitive; 1 ng of 9,10,18-trihydroxyoctadecanoic acid in the form of methyl-9,10,18-*tris*-trifluoroacetyloctadecanoate can be detected.

3

Early Work on the Plant Cuticle

Remarkable advances have been made since the Second World War in knowledge of the morphology and chemistry of the plant cuticle. The advances have been due to the development of highly sophisticated investigational techniques described in the previous chapter, including particularly electron microscopy and chromatographic analysis. In a discussion of early work on the cuticle, it is therefore appropriate and convenient to confine it to investigations conducted during the pre-war period.

MORPHOLOGY

The cuticular membrane

The presence of a superficial membrane on plants has been known for about 200 years. First suggested by Ludwig in 1757, its existence was recognised by de Saussure in 1762 and by Hedwig in 1793 (Barthélemy, 1868). Some workers, including Sprengel and Mirbel, denied its presence, but doubts as to its existence were resolved by Brongniart (1830-34) who isolated it by allowing cabbage leaves to disintegrate in water over a long period. He showed it to be a continuous, non-cellular structure moulded on the epidermal cells and sheathing the epidermal hairs, and named it the *cuticle*. The membrane was described as colourless, transparent and of simple structure; it was marked by reticulate lines and showed oval blemishes which had the appearance of button-holes. Examination of the blemishes—the stomata—revealed their openings. Brongniart concluded that the membrane was probably little permeable to liquid or gas, and that its essential function was to protect the plant against excessive evaporation of water. It was therefore not needed on water-submerged plants. Brongniart's 'cuticle' was the outermost, cellulose-free layer as distinct from other layers making up the epidermis that might occur beneath it; he suggested that it was composed of 'cutine', fatty material taking Sudan stain. He found that many plants have a thin skin covering the external surface of the cellulosic layer of the epidermis. Henslow (1831) isolated a similar skin from the corolla, stamens and style of foxglove (*Digitalis purpurea*) thus confirming 'the existence of a delicate membrane investing the epidermis'. Henslow obtained his cuticles by digesting the plant material with dilute nitric acid. Some of the earliest references to cuticles occur, oddly enough, in work reported by Brodie (1842) and others on fossilised plant residues; Brongniart himself was led to the study of cuticles by his interest in fossilised vegetation. The early work on cuticles in fossils is reviewed by Stace (1965).

After Brongniart, the interest of cytologists in the nature and properties of the cuticle grew rapidly. Chemical macerating agents such as Schultze's solution, which consisted of nitric acid with added crystals of potassium chlorate, were used to destroy the cellular tissues and liberate the membranes. Cuticles were found on plants by Treviranus, Link, Schleiden, von Mohl and others. Von Mohl (1845) reported that the surface membrane extended within the respiratory chambers of the stomata; this had been observed by Payen in 1840. Trécul (1856) confirmed that a cuticle, similar to that outside, lined the substomatal cavity and showed by staining tests that the cuticle was composed of superimposed layers and traced the participation of small granules in its development. He further suggested that the formation of the cuticle was due to a physiological process rather than to the influence of atmospheric agents.

The role played by the cuticle in gaseous exchange processes became the subject of much discussion. Some workers, including Dutrochet (1832), regarded the cuticle as impervious and considered that the open stomata provided the sole pathway; others, e.g. Duchartre (1856), believed that the epidermal cells participated in the respiratory process. From experiments with *Nerium oleander* leaves, subjected to an atmosphere exceptionally rich in carbon dioxide (30%), Boussingault (1864, 1865) concluded that diffusion occurred through the adaxial surface despite the presence of stomata on the abaxial. Barthélemy (1868) suggested that 'an all too popular habit of generalising and of making conclusions on the basis of comparisons' had led scientists 'to regard the stomata as the only organs which introduce and expel gas'. He determined the rates of diffusion, under pressure, of air, carbon dioxide, oxygen, ozone and nitrogen through leaves of Indian bean (*Catalpa* sp.), *Magnolia* sp., virginia creeper (*Ampelopsis* sp.) and maple (*Acer* sp.), and analysed the diffused gases. Carbon dioxide and ozone diffused more rapidly than oxygen and nitrogen, and air was enriched in oxygen during its passage. Barthélemy concluded that the carbonic acid of the air dissolves in the cuticle and passes in noticeable quantities, in solution, into the interior parenchyma. He ascribed to the cuticle a definite role in the absorption and exhalation of carbonic acid and suggested that the nitrogen of the air was introduced and exhaled principally by the stomata. His theory supported the view of de Saussure and Brongniart that the quantity of carbon dioxide decomposed by a leaf was proportional to its surface area and not to its volume since it was dependent only on the quantity of gas which passed into the cuticle in solution. The theory was accepted, but later rejected. Blackman (1895) concluded that, at normal atmospheric concentration, practically the sole pathway of carbon dioxide into or out of the leaf is provided by the stomata; since then reports have been made that cuticular absorption can indeed occur (Chapter 7).

The cuticular wax

The cuticle was believed to be formed *in situ* by a process of modification of the cellulosic constituents of the epidermal cell wall. This view, the 'metacrase' theory, was put forward by de Saussure and others. Controversy arose about the origin of the epidermal wax. Some workers including Wiesner (1871) suggested that the wax was secreted to the surface in a volatile solvent. Karsten (1857, 1860), on the other hand, believed that the wax was formed from epidermal cell wall substances. Uloth (1867) proposed an empirical formula of $C_{26}H_{16}O_4$ for the wax of *Acer striatum* and suggested that it was a product of the breakdown of cellulose. The position was clarified by de Bary (1871) who showed that the appearance of wax was not accompanied by a decrease in cellulose, and that the wax could be removed with hot alcohol to leave the cuticle unchanged in appearance, properties and reactions. No changes were observed in the structure or appearance of the epidermal cells or the cuticle while the wax was in the process of formation. In the Gramineae, with the exception of *Saccharum*, the cell walls on which the wax was formed were fully silicified by the time it began to appear. De Bary confirmed that the cuticle could be regarded as a superficial deposit covering the outermost layer of cells, but fairly sharply delimited from it and that it was composed of superimposed cuticularised layers impregnated with wax. In some species, among which *Acer striatum* was a good example, a large amount of wax was embedded in the intermediate layer while giving rise to little or no secreted wax on the surface; Uloth (1867) had designated this as 'stratified cerification' and believed that it indicated a progressive transformation of the epidermal tissues into wax.

De Bary believed that the wax was a product of active secretion and postulated the presence of canals within the cuticle. Observations of the development of single-layer granular wax films on monocotyledon leaves (*Galanthus nivalis* and *Tulipa sylvestris*) and of rod formations on *Benincasa hispida* confirmed that both these types of wax structure arose without any apparent alteration in the underlying cuticle. When the wax bloom was wiped off it was renewed, again without any structural changes being observable in the epidermal tissues. Thus for granular and rod formations of wax the secretion theory seemed to be the valid one. The same was found to be true for the aggregate coatings developed in such species as *Kleinia*, *Eucalyptus*, *Lonicera*, and *Secale*. The origin of the secreted wax was studied mainly in *Heliconia farinosa* because of the high transparency of the epidermal cells. Neither in the outer chlorophyll-free layer of parenchyma cells nor in the inner chlorophyll-containing parenchyma cells was there any visible trace of wax among the cell contents. The same was true for epidermal cells of *Strelitzia*, *Galanthus*, *Tulipa* and *Cotyledon orbiculata*. Boiling with alcohol made no difference to the granular inclusions visible in the epidermal cell contents of *Myrica* fruits, *Saccharum*, *Chamaedora*, *Coix*, *Sorghum* or *Ceroxylon* (*Klopstockia*). De Bary concluded that if wax was present in the cell protoplasm at all it must occur in an undetectably fine state of division.

The major contribution of de Bary (1884) was the classification and description of the waxes of many species of plant. He defined the physical properties of the waxes and described four main types of waxy coverings (see, however, p. 109). The first type, the aggregate wax coating, consisted either of a dense accumulation of fine rods or needles, less than 1 μm in diameter and rising vertically or obliquely from the epidermal surface, or of several superimposed layers of extremely small granules. The length of the rods or needles did not exceed the thickness of the underlying epidermal cell-wall. The second type was a single-layer granular coating, in which the individual elements reached or exceeded 1 μm in diameter. In some species the granules occurred closely packed together to produce a continuous wax film; in others there were interspaces between the individual particles. The third type of wax coating was made up of a large number of fine rod-like structures. In the typical form found on the leaves of *Heliconia farinosa* they were about 1 μm in diameter and up to 50 μm or more long, standing out vertically from the epidermal surface. The lower portions of the rods were straight, and the upper portions curled into various shapes which interlaced with one another. They covered most of the epidermal cells apart from those immediately around the stomatal openings. Other species in which this type of waxy coating occurred included *Musa* (in which the rods tended to be distributed along the edges of the epidermal cells), *Strelitzia ovata* (in which the rods close to the nodes were 2–4 μm thick and up to 150 μm long), *Eulalia* (*Miscanthus*) *japonica*, *Coix lacryma-jobi* and *Sorghum*. The rods were generally cylindrical or somewhat flattened in shape, colourless and rather brittle The smaller ones were uniformly transparent without any fine structure, but the larger ones often exhibited a striated structure indicating the presence of layers of different transparencies. The fourth type of wax coating took the form of a continuous film or incrustation covering the whole cuticle, apart from the stomatal openings, like a brittle transparent glaze. It varied between 1 μm in thickness (*Sempervivum tectorum*) and 70 μm (older twigs of *Euphorbia canariensis*). Fruits of *Myrica* species had a wax coating up to 50 μm thick, which could be lifted off in pieces to show an impression of the epidermal surface. It appeared in section to be formed of two non-separable layers of somewhat different structure and refractive index. The coating was a mixture of fatty and waxy substances. A similar wax incrustation observed on the vegetative organs of *Panicum turgidum* formed a brittle layer about 30 μm thick. In section it was seen to be made up of vertical fused rods with numerous planes of cleavage.

De Bary also examined waxes of commercial importance including carnauba wax from *Copernicia cerifera*. Indian wax palms had wax layers on their trunks up to 5 mm thick. Samples studied contained a resin-like component readily soluble in cold alcohol in addition to the main mass of insoluble wax. The coating was made up of a large number of elongated prisms, each fitting over one epidermal cell. The prisms exhibited three-directional striations, and were cemented together by a clear glass-like material.

Ambronn (1888) examined the cuticle with the polarising microscope; the outer layers were optically negative with reference to the tangential direction of the cell wall, whereas the inner cellulosic layer was optically positive. The negative birefringence was attributed to the optical properties of the impregnating wax; on melting the wax, the layer was isotropic. Van Wisselingh (1895) confirmed that the cuticle consisted of a layer of cutin lying over a so-called 'cutinised lamella' in which cutin-like substances were deposited in a layer of cellulose. The outer layer, or cuticle proper, was fatty in nature and was thought to be formed by the oxidation of oily materials which were presumed to be the products of cell metabolism and to have diffused through the cellulose-pectin component of the wall of the epidermal cell. Géneau de Lamarlière (1906) proposed staining tests which are used today for the identification of cuticle components; the dye Sudan III in ethanol or safranin in aqueous ethanol was used for cutin, ruthenium red for pectin and iodine/phosphoric acid for cellulose. He examined the epidermal structure of many aquatic and aerial plants and observed that cells bordering air spaces, including those lining the substomatal cavities, are cutinised.

The secretion of waxes to the surfaces of plants was further studied by Dous (1927), Ziegenspeck (1928) and Pohl (1928). From their work the consensus of opinion was that the waxes were formed in the cellular tissues and passed through micropassages to the surfaces where they hardened. Martens (1934) concluded that the cuticle as a whole is secreted in a fluid form and then coagulates in the air; the acids forming cutin probably migrate, in a low molecular state, into the wall of the epidermal cell and there polymerise. He thought it possible that some material with solvent power aids the movement of the wax, or alternatively that the components of the wax migrate in an uncombined form. Weber (1942) failed to find pores in the cuticle and also believed that the wax is secreted in a liquid form which infiltrates through the epidermal wall and cuticle and later hardens into the forms described by de Bary. Anderson (1934) obtained evidence that a cutinised cellulose lamella could be deposited on the inner surface of the outer wall of the epidermal cell, which suggested that cutinisation may not be dependent upon exposure to external factors. This evidence supported much earlier work by Damm (1901) who described cases in which epidermal cells were cutinised on all sides so that a so-called 'cuticular epithelium' was formed. These observations were of interest in that they implicated the epidermal cells as the site from which the cutin precursors arose. Martens (1933) in a study of the origin of superficial corrugations on petals of *Tradescantia virginiana* found that the folds resulted from an excessive secretion of cuticular substances. His work emphasised the role played by secretion in the formation of the cuticle and led finally to the lapse of the 'metacrase' theory that it was formed *in situ*.

The formation and structure of the cuticle

Priestley (1921) recognised the protective character of the cuticle and put forward the enlightened view that it should be taken into account by plant pathologists when considering the entry of parasitic fungi through the uninjured plant surface or when discussing the possible effects of spray fluids on the plants they protect. He also suggested in connection with the invasion of plants by fungi that the cuticle was chemically so composed that it was unlikely to be decomposed by ordinary hydrolysing enzyme action, and that the difference between cutin and suberin was due to different constituent organic acids and to the differing conditions under which they were transformed into the mature materials.

Lee and Priestley (1924) pointed out that, in botanical literature, the cuticle was usually said to consist of 'cutin'; they believed this to be not a chemical entity, but an aggregate of substances varying in composition but occurring always in the same location on the plant and having the same general characters. They described the progressive development of the cuticle in the shoots of angiosperms. As differentiation began, fatty substances moved freely along the cell walls until they reached the surface where they formed a film which underwent physical changes to form a rigid layer. Many of the fatty substances contained unsaturated fatty acids which, in the presence of oxygen, oxidised and condensed to give varnish-like substances insoluble in water and fat solvents. The view that the cuticle resembled a varnish had long been held (see quotation, chapter 1, p. 1). An intermediate lamella, the cutinised layer, could be distinguished between the outermost continuous fatty deposit and the cellulose wall within.

Lee and Priestley confirmed the views of de Bary and van Wisselingh that the cuticle could be regarded as a superficial cellulose-free fatty deposit covering the outermost layer of cells. They also showed that light and humidity affected the thickness and consistency of the cuticle and ascribed this to the influence of the environmental factors upon the oxidation and condensation of the fatty acids. The thickness of the cuticle of some plants increased as the quantity of available moisture fell and as light intensity rose. They examined the cuticles of leaves of *Ribes* and *Prunus* spp. from a fertiliser trial and found that a high level of nitrogen gave a thicker cuticle and a high level of calcium a thinner one. These effects, they suggested, were due to the influence of potassium and calcium upon the outward diffusion of the fatty acids, the soluble potassium salts being more mobile than the insoluble calcium. On the other hand, the occurrence of a high proportion of potassium to calcium in a submerged water plant produced a thin cuticle because, it was suggested, the soluble potassium salts of the acids were leached into the surrounding water. Lee and Priestley recognised the presence of a membrane on the submerged roots of water plants, but believed that anything more than a tenuous structure was unlikely because, in the absence of exposure to air, the oxidation and condensation of fatty acids to form a continuous, varnish-like covering could not be expected.

The function of the cuticle also received attention from Lee and Priestley. Loss of water from the plant decreased as the cuticle thickened. The cuticles of plants subsisting in a moist atmosphere, or of etiolated plants, were soft although not necessarily thin. In an interesting diversion, they suggested that the cuticle might influence the final anatomy of the leaf. The configuration of the palisade parenchyma, perpendicular to the surface, was usually regarded as being connected with the incidence of light. During the early growth of the cuticle, a point was reached when it was no longer capable of stretching; subsequent elongation of the palisade cells could then take place only at right angles to the inelastic surface layer.

Lee and Priestley's theory that the cuticle arises from lipid materials which are formed or mobilised in the epidermis, migrate to the surface and are oxidised, followed an earlier suggestion by Priestley and Woffenden (1922) that early suberisation, for example in the formation of wound cork, depends primarily upon the presence of air. Artz (1933) and Frey-Wyssling and Häusermann (1941) reported that a tenuous cuticle occurs on the surface of mesophyll cells where they are exposed to intercellular air passages and this observation has been widely quoted. Priestley (1943) in a review article discusses the formation, patterns and functions of the cuticle in angiosperms.

Polarised light microscopy, in the hands of Frey (1926), Anderson (1928, 1935), Frey-Wyssling (1930), Meyer (1938) and Roelofsen (1952) revealed the complex interrelations of cutin, pectic material and cellulose within the cuticle. Anderson showed that the epidermis is bounded by an optically isotropic cuticle and that a fairly wide pectic layer, also isotropic, is interposed between the cuticular and cellulosic layers. Frey and also Anderson examined the distribution of cutin in the cuticle of leaves of *Clivia nobilis* and presented a picture of the outer wall of the epidermal cell and overlying cuticle which accords well with present-day knowledge (see p. 80). A cellulose-pectin layer adjacent to the cell lumen merges into a zone in which pectic material predominates; superimposed on this is a layer composed of pectin, cellulose and cutin which shows decreasing concentrations of pectin and cellulose and an increasing concentration of cutin as it extends to the outermost zone of cutin, or cuticle proper. The negative birefringence shown by a cuticular layer was found by Meyer to be due to the presence of a fusible wax, the wax-free cutin framework being isotropic. By this time much information had been obtained on the chemical and physical nature of the various components of the cuticle; the optical characteristics could therefore be interpreted in the light of the chemical and physical data.

The work laid the foundation of the present concept of the structure of the cuticle and is admirably summarised by Frey-Wyssling (1948). The waxes contained compounds of comparatively low molecular weights in a non-polymerised condition and occurred as rods of less than 10 μm length, the cellulose was in the form of long-chain molecules, the pectin existed as long chains and the cutin was composed of acids linked together through hydroxyl and carboxyl groups in the nature of an 'estolide' as defined by Bougault and Bourdier (1908). Cutin had some of the characteristics of an acid or polymeric anion (Brauner, 1930), and so probably not all of the carboxyl groups of the acids were esterified. The optically isotropic nature of cutin indicated that the linkage of the acids was not in the form of a linear chain, but was reticular in spatial directions. From the optical analysis, a scheme was suggested by Frey-Wyssling showing the relative positions of the four major components in the cuticular layer. On the outside occurred layers of wax molecules in radial arrangement, on the inside lamellae of cellulose and pectin, and between them amorphous cutin in random orientation. The cutin in the intermediate zone contained hydrophobic methyl groups orientated towards the surface, and hydrophilic hydroxy and carboxyl groups orientated towards the inner cellulosic layer. The outer layers probably consisted of wax and cutin only; towards the inside, the cellulose was admixed with cutin and could be liberated only by removing the cutin by hydrolysis. The picture was built up of an outer, true cuticle consisting of cutin carrying occluded and surface wax, and below this, a cuticular layer in which the proportion of cutin in the layer progressively decreased and that of cellulose and pectin increased with approaching nearness to the cellulosic outer wall of the epidermal cell. Frey-Wyssling (1948) compared the surface wax extrusions with the fine protruding threads formed on the addition of water to a lecithin surface. Kreger (1948) qualitatively examined the surface waxes of many plants by a micro X-ray diffraction method. The waxy coverings were classified into groups according to composition as indicated by the diagrams obtained. Kreger compared his classification with that proposed by de Bary, and suggested that the different forms assumed by the wax covers on leaves are probably caused by differences in chemical composition.

CHEMISTRY

Waxes

Chemical work on the waxy components of the plant cuticle effectively dates from the beginning of this century. Information on waxes known to be derived exclusively from the surfaces of plants was first obtained from the examination of materials of commercial importance. Stürcke (1884) noted a preponderance of higher alcohols and a low content (<1%) of paraffins in carnauba wax, and isolated from it an isomer of lignoceric acid (possibly 2-methyltricosanoic) which he named carnaubic acid and a compound which he thought was a lactone. Leys (1913) found that the wax contains about 50% of alcohols, a value confirmed by Murray and Schoenfeld (1951) and others. Heiduschka and Garies (1919) identified melissyl alcohol and melissic acid in the wax, and Gottfried and Ulzer (1926) considered the small amount of paraffin present to be heptacosane. By contrast, candelilla wax was shown to contain about 50% of paraffins (Sanders, 1911; Berg, 1914; Buchner, 1918). Hentriacontane was isolated from the wax by Fraps and Rather (1910) and Sanders (1911), and dotriacontane by Meyer and Soyka (1913). A preponderance of a C_{30} alcohol in the wax esters was noted by Sanders (1911). An alcohol named raphia alcohol $C_{20}H_{41}OH$, possibly an isomer of arachidyl alcohol, was obtained by Haller (1907) from raphia wax (*Raphia pedunculata*) and an ester was identified by Greshoff and Sack (1901) as the chief component of Pisang wax (*Musa paradisiaca*). Legg and Wheeler (1929a) detected montanyl and melissyl esters in *Agave* wax.

Work on waxes obtained by extracting plant tissues, and therefore derived not solely from the surface, confirmed that paraffins and alcohols, both free and as esters, are important components. The wax of American cotton (*Gossypium*) received considerable attention; it was one of the first waxes to be examined chemically. Knecht and Allan (1911) obtained from it hentriacontane and dotriacontane, and palmitic, stearic and another acid believed to be melissic; Fargher and Probert (1923, 1924) isolated paraffins including hentriacontane, alcohols including the C_{30} compound named gossypyl (melissyl) alcohol, and *n*-fatty acids in the range C_{16}–C_{32} including the C_{30} member called gossypic (melissic) acid. From Japan wax, a vegetable tallow (*Rhus succedanea*) Schaal (1907) obtained dicarboxylic acids identified as nonadecanedioic, eicosanedioic and heneicosanedioic; the latter was given the name japanic acid. In 1911, Tassilly showed that Japan wax is of a glyceride nature, composed chiefly of palmitin and palmitic acid. Power and Moore (1910) isolated hentriacontane, pentatriacontane, ceryl alcohol, palmitic, stearic and linoleic acids, and prunol (ursolic acid) from the wax of leaves of the black cherry (*Prunus serotina*).

Bougault and Bourdier (1908–9) showed that the waxes of conifers (*Juniperus*, *Picea*, *Pinus*, *Thuja*, spp.) contain high proportions of 16-hydroxyhexadecanoic (juniperic) acid and 12-hydroxydodecanoic (sabinic) acid. They suggested that in the natural wax these compounds are polymerised in chains, the acid group of one constituent being esterified with the alcohol group of another. The polymeric compounds constituted a new class of natural products and were named ***'etholides'***. Bougault and Bourdier suggested that the etholides of conifer waxes differ chiefly in the numbers of the associated **ω-hydroxy acids** and in the order of their association.

Much work was done in the early days on the wax obtained from apple (*Malus*) peel. There was some assurance that the wax was derived largely from the cuticle; the cellular tissue comprising part of the peel probably contributed comparatively little fatty material. Power and Chesnut (1920) isolated from the peel a hydrocarbon which they considered was *n*-triacontane. Sando (1923) isolated three principal components of apple wax; a hydrocarbon fraction which he believed was composed chiefly of *n*-triacontane, an alcohol thought to be *n*-heptacosanol and a petroleum ether-insoluble triterpenoid fraction for which the name malol was proposed. This fraction was later re-named ursolic acid (Sando, 1931) since it was identical with urson (ursolic acid) previously isolated from the leaves of the bearberry (*Arctostaphylos uva-ursi*) and with prunol from the leaves of the black cherry (van der Haar, 1924). That Sando's alcohol was *n*-heptacosanol was accepted by Rivière and Pichard (1924).

Identification of the paraffins and alcohols obtained from waxes was not easy; reliance was placed chiefly upon melting points and transition temperatures, but there was little assurance that the substances were not mixtures. An important step forward was made in the classic work of Chibnall and Piper and their colleagues. In 1931, Piper, Chibnall, Hopkins, Pollard, Smith and Williams synthesised long-chain compounds of possible biochemical importance and determined their melting points, transition temperatures and crystal spacings. Chibnall, Piper, Pollard, Smith and Williams (1931) then utilised the data in an examination of the unsaponifiable fraction of apple wax. The crude wax was obtained by extracting dried, powdered peel with petroleum ether, and purified by precipitation with an excess of acetone. A paraffin fraction was found to contain *n*-nonacosane and *n*-heptacosane; the alcohols *n*-hexacosan-1-ol, *n*-octacosan-1-ol, *n*-triacontan-1-ol and *n*-nonacosan-10-ol were also identified. Markley, Hendricks and Sando (1932) confirmed the presence in apple peel of *n*-nonacosane and nonacosan-10-ol. Markley and Sando (1934, 1937) showed that *n*-nonacosane, *n*-hentriacontane, ursolic acid and fatty acids occurred in the wax obtained from cranberries (*Vaccinium oxycoccos*), and that *n*-nonacosane, ursolic acid and fatty acids were present in the wax of the cherry (*Prunus avium*). Pear cuticle wax contained nonacosane, primary alcohols C_{20}–C_{30} with tetracosanol and hexacosanol predominating, acids within the range C_{15}–C_{24}, and ursolic acid (Markley, Hendricks and Sando, 1935). Components of grape (*Vitis labrusea*) wax were nonacosane, hentriacontane, alcohols C_{22}–C_{28} and oleanolic acid (Markley, Sando and Hendricks, 1938), and of grapefruit (*Citrus grandis*) wax nonacosane, hentriacontane, umbelliferone $C_9H_9O_3$ and a cyclic ketone $C_{30}H_{52}CO$ (Markley, Nelson and Sherman 1937). *n*-Hentriacontane was isolated from spinach (*Spinacia oleracea*) leaf wax and its identity confirmed by X-ray analysis (Clenshaw and Smedley-Maclean, 1929; Collison and Smedley-Maclean, 1931). The alcohols *n*-tetracosan-l-ol and *n*-hexacosan-l-ol were also found in this wax (Heyl and Larsen, 1933).

Chibnall and Piper and their colleagues reported the analysis of many other plant waxes. Their work was greatly assisted by the availability of an extended range of compounds which could be used for reference purposes. Piper, Chibnall and Williams (1934) prepared alcohols, acids, acetates and ethyl esters in the range C_{26}–C_{36}, determined melting points and crystal spacings, and used the data to interpret the composition of mixtures of unknown alcohols and acids. In addition to paraffins, alcohols and fatty acids already referred to, ketones were found in some waxes. The findings of

Chibnall and Piper and their colleagues are summarised in Table 3.1; included are the results of the re-examination of some waxes investigated earlier by other workers.

Table 3.1 Constituents of plant waxes

Source of wax	Constituents and references
American cotton (*Gossypium hirsutum*)	*n*-primary alcohols C_{24}, C_{26}, C_{28}, C_{30}, C_{32}, C_{34}; *n*-fatty acids C_{24}, C_{26}, C_{28}, C_{30}, C_{32}, C_{34} Chibnall, Piper, Pollard, Williams and Sahai (1934)
Banana (*Musa* spp.)	*n*-primary alcohols C_{28}, C_{30}, C_{32} Chibnall, Piper, Pollard, Williams and Sahai (1934)
Brussels sprout (*Brassica oleracea* var. *gemmifera*)	*n*-nonacosane; *n*-hentriacontane; *n*-hexacosan-1-ol; *n*-octacosan-1-ol; *n*-nonacosan-15-ol; *n*-nonacosan-15-one Sahai and Chibnall (1932)
Cabbage (*Brassica oleracea* var. *capitata*)	*n*-nonacosane; *n*-hentriacontane; *n*-nonacosan-15-one Channon and Chibnall (1929)
Cactus (*Opuntia sp.*)	*n*-tritriacontane; *n*-pentatriacontane; *n*-heptatriacontane; *n*-octacosan-1-ol; *n*-triacontan-1-ol Chibnall, Piper, Pollard, Williams and Sahai (1934)
Candelilla (*Euphorbia cerifera*)	*n*-paraffins C_{31}, C_{33}; *n*-primary alcohols C_{28}, C_{30}, C_{32}, C_{34}; *n*-fatty acids C_{30}, C_{32}, C_{34} Chibnall, Piper, Pollard, Williams and Sahai (1934)
Carnauba (*Copernicia cerifera*)	*n*-primary alcohols C_{26}, C_{28}, C_{30}, C_{32}, C_{34}; *n*-fatty acids C_{30}, C_{32}, C_{34} Chibnall, Piper, Pollard, Williams and Sahai (1934)
Clover (wild white) (*Trifolium repens*)	*n*-octacosan-1-ol; *n*-triacontan-1-ol; *n*-fatty acids C_{26}, C_{28}, C_{30} Chibnall, Piper, Pollard, Williams and Sahai (1934)
Cluytia similis	*n*-hexacosan-1-ol; *n*-octacosan-1-ol Chibnall, Piper, Pollard, Williams and Sahai (1934)
Cocksfoot and perennial rye-grass (*Dactylis glomerata* and *Lolium perenne*)	*n*-tetracosan-1-ol; *n*-hexacosan-1-ol Pollard, Chibnall and Piper (1931)
Hemp (*Cannabis indica*)	*n*-nonacosane; *n*-hentriacontane Chibnall, Piper, Pollard, Williams and Sahai (1934)
Lucerne (*Medicago sativa*)	*n*-triacontan-1-ol Chibnall, Williams, Latner and Piper (1933)

Table 3.1 (*continued*)

Source of wax	Constituents and references
Mustard (white) (*Brassica hirta*)	*n*-nonacosane; *n*-hentriacontane; *n*-octacosan-1-ol; *n*-triacontan-1-ol; *n*-dotriacontan-1-ol; *n*-fatty acids C_{26}, C_{28}, C_{30} Chibnall, Piper, Pollard, Williams and Sahai (1934)
Raphia (*Raphia vinifera*)	*n*-primary alcohols C_{28}, C_{30}, C_{32} Chibnall, Piper, Pollard, Williams and Sahai (1934)
Rose (*Rosa* sp.)	*n*-paraffins C_{21}–C_{35} Chibnall, El Mangouri and Piper (1954)
Sandal (*Santalum album*)	*n*-hentriacontan-16-one; *n*-hentriacontan-10-ol-16-one; *n*-octacosan-1-ol; *n*-triacontan-1-ol Chibnall, Piper, El Mangouri, Williams and Iyengar (1937)
Swede Turnip (*Brassica rapa*)	*n*-nonacosane; *n*-hentriacontane; *n*-nonacosan-15-ol Chibnall, Piper, Pollard, Williams and Sahai (1934)
Tobacco (*Nicotiana tabacum*)	*n*-paraffins C_{29}, C_{31}, C_{33} Chibnall, Piper, Pollard, Williams and Sahai (1934)
Wheat (*Triticum vulgare*)	*n*-octacosan-1-ol Pollard, Chibnall and Piper (1933)

Some samples of *n*-triacontane, isolated by others and sent to Chibnall for examination, were found to be mixtures of *n*-paraffins. The detection of a new type of constituent, the ***ketol*** *n*-hentriacontan-10-ol-16-one, in sandal wax (*Santalum*) was of special interest. The outstanding features of the results, however, were that the compounds found were saturated, that the paraffins, secondary alcohols and ketones contained odd numbers of carbon atoms and that the primary alcohols and fatty acids even numbers. Chibnall and Piper and their co-workers expressed the view that all waxes are essentially of the same general type and contain odd carbon-numbered *n*-paraffins from C_{25} to C_{37} and even carbon-numbered *n*-primary alcohols and *n*-fatty acids from C_{24} to C_{36}. The possible presence of higher or lower odd carbon-numbered paraffins or of even carbon-numbered alcohols and acids, however, was not excluded.

Of the many plant waxes examined by Chibnall and Piper and their colleagues, only those of apple, carnauba and candelilla were probably solely of cuticular origin. Their usual procedure was to extract plant tissues, such as whole leaves, with petroleum ether, precipitate the wax by the addition of an excess of acetone to the concentrated extract and saponify the wax before the isolation of its constituents. The petroleum ether used for extraction may have acted selectively upon the cuticular waxes so that all their components were not necessarily obtained; both Gane (1931), by immersing apple fruits in ether, and Markley and Sando (1931), by extracting apple peel with ether after petroleum ether, obtained an acetone-insoluble 'hard' wax, an acetone-soluble 'soft' wax or oil, and ursolic acid. Gane's oil consisted chiefly of unsaturated esters. Chibnall and his colleagues examined only the acetone-insoluble fraction and so some of the constituents of the cuticle may have been missed. Furthermore, the saponification of this fraction obscured the extent to which the alcohols and fatty acids occurred in combination as esters or in the free state. In their work on the apple cuticle, they recognised that a selective extraction had been made and that the constituents found represented only a part of the wax. The yield of fatty acids from the saponified acetone-insoluble fraction of apple wax was small, showing clearly that the alcohols in this fraction were largely free.

Sahai and Chibnall (1932) found that the wax of Brussels sprout comprised 0·18% of the fresh leaves, and concluded that the wax was synthesised continuously and did not alter in composition throughout the life of the plant. These workers and Jordan and Chibnall (1933) believed that the waxes and unsaponifiable materials of plants were the end-products of metabolism. Chibnall and Piper and their co-workers concluded that the waxes of different plants vary in chemical composition only in the proportions in which the constituents are present, and that the physical properties of the waxes are determined not only by the amounts and chain lengths of the paraffins, free primary alcohols, free acids and esters present, but also by the chain lengths of the components of the esters. The chemical relationships found between the wax constituents led Channon and Chibnall (1929) and Chibnall and Piper (1934) to propose a general scheme for the metabolism of waxes (see Chapter 6).

The acetone precipitation procedure for the initial fractionation of waxy materials was retained by later workers, but attention was given to the nature both of the acetone-insoluble 'hard' or 'true' wax and the acetone-soluble 'soft' wax or 'oil'. Kurtz (1950) examined the relationship between the characteristics and yields of waxes from different species and plant age. Air-dried leaves were extracted with petroleum ether and hard and soft wax fractions obtained; the relative proportions of these varied with the species of plant. Blair, Mitchell and Silker (1953) precipitated the wax from alfalfa (lucerne) (*Medicago sativa*) with acetone. The paraffins, isolated by chromatography on a magnesium oxide column, approximated to a mixture of C_{29} and C_{31} compounds. Esters were also present and gave on saponification a mixture of *n*-octacosan-1-ol and *n*-triacontan-1-ol; these alcohols also occurred in the free state in the wax. The leaf wax of *Anona senegalensis* was investigated by Mackie and Misra (1956); *n*-hentriacontan-16-one (palmitone), the C_{28}, C_{30} and C_{32} *n*-primary alcohols and higher saturated *n*-fatty acids were isolated from the hard wax fraction, and palmitone, unsaturated and saturated fatty acids and sesquiterpene oils from the soft. The soft wax fraction possessed ***anthelminthic*** properties.

Cutin

The chemistry of cutin, which provides the framework of the delicate skin of soft tissues, is intimately bound up with that of suberin, which is the basic component of corky deposits found within and on the surface of plants. Cutin and suberin are clearly similar. They resist natural decay, are attacked only slowly by cold strong acids, react similarly to staining reagents and have long been referred to as acidic, fatty substances. Attempts to unravel their chemistry were first made at the beginning of the nineteenth century. Interest in the nature of cuticles and cutin arose from the work of plant physiologists who were investigating the respiratory functions of leaves; Brongniart (1830) drew attention to the need for more information on the structure of their cuticles to aid an understanding of the observations that had been made.

Cuticles were isolated from the cellular tissues by digesting leaves in strong, hot acids (nitric, chromic) or Schultze's reagent. Von Höhnel (1878) treated leaves of *Cereus* spp. with nitric acid and observed the 'true' cuticle as a thin dense lamella which formed the outermost layer of a thick cuticular complex. The action of warm caustic alkali also revealed the outermost lamella. By prolonging the action of the reagents the 'true' cuticles were isolated in the form of thin membranes which eventually dissolved to give a whitish waxy substance called cerin acid.

Important advances were made from 1859 onwards by Fremy who found that cuticular membranes were readily attacked by alcoholic alkali solutions. Fremy and Urbain (1885) gave the name cutose to the material which formed part of the epidermis of leaves and which resisted the action of strong acids. They recognised that the epidermis was composed of three layers; a surface waxy layer soluble in boiling alcohol, an inner cellulosic layer which from its solubility in ammoniacal copper solution after treatment with hydrochloric acid was designated paracellulose, and an intermediate membranous layer composed of cutose. The epidermal membranes were isolated by macerating leaves in warm water or by boiling them in hydrochloric acid, the waxy fraction was removed by extraction with alcohol and then ether, and the cellulosic substance was dissolved away by ammoniacal copper solution or sulphuric acid trihydrate. Cutose, isolated in this way, was obtained in quantity from the leaves of *Agave* sp. It dissolved in boiling caustic alkali or even boiling carbonate solution to give a water-soluble soap, from which two new fatty acids were isolated. One, a solid of m.p.76°, was named stearocutic acid and the other, a liquid, oleocutic; they occurred in cutose in the proportion of 1 part of stearocutic acid to 5 parts of oleocutic. The acids were isolated also from the cutose of apple and ivy (*Hedera helix*). Under certain influences, e.g. of boiling alcohol, they combined to form a double acid and this also could be saponified by alkali. Oleocutic acid could itself undergo change to a membrane-like form. Fremy and Urbain considered therefore that the cutosic membrane was built upon a combination of the two acids. From its chemical stability, they concluded that its principal role was the protection of delicate plant organs, and observed that it was found not only at the surface of aerial tissues, but often penetrating their interior. Some 40% of a substance similar to cutose occurred in cork; both Fremy and von Höhnel believed that cutose and suberin, although chemically related, were individual substances. Fremy's 'cutose' later became known as 'cutin'.

Legg and Wheeler (1925) examined the cutin of leaves of *Agave americana*. The cuticle was peeled off as sheets with cellular tissue attached, and extracted with alcohol to remove the wax. After additional extractions with benzene and chloroform, the cuticle was freed from cellulose by treatment with cuprammonium solution, washed with acid and dried. The material obtained was regarded as cutin; this on saponification with 3% alcoholic potash solution for 24 hr yielded a mixture of acids which was resolved by utilising the differential solubilities of the potassium or copper salts in water or alcohol. The chief products obtained were semi-liquid acids named cutic and cutinic; cutic acid $C_{26}H_{50}O_6$ comprised 65% and cutinic acid $C_{26}H_{44}O_6$ 10% of the total weight of acids liberated from the cutin. Another acid, $C_{19}H_{38}O_6$, was obtained as yellow crystals m.p. 107–108 °C. Legg and Wheeler (1929a) showed that the cuticle of *A. rigida* consisted of 55% cutin, 20% wax, 15% cellulose and 10% water-soluble material, and that its cutin contained the same kinds and quantities of acids as that of *A. americana*. They also obtained (1929b) sebacic (decanedioic) acid and an acid $C_{15}H_{28}O_3$ m.p 81–82 °C from cutin isolated from 'Papier Kohle' and bituminous coal, and supported Fremy's conclusion as to the chemical similarity of cutin and suberin.

Lee (1925) worked on the cuticles of chrysanthemum, rose and bluebell (*Endymion nonscriptus*) flower petals and of rhubarb (*Rheum rhaponticum*) petioles. She considered cutin to be a complex mixture of fatty acids, both free and combined with alcohols, that had undergone condensation and oxidation. Fatty acid soaps, unsaponifiable material probably containing higher alcohols, resinous substances and a compound giving reactions for tannin were also thought to be present. The cutin always contained larger proportions of oxy-fatty acids than fatty; this was ascribed to oxidation processes taking place during the formation of the cuticle, in accordance with the view of Lee and Priestley (1924) that the cuticle arose from the migration of lipoidal substances (unsaturated acids) to the surface and their oxidation to give a varnish-like covering. Lee considered that an essential difference between cutin and suberin lay in the nature of their constituent oxy acids. Early recognition of the structure of cutin came from the work of Markley and Sando (1933) on the skins of apple fruits. Cuticles were isolated from the surface layer by treatment with hydrochloric acid and extracted with solvents to remove the wax; the residue consisted primarily of cutin admixed with impurities derived from the cellular walls and with adhering remnants of the epidermal and sub-epidermal tissues. From their observations on treating this with alcoholic potash solution, they concluded that cutin consists of an ester-like substance which on saponification yields solid or semi-liquid acids. Some earlier workers believed that the cutin was bound to cellulose as a 'cutocellulose'; the results of Markley and Sando indicated that it existed in the membrane uncombined rather than as a definite chemical entity.

Suberin

The name 'suberin' was used by Chevreul (1815) to describe the substance, insoluble in water or alcohol, which comprises a large proportion of bottle cork. The name became applicable by common consent to the substance which appeared to give cork tissue or periderm its special properties, in particular, its impermeability to water. The relationship between suberin and cutin has long been debated; Priestley (1921) referred to the lack of precision in the use of the terms and suggested that the substances were perhaps better differentiated by their positions in the plant, suberin being formed in the periderm and cutin in the outermost walls of epidermal cells.

Investigations on the chemical nature of cork preceded those on the cuticle. Brugnatelli as early as 1878 obtained an acid named suberic acid by treating cork with nitric acid. La Grange in 1797 obtained the same material by oxidising cork. Suberic acid $COOH(CH_2)_6COOH$ and azelaic acid $COOH(CH_2)_7COOH$, also obtained by treating cork with nitric acid, were later believed to be decomposition products of suberin. Chevreul (1807, 1815) examined an alcoholic extract and the residual solids of cork; from the extract he obtained a white crystalline material which he believed to be a wax and named 'cerine'. Kügler (1884) and other workers also isolated cerine but still in impure condition. Istrati and Ostrogovich (1899) resolved cerine into two compounds by fractional crystallisation from chloroform; the less soluble they called cerin and the more soluble friedelin. Drake and Jacobsen (1935) showed the empirical formula of cerin to be $C_{30}H_{50}O_2$ and that of friedelin to be $C_{30}H_{50}O$. Both are classified as triterpenoids; cerin is 2-β-hydroxy-friedelin (Corey and Ursprung, 1956). They occur in cork to the extent of about 3%.

The close chemical similarity of suberin to cutin was recognised by von Höhnel (1877) who showed that cork, like the cuticle, could be hydrolysed by alkali to release acids. Cork was more susceptible to degradation than the cuticle, but the cerin acid he obtained from cork was considered to be identical with that from the cuticle. Fremy and Urbain (1882, 1885) believed that cork, like the cuticle, contained cutose and obtained from it products similar to their cuticle stearocutic and oleocutic acids. Differences in the results of staining tests, however, indicated that suberin and cutin were not identical in nature. Kügler (1884) also obtained a mixture of acids by the hydrolysis of cork, and isolated one member which he named phellonic acid, melting at 96°C.

An important step forward was made by Gilson (1890). The powdered cork of *Quercus suber* L. was saponified with alcoholic potash and three acids were isolated; phellonic acid m.p. 95–96°C, an acid known as phloionic acid and a semi-liquid suberinic acid, each with the properties of an oxy-fatty acid. Phellonic acid was also obtained by van Wisselingh (1893). Von Schmidt (1904, 1910) showed that it was a saturated hydroxy-fatty acid, and also isolated another acid which he named phellogenic acid. Scurti and Tommasi (1913, 1916) thought phellonic acid was α-hydroxybehenic acid ($CH_3(CH_2)_{19}CHOHCOOH$) and this view was accepted by Zetzsche and Rosenthal (1927) and Zetzsche, Cholatnikow and Scherz (1928). The phellonic and phellogenic acids of cork were further examined by Zetzsche and Bähler (1931a).

Knowledge of the hydroxy acids present in oak suberin was much advanced by Zetzsche and his collaborators. Phloionic acid melting at 124°C and phloionolic acid melting at 107–108°C were obtained from cork and identified by Zetzsche and Bahler (1931b). Their identities were confirmed by Zetzsche and Weber (1938); phloionic acid was 9,10-dihydroxyoctadecanedioic $COOH(CH_2)_7(CHOH)_2(CH_2)_7COOH$ and phloionolic acid was 9,10,18-trihydroxyoctadecanoic $CH_2OH(CH_2)_7(CHOH)_2(CH_2)_7COOH$. The constitution of phloionic acid was confirmed by synthetic work by Ruzicka, Plattner and Widmer (1942) and Hunsdiecker (1944). Phloionic acid was also synthesised by Gensler and Schlein (1955) who confirmed its *threo* configuration. Zetzsche and Sonderegger (1931) obtained cerin and friedelin from cork, which was then hydrolysed to yield phloionic, phloionolic and suberic acids. The suberic acid was not identical with an hydroxyoleic acid (an isomer of ricinoleic) isolated by Scurti and Tommasi from elder (*Sambucus nigra*) and named by them suberinic acid.

Zetzsche (1932) gave detailed descriptions of the isolation, properties, and products of hydrolysis of cutin and suberin. The suberin content of *Q. suber* bark was 35–44%, and the suberin contained 14–15% of phellonic acid, 1·4% of phloionic acid, 1·7% of phloionolic and 1% of eicosanedioic. He suggested that, from its properties, the acid of m.p. 107–108°C isolated by Legg and Wheeler (1925) from *Agave* cutin was identical with phloionolic acid obtained from suberin. Lee (1925) had concluded that an important difference between suberin and cutin was the presence of phellonic acid in suberin but not in cutin. Suberin was known to yield hydroxy-fatty acids which had been identified; cutin had yielded oxy acids but none had been identified. The chemical similarity between suberin and cutin was indicated by the isolation, by Matic (1956), of phloionolic acid from *Agave* cutin. Lee's conclusion is still valid; phellonic acid has not been detected in any cutin so far examined. Zetzsche (1932) described reactions which showed that cutin could be differentiated from suberin but which gave no information on whether cutin was identical in composition in different plants. The cutin of different plants could be composed of different acids or of the same acids in different proportions; the second of these alternatives has proved to be the correct one (Chapter 5).

4

The Anatomy and Morphology of Cuticles and Barks

CUTICLES

The word epidermis is used to describe the outermost layer of cells on a plant body. The epidermis of leaves and other parts of the plant varies in the number of its layers, its form, structure, the number of stomata that penetrate it, the number and form of the trichomes that emerge from it and in its possession of specialised cells.

The epidermal cells

Epidermal cells form a continuous layer over the surface of the plant body in its primary state. Most of the epidermal cells are similar within a single plant; they vary from species to species in detail, but are basically tabular

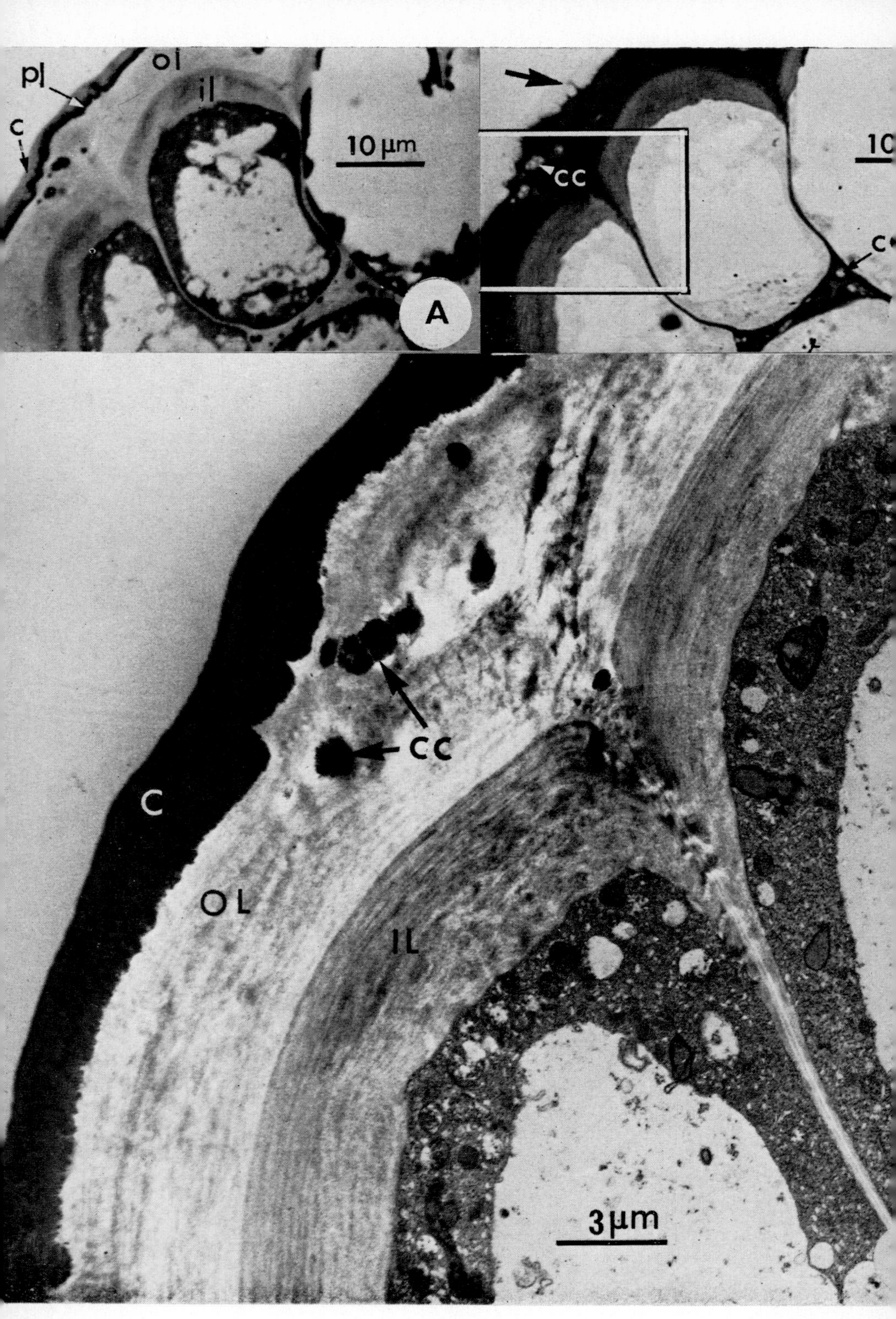

FIG.4.1

in shape (see Fig. 1.1). A few, however, show various degrees of specialisation related to their superficial position on the plant. Some form the guard cells of stomata (see Fig. 1.2) and hydathodes, some form trichomes and root hairs and a few may be secretory or sclerenchymatous.

Comparatively little work has been done on the fine structure of the epidermal cells of leaves. Hallam (1967) has shown that in *Eucalyptus* the epidermal cells of the leaves are not normally as extensively vacuolated as those of the palisade or the mesophyll within (see Fig. 1.1). In the leaves of *Ligustrum* and *Eucalyptus* and the coleoptile of *Avena* (Fig. 4.1) some, but not all, of the epidermal cell vacuoles are filled with granular material, apparently strongly osmiophilic, a reaction within the vacuole that is not noticed in the photosynthesising cells within the leaf. Epidermal cells are widely assumed to contain leucoplasts, but, with the possible exception of the guard cells, not chloroplasts. Amongst epidermal cells that are known to contain chloroplasts are those of some water plants and ferns and a few higher plants (Esau, 1965; Meyer, 1962). However, the leucoplasts of epidermal cells are sometimes difficult to find. In an extensive investigation into the fine structure of the leaves of many species of *Eucalyptus*, Hallam (1967) failed to find any obvious **plastids** in the epidermal cells and they have not been seen in *Ligustrum* and *Phaseolus* leaves. In every case well-preserved chloroplasts were seen in the mesophyll, palisade cells and guard cells of these leaves (see Fig. 1.1). However, Drawert and Mix (1963) in their study of the epidermal cells of the scale leaves of *Allium cepa* found within them numerous leucoplasts with the usual vestigial internal development. The distribution and structure of the plastids in epidermal cells obviously requires further investigation. Epidermal cells contain numerous normal ***mitochondria, endoplasmic reticulum***, ***spherosomes*** and ***Golgi bodies***. It is also believed that material secreted from the Golgi bodies is incorporated into the epidermal walls, particularly into the outer walls (Hallam, 1967) in ways very similar to those described in many other tissues (Clowes and Juniper, 1968). The cytoplasm of the epidermal cells in petals, some leaves, stems and petioles may contain anthocyanins and other pigments.

The walls of epidermal cells vary strikingly in thickness in different plants and in different parts of the same plant. Frequently the outer wall is the thickest (see Fig. 1.1). Often the outer wall of each epidermal cell protrudes

---

FIG. 4.1 Light and electron micrographs of sections through the cuticle of a coleoptile of *Avena sativa* var. *Victory*. ***A***, A section photographed under the light microscope stained with methylene blue azure II. The cuticle is indicated by the arrow c; the pectin layer by the arrow pl and two layers of the wall are recognised, an outer layer (ol) and an inner layer (il). ***B***, as in ***A***, but stained with PAS (periodic acid-Schiff). Note the thin strands of material which penetrate into the cuticle (heavy arrow) and the cutin cystoliths (cc) just under the cuticle. ***C***, An electron micrograph of a section adjacent to the area outlined in section ***B***. The same general features are visible as in the light micrographs except for the pectin layer. Glut/Osm/Pb/Ur. (T. P. O'Brien, 1967, Observations on the Fine Structure of the Oat Coleoptile. I. The Epidermal Cells of the Extreme Apex. *Protoplasma*, **63**, 393, Figs. 1–3)

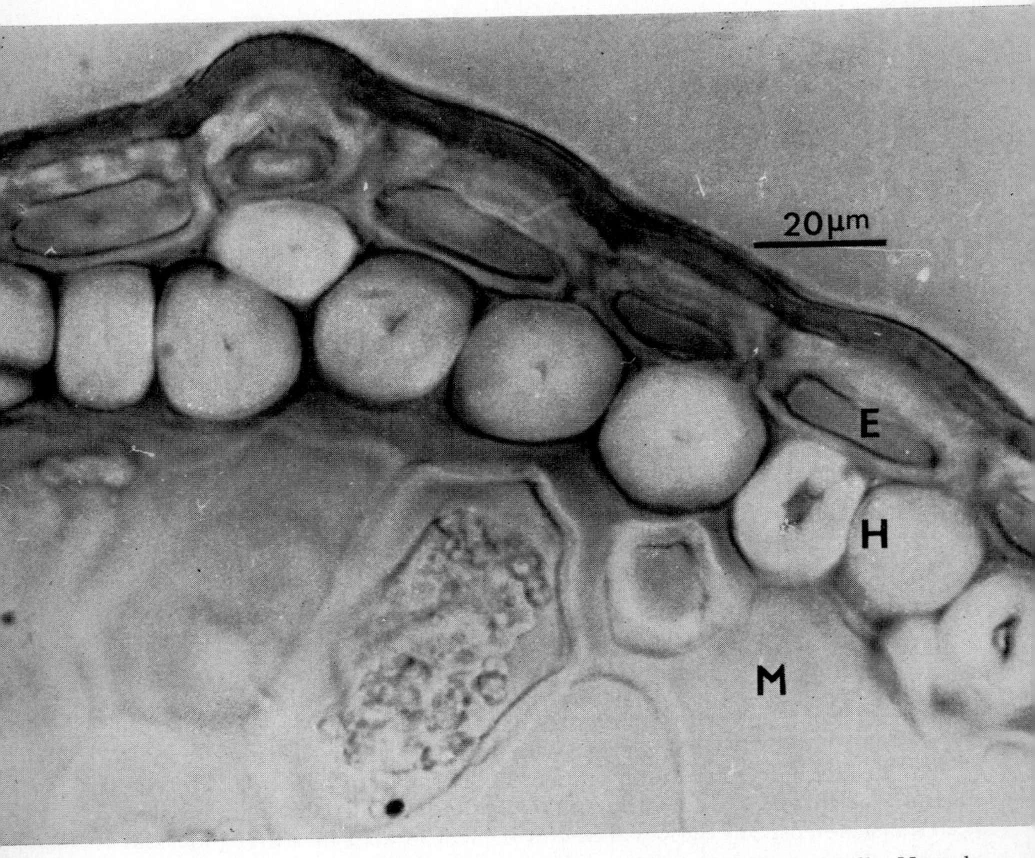

FIG. 4.2 Light micrograph of a section through a spruce (*Picea*) needle. Note the very heavy cuticle, the thick outer tangential walls of the epidermal cells ***E***, and the heavily thickened walls of the hypodermis ***H***. ***M*** is the mesophyll. (Courtesy von O. L. Lange)

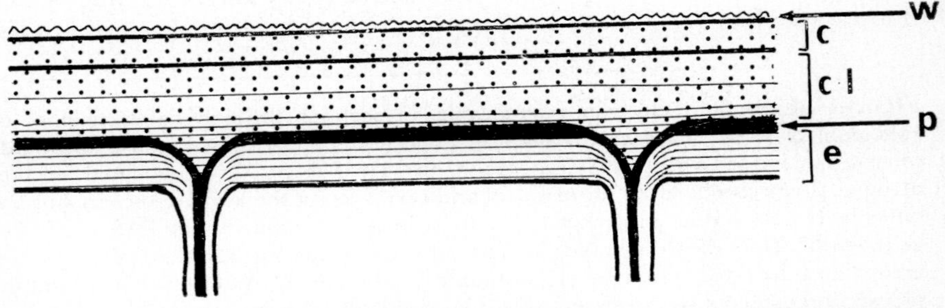

FIG. 4.3 A hypothetical cuticle: w, the epicuticular wax; c, the cuticularised layer; cl, the cutinised layer of the cell wall; p, the 'pectin' layer, and e, the epidermal cell wall

as a small papilla. In many shade plants with velvety leaves such as *Ficus barbata*, *Begonia* and *Anthurium* species the outer walls are carried up into high conical papillae (see Fig. 7.7). Sometimes these papillae are the result of excessive development of the pectin layer, of the cellulose layer, of the cuticle itself or in *Petraea volubilis* of a concretion of silica (Haberlandt, 1914). Haberlandt believed that these papillose epidermal cells may have a function in concentrating as a lens the limited light below the canopy of a tropical forest. This hypothesis will be considered in Chapter 7. In conifers all of the epidermal walls are thick, sometimes so extensively thickened as almost to obliterate the lumen (Fig. 4.2). The walls may be ***primary*** or ***secondary***, although the problem of primary and secondary deposition of wall material in epidermal cells has scarcely been investigated. The radial and the inner tangential walls of epidermal cells have primary pit fields (Esau, 1965). They also have ***plasmodesmata*** visible under the light microscope (Franke, 1962) and under the electron microscope. Clowes and Juniper (1968) suggest that the plasmodesmata revealed by the electron microscope should be distinguished by the name 'microplasmodesmata' from those seen under the light microscope since they differ markedly in thickness.

An epidermis may consist of more than one layer of cells. Such extra layers are usually morphologically and physiologically distinct from the ground tissue, i.e. palisade and mesophyll, that lies within the leaf. Such sub-epidermal layers are sometimes called the hypodermis and the number of extra layers that this may contribute to the epidermis varies from one to sixteen (Esau, 1965). When a leaf possesses a hypodermis, the outermost layer resembles the normal epidermis in having a cuticle and the inner layer or layers commonly lack chlorophyll and are differentiated as storage tissue or structural tissue. Examples of leaves with a single hypodermal layer are shown in Figs. 2.1 and 4.2. In the micrograph of the spruce (*Picea*) needle, the hypodermal cells have differentiated into fibre-like cells and the lumen of these cells is almost completely obliterated (von Lange and Schulze, 1966).

The conventional view of the epidermis and cuticle of a mesophytic leaf is summarised in Fig. 4.3. The walls of the epidermal cells (e) are thought to have a normal architecture and to consist of cellulose embedded in a matrix of pectin and hemicellulose. Outside this layer there is generally believed to lie a band of more or less pure pectin continuous with that of the anticlinal walls of the epidermal cells (p). Outside this again occurs the so-called cutinised layer (cl) in which cellulose and pectin are incrusted with cutin and on the outside of the leaf is a more or less pure layer of cutin (c). Waxes are supposed to occur within the cutinised layer, but are often assumed to be absent from the pure layer of cutin. A thin layer of more or less pure wax also occurs outside the pure cuticle layer (w). This view has, if at all, only a limited application and more recent work with the electron microscope casts some doubt even on these general conclusions.

Cellulose

In vascular plants the basic skeleton of almost all cell walls including those over which a layer of cutin develops is cellulose. Cellulose is a β-1,4 linked glucan in which each glucose molecule is rotated at 180° to its neighbour in the chain. The hydroxyl groups of a cellulose chain face outwards, rendering linkage with other adjacent molecules possible. The basic structural unit of cellulose in the wall is the microfibril which consists of 300–500 individual cellulose molecules arranged strictly parallel to one another. This microfibril is probably rectangular in cross-section rather than square which may account for some of the discrepancies in measurement given; it seems most likely that its dimensions are about 8 nm by 25 nm. The surface of the microfibril possesses a number of hydroxyl groups available for linkage with other carbohydrates and the boundary between the microfibril and the matrix material in between is probably not an abrupt one, but a gentle transition from the crystalline to the non-crystalline state. Links with proteins probably also occur (Lamport, 1965). Isolated and shadowed preparations of microfibrils under the electron microscope give the impression of a textile such as linen, in which each thread is in contact with its neighbours. This is a misleading impression because in the unmacerated state each microfibril is separated from its neighbour by about 40 nm which is about twice the microfibrillar diameter, and moreover the cellulose fibrils comprise in a growing cell wall only about 2·5% of the weight of the wet cell wall as a whole (Houwink and Roelofsen, 1954).

Other components of the epidermis

Between the cellulose microfibrils lie the non-crystalline polysaccharides. These, by some authors, are distinguished as separate groups such as pectin, protopectin and hemicellulose A and B. These fractions are obtained from cell wall preparations by extraction with aqueous solutions of increasing alkalinity. This process progressively dissolves given sugars by breaking ***covalent bonds***. When a sufficient number of bonds are broken the polysaccharides in turn pass into solution. There is, however, no fundamental chemical distinction between any of the so-called fractions; the process degrades a spectrum of polysaccharides from those with a high content of uronides and a low content of xylan and gluco-mannan to those with a low or non-existent uronide fraction and a high level of xylan and gluco-mannan (Barrett and Northcote, 1965). The variability of the non-crystalline fraction of the cell wall also leads to confusion. For example, the so-called pectic fraction from sunflower (*Helianthus annuus*) heads contains over 90% polyuronic acid, whereas the equivalent fraction from oat (*Avena sativa*) seedlings has only 23% of galacturonic acid (Bishop, 1955; Bishop, Bayley and Setterfield, 1958). Efforts to extract pure polyuronide fractions from plant cell walls, with the possible exception of the *Helianthus* heads, have, except where extensive degradation of the polysaccharide is carried out, failed completely, presumably due to extensive bonding between the neutral and acid sugars

(Albersheim, 1965b). The matrix is amorphous and highly hydrated in the normal state.

The cell walls of leaves, in particular those of the epidermal cells, are comparatively thin. However, the walls of the epidermal cells are generally thicker than those of the palisade cells and mesophyll, and the outer radial walls are generally thicker than anticlinal or inner radial walls (see Fig. 1.1). Very little secondary cell wall deposition is likely to occur in the epidermal cell of a normal mesophytic leaf. However, in xeromorphic plants, the epidermis is believed to have lignified walls. Cell walls believed to be incrusted with lignin have all the properties associated with wood. Only a few epidermal walls of leaves undergo this incrustation with lignin and, as a result, very little is known about their particular pattern of lignification, but presumably it follows a course similar to that which occurs in structural tissues. Lignification normally begins in the middle lamella region. This is the region of the cell wall laid down at mitosis and consisting exclusively of noncrystalline polysaccharides usually with a high uronic acid fraction (Albersheim, 1965a). This concentration of lignification may occur because the density of material in the middle lamella is lower than in the primary and secondary walls deposited on either side, hence there is more room for the polymers of the lignin to penetrate. Two potential lignin precursors, coniferin and syringin, are found in meristematic tissues. Labelled coniferyl alcohol fed into the cambial zone is incorporated into the lignin of mature wood. Therefore it is possible that areas of lignification found in leaves are derived from precursors that originate in the meristematic tissues. In *Eucalyptus* species and some other xeromorphic plants lignification of the outer radial walls is irregular and forms a domed structure to the outer surface of the cell over which the cuticle lies (Hallam, 1967).

Between the cellulose-rich layer of the epidermal wall of *Avena* **coleoptile** O'Brien (1967) recognises an additional layer (***ol*** in Fig. 4.1). This outer layer is rich in hemicellulose, poor in cellulose and continuous with the anticlinal primary walls of the epidermal cells. He believes that this layer is analogous to the primary wall. However, no clear division between an inner layer and an outer layer is apparent in for example the epidermal wall of celery (*Apium*) (Fig. 4.4). The range of plant material upon which electron microscope examinations have been carried out is so limited that judgement must be reserved as to which is the more common condition.

The pectin layer

Between the layer or layers of the epidermal cells and the cuticle above there is believed in many plants to be a layer of pectin, which is believed to be continuous with the middle lamella of the anticlinal walls of the epidermal cells (Fig. 4.3). This layer stains with ruthenium red and is isotropic under polarised light (Roelofsen, 1959). Neither the existence of a ruthenium red positive band nor its isotropy is positive proof that this layer is pectin. Ruthenium red is specific for carboxyl groups (carboxyl groups are found in cuticular constituents other than pectin) and isotropy is also found elsewhere in the cuticle (see Fig. 2.1). The view that there exists a layer of more or less

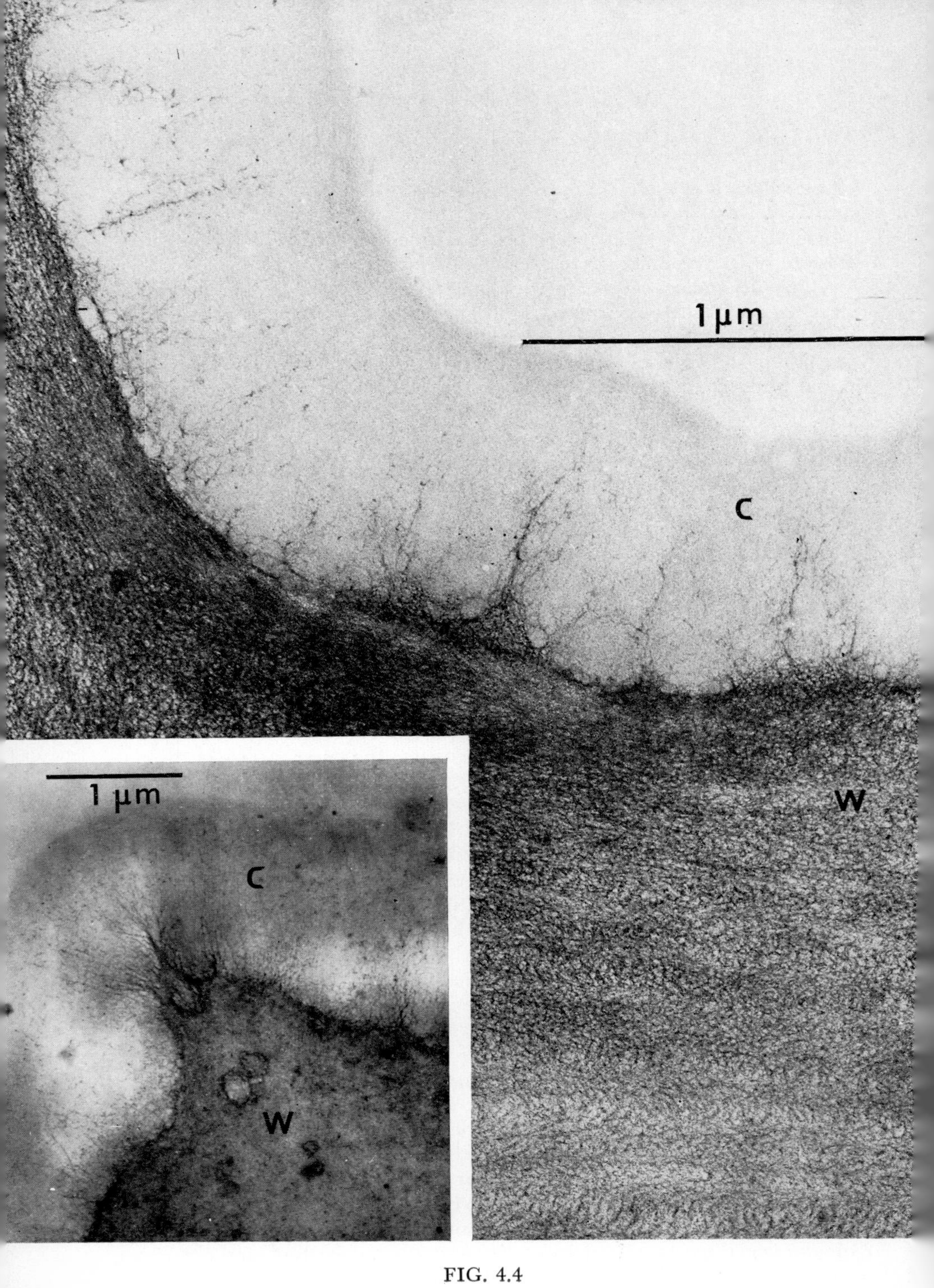

FIG. 4.4

pure pectin is, however, supported by observations made during the isolation of cuticles from leaves (see p. 47). Orgell (1955) and other workers found that various species respond quite differently to treatment with pectinases. Some cuticles are detached intact within 3–12 hr, e.g. *Convolvulus arvensis*, *Peperomia* spp., *Philodendron* spp. and *Vinca major*. Others are detached rapidly, but come away only as flecks of cuticle and not as intact sheets. Examples of this group are *Lactuca scariola*, *Phaseolus vulgaris*, *Rumex crispus* and *Sonchus oleraceus*. Others such as *Citrus* spp., *Nicotiana* spp. and *Prunus* spp. give intact cuticles, but they are detached more slowly, between 12 hr and 3 days. Pear leaf cuticle, studied by Norris and Bukovac (1968) also seems to come into this category. The cuticles of some species, amongst them *Vitis* sp., *Ficus elastica* and *Zea mays*, separate very slowly or fail to separate at all. After treatment with pectinase the tissues lose their affinity for ruthenium red. Skoss (1955) isolated cuticles using a preparation of *Clostridium roseum*, a micro-organism obtained from the gut flora of a herbivore. His results are very similar to those above and the plants he selected separated into the three groups as before. These groups suggest that the distribution of the pectin layer is very variable in amount when it is present and in a very large number of species may be absent altogether or traversed by so much pectinase-resistant material that it fails altogether to act as a separating layer between the epidermal cells and the cuticle. The fragmentation of the cuticle, noticed by both Orgell and Skoss in many species, may be due either to a cuticle so thin and fragile that it breaks up spontaneously or to the interleaving between the cuticle of material susceptible to pectinase.

Holloway and Baker (1968b) have adopted a different approach and use, instead of a pectinase preparation, zinc chloride-hydrochloric acid solution. They claim that not only does this detach a wider range of cuticles, but it also does so quicker and with less fragmentation than the other methods described above. Therefore some of the earlier work and the interpretations derived from it should perhaps be re-examined.

Recent work with the electron microscope by O'Brien (1967) shows that polyuronides, the principal constituents of pectin, are present throughout the width of the epidermal wall, but are especially rich in the layer designated **pl** in Fig. 4.1. This is in general agreement with the cuticle isolation experiments and with the polarising light experiments of Meyer (1938) (Fig. 4.5) except that neither the electron microscopy nor the periodic acid–Schiff (PAS) (see p. 82) and methylene blue-azure II tests with light microscopy suggests that this layer is continuous with the middle lamella of the anticlinal walls of the epidermal cells.

If then, as the evidence suggests, there frequently is a layer or possibly a series of layers of pectin between the epidermal walls and the cuticle, as the

FIG. 4.4 Electron micrographs of sections through the cuticles of a celery (*Apium*) petiole and (inset) of a leaf of *Eucalyptus cinerea*. The projections from the wall (***W***) into the cuticle (***C***) have not yet been positively identified, but see p. 81 and Fig. 4.1 ***B***. *Apium* section, Glut/Osm/Pb (Courtesy M. C. Ledbetter). *Eucalyptus* section, Glut/Osm/Pb/Ur (Courtesy N. D. Hallam)

leaf expands there must be a continuous provision of pectin into this space. Therefore some mechanism exists to supply not only the cellulosic and other matrix materials of the epidermal wall and the cutin and waxes of the cuticle, but also a more or less homogeneous layer of pectin in between. It is unlikely that the layer of pectin represents the remains of the middle lamella

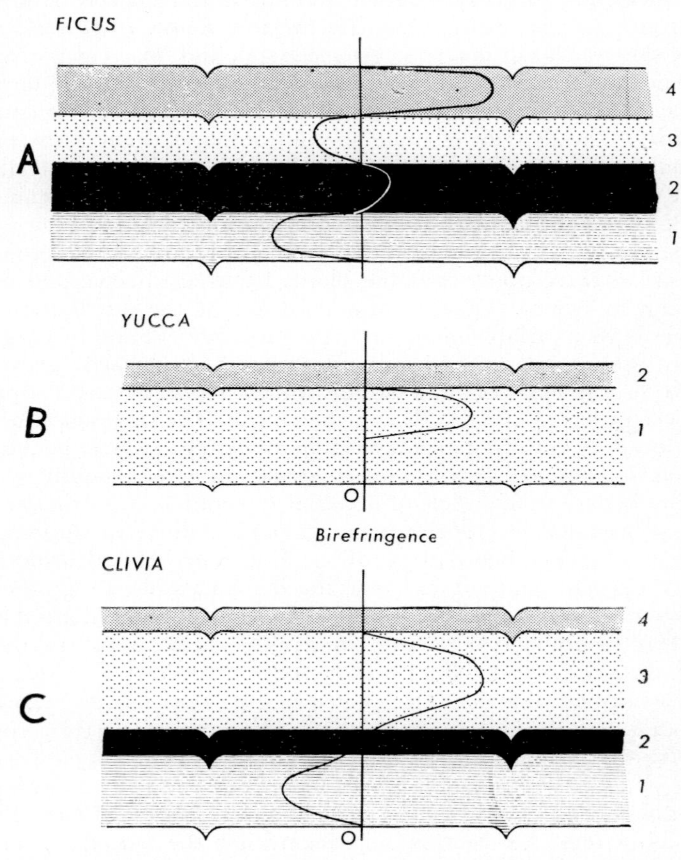

FIG. 4.5 The distribution and optical properties of different components of three cuticles. The line 'O' indicates the position of minimal optical activity and deviations of the line to either side, positive or negative birefringence. In *Ficus* (***A***), the epidermal cell wall (1) is strongly birefringent and the 'pectin' layer (2) above it weakly birefringent, but of opposite sign. The cutinised layer (3) shows the same direction of birefringence as the cellulose cell wall, but the cuticle (4) is anomalous in being strongly birefringent in the opposite direction. *Yucca* (***B***) has a conventional isotropic cuticle (2), no obvious pectin layer and a weakly birefringent cutinised layer (1). *Clivia* (***C***) has the most common structure. The cuticle proper (4) is isotropic, the cutinised layer (3) is birefringent, the pectin layer (2) is isotropic again and the epidermal wall (1) is birefringent at opposite sign to the cutinised layer. (Redrawn from Meyer, 1938)

laid down when the cell was formed since the surfaces of the epidermal cells, over which this layer is developed, enlarge greatly after their pattern is established. The developmental sequence of the pectin layer from the primordial to the mature leaf state has not been followed, so it is not known for certain whether it is deposited before or after cell extension has ceased. There are precedents in other tissues for the deposition of pectin both during extension and in the mature and fully elongated state. Majumdar and Preston (1941) have shown that alternate layers of cellulose and pectin are laid down predominantly during extension growth in the development of collenchyma cells of *Heracleum sphondylium*. What controls the alternating pattern of deposition is not known and a study of the activity of the Golgi bodies and the distribution of the microtubules on the lines of the work done by Wooding and Northcote (1964) would be very instructive. The deposition of pectic materials in cells which have ceased to grow is reported by Carlquist (1956); here warts of pectic material are intruded into the intercellular spaces of the collenchyma of some Compositae. This intrusion takes place after the collenchyma cells have attained their mature shape, and hence presumably after deposition has ceased in the collenchyma cells themselves.

According to O'Brien (1967) the pectin layer could originate from the anticlinal walls of the epidermal cells to unite in a continuous layer over the surfaces of the epidermal cells. Because of relatively unsophisticated isolation studies using mixed enzyme preparations and the limited use of electron microscopy, the structure of the transition zone from the epidermal wall to the cuticle is imperfectly understood. In leaves from which the cuticle cannot be removed or can be removed only with difficulty, cutinisation may penetrate far down in the anticlinal walls of the epidermal cells; at least a third of the way down according to Scott, Shroeder and Turrell (1948) and Meyer (1938). The penetration of cuticular pegs down the anticlinal walls of the epidermal cells can continue so far that they are found on isolation to have been penetrated by the plasmodesmata joining epidermal cells (Skoss, 1955).

The cutinised layer

The cutinised layer, conventionally thought of as that region outside the layer of pectin (see Figs. 4.1 and 4.5) contains cellulose and presumably matrix materials as well. However, the cellulose of this cutinised layer is so completely enveloped by the cutin that it no longer gives the normal chemical reactions. According to Roelofsen (1959), in sections cut from walls from which the wax has been extracted, the cutinised layer containing cellulose becomes positively birefringent with respect to the tangent plane of the wall, whereas the pure cuticle above stays isotropic or nearly so. However, as Roelofsen points out, the amount of cellulose is so small that its birefringence may be masked by traces of residual wax. Moreover, as Hülsbruch (1966), O'Brien (1967) and Hallam (1967) have shown, and as can be seen in Figs. 4.1 and 4.4, cellulose fibres appear to extend into the cutinised layer, but not necessarily in the same plane as those of the epidermal walls. Figures 4.1 and 4.4 suggest that they may even form a reticulum within the cutinised layer. According to O'Brien this fibrillar material gives a positive periodic acid–Schiff reaction under the light microscope. The reaction may indicate the presence of cellulose or of certain hemicelluloses or pectin. However, the obvious fibrillar nature of this reticulum suggests that the fibres are, at least in part, cellulose. In *Eucalyptus* this fibrillar material forms small tufts projecting up into the domes of cuticle over the radial walls of the epidermal cells (Hallam, 1967) (see Fig. 4.4 insert). In *Ilex integra* (Hülsbruch, 1966) these radial striations are so well developed in older leaves they can clearly be seen under the polarising microscope. Roelofsen (1959) suggests that the appearance of positive birefringence (with reference to the tangent plane of the wall) on removing the wax from the cutinised layer is due to the presence of orientated cellulose microfibrils previously masked by waxes The electron microscopy of four unrelated plant cuticles indicates that what is thought to be the cellulose in the cutinised layer is not orientated and hence cannot be responsible for the reappearance of birefringence.

Under the electron microscope the fibrillar material extends throughout the cutinised layer and is often well developed around the cutin ***cystoliths*** (see p. 114 and Fig. 4.1), but does not quite reach the surface of the cuticle. It may be that the cellulose-free layer corresponds to the cuticle proper (as defined by Esau).

Deposits of silica (plant opals) are found in the epidermal walls of many plants, for example *Equisetum*, many Gramineae, Cyperaceae and Palmae (Metcalfe and Chalk, 1950). Deposits of calcium carbonate, in the form of cystoliths, are found in outgrowths of some epidermal cell walls (Esau, 1965). Other cystoliths, but of an unknown constitution, occur in the cuticle itself (O'Brien, 1967). Others are described by Fritz (1937).

The whole of the carbohydrate moiety of the epidermal cell wall and the cuticle are completely bathed in water (this is the 'apoplast' of Crafts, 1961). Gaff, Chambers and Markus (1964) have shown that *Helxine* leaves induced to take up colloidal gold in the transpiration stream move the small particles about readily in the epidermal walls. They obtained some evidence that these particles move not randomly in the walls, but in channels up to 30 nm in diameter.

Ectodesmata

Through the outer tangential walls of the epidermal cells of a wide range of plants run canals, or more precisely different regions within the walls, now known as ectodesmata (Figs. 4.7 and 4.8). They were discovered by Schumacher and Halbsguth in 1939, but were called by them plasmodesmata. There seems no doubt now that they are not plasmodesmata in the usual sense, i.e. they are not cytoplasmic continuities. Schnepf (1959) showed that under the electron microscope there is no sign of a ***plasmalemma*** lining the canal as is seen in ordinary plasmodesmata (Clowes and Juniper, 1968). However, Schnepf's electron micrographs do reveal differences in regions of the wall corresponding to the ectodesmata (Fig. 4.6). The differences are apparent both in the electron microscope and the light microscope only if advantage is taken of the capacity of ectodesmata to reduce silver salts or, as

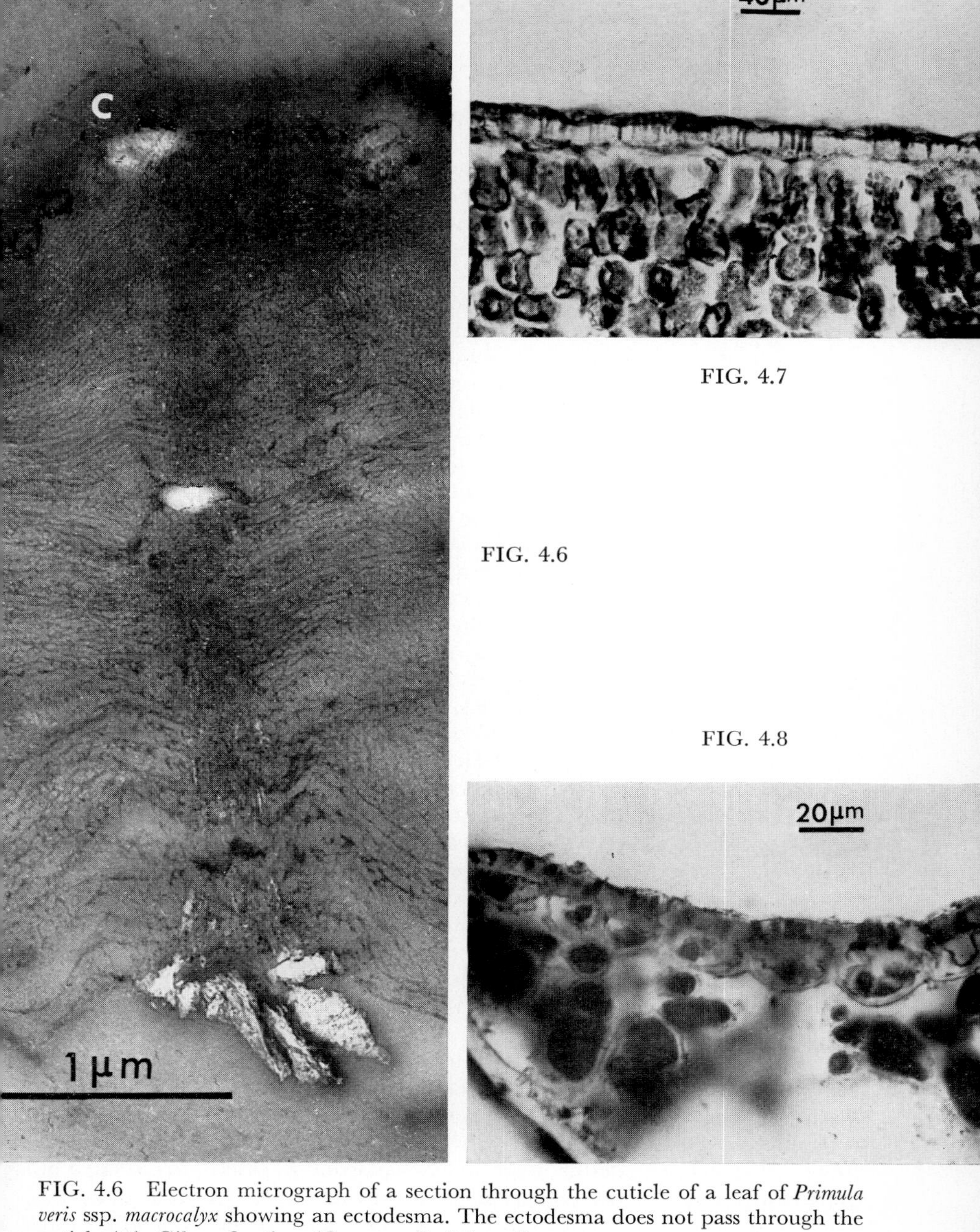

FIG. 4.7

FIG. 4.6

FIG. 4.8

FIG. 4.6 Electron micrograph of a section through the cuticle of a leaf of *Primula veris* ssp. *macrocalyx* showing an ectodesma. The ectodesma does not pass through the cuticle (*C*). Gilson fixative. (Courtesy E. Schnepf)

FIG. 4.7 Light micrograph of a section through the adaxial epidermis of a leaf of *Plantago major* showing ectodesmata. (Franke, W., 1967, Ektodesmen und die peristomatäre Transpiration. *Planta*, **73**, 138–154. Springer, Berlin–Heidelberg–New York)

FIG. 4.8 Light micrograph of a section through the adaxial epidermis of a leaf of *Helxine soleirolii*. Note that the ectodesmata in this leaf sometimes look like inverted mushrooms (See Fig. 4.9.) (Franke, W., 1967, Ektodesmen und die peristomatäre Transpiration. *Planta*, **73**, 138–154. Springer, Berlin–Heidelberg–New York)

in the so-called ***Gilson fixative,*** mercuric chloride. Ectodesmata are usually invisible under ordinary conditions of fixation and staining for the light and electron microscope. Arens (1968) has reported that there does seem to be one situation in which ectodesmata may be seen without the use of any special staining techniques. A radial structure originating from the stomata of *Ouratea spectabilis* proved to be rows of ectodesmata which were visible because they were so large. These ectodesmata could also, under experimental conditions, be made to secrete fluid droplets. Thus Arens thinks that they may be pathways of peristomatal transpiration.

Ectodesmata, in contrast to plasmodesmata, do not connect one cell to another, but pass from the plasmalemma of the epidermal cells through the outer epidermal cell walls or through the walls of guard cells and some trichomes (Fig. 4.9). They do not extend to the outside of the plant and are always covered by the cuticle. In the epidermal cells of some plants such as *Helxine soleirolii* they have the shape of an inverted mushroom (Fig. 4.8 and 4.9). In other plants the ectodesmata extend only part of the way through the epidermal walls or are apparently segmented (Franke, 1964). A diurnal rhythm in the numbers of ectodesmata per unit area of epidermal wall has been detected. For example, in *Passiflora incarnata* a greater number of ectodesmata is observed by night than by day (Lambertz, 1954). It seems unlikely that they are retracted and reformed by the cytoplasm and more likely that enzymic changes in the epidermis from day to night bring about the differences in number observed. This view is supported by the observation that leaves of *Plantago major* floated on solutions of amino acids, sugars or caffein have more ectodesmata than untreated leaves, whereas inorganic salts reduce their frequency (Schumacher and Lambertz, 1956). An increase in the number of ectodesmata was noticed by Brants (1964) after she had applied pressure to the leaves to assist infection by virus. Suchorukov and Plotnikova (1965) found that changes in light or humidity, contact with poisons, or the action of toxins produced by infection may cause them to disappear or change their shapes. Ectodesmata are not evenly distributed. In *Plantago* leaves the areas that are especially rich in ectodesmata are the epidermal cells above, on both sides and beneath the larger of the leaf veins, the guard cells, the conical hairs and the cells surrounding the capitate hairs (Franke, 1961) (Fig. 4.10).

In surface view ectodesmata appear as points. Frequently, as in *Plantago* leaves, they are crowded, often in straight lines, above the anticlinal walls. Often they encircle leaf hairs, and most frequently they are found in heavy concentrations over the guard cells of the stomata (Fig. 4.10). On the present evidence, ectodesmata are not cytoplasmic in origin, are relatively ephemeral regions of the epidermal cell walls and their frequency can be raised or depressed by changes in the environment. They are not structural features in the normal sense, but are probably local concentrations of reducing substances. Their possible role in the adsorption and loss of substances through the cuticle will be discussed in Chapters 7 and 8.

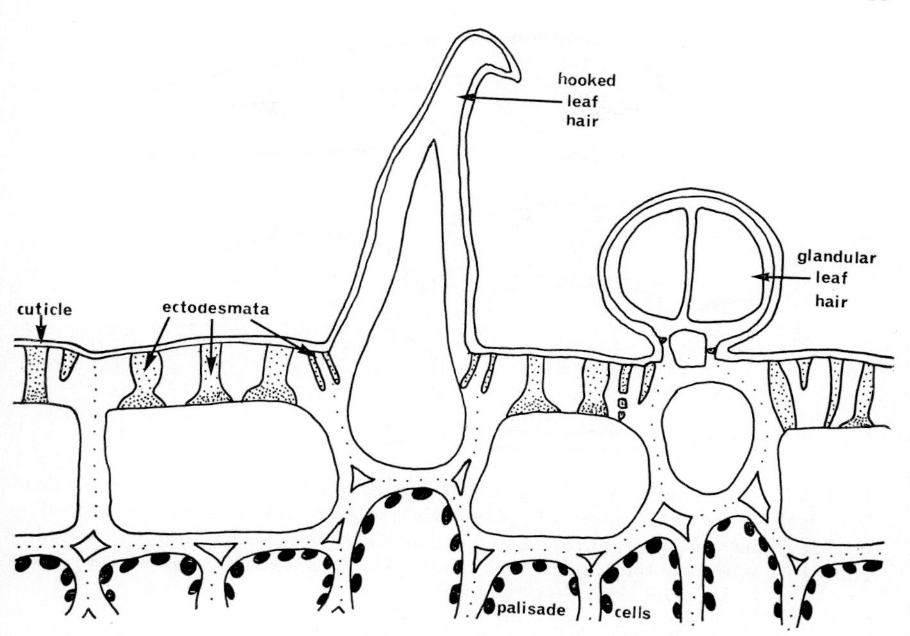

FIG. 4.9 Drawing of ectodesmata and leaf hairs in *Helxine soleirolii*. (Redrawn from Franke, 1960)

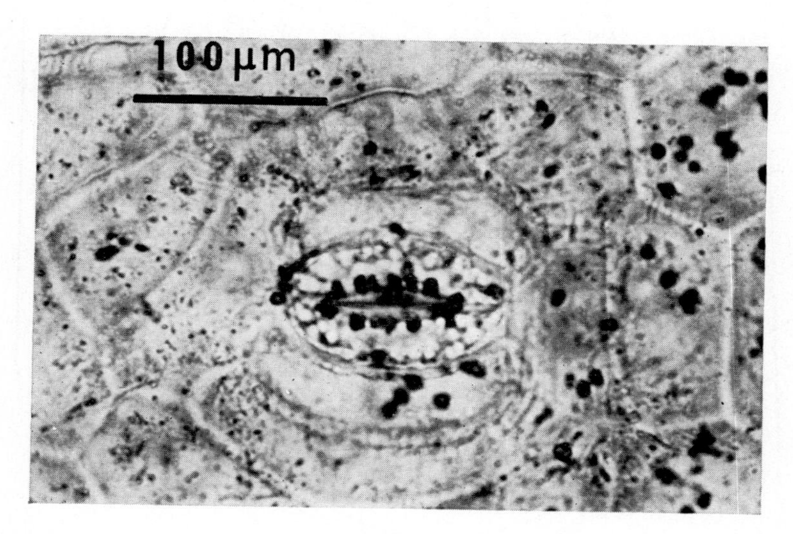

FIG. 4.10 Light micrograph of the adaxial epidermis of *Zantedeschia aethiopica* showing a stoma surrounded by ectodesmata. (Franke, 1967, Ektodesmen und die peristomatäre Transpiration. *Planta*, **73**, 138. Springer, Berlin–Heidelberg–New York)

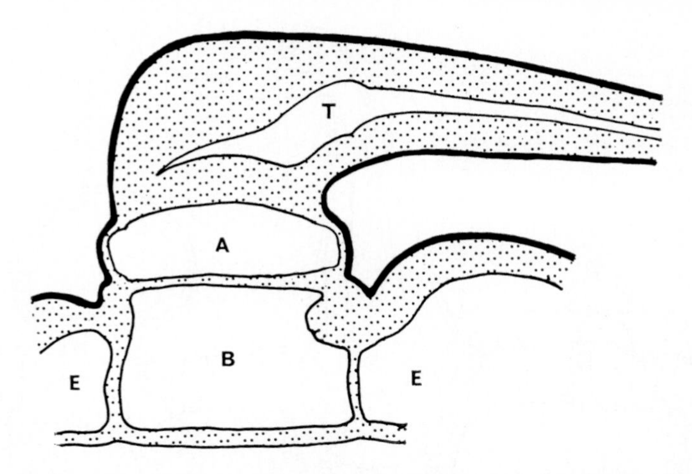

FIG. 4.11 Drawing of a trichome on the adaxial surface of *Convolvulus cneorum* leaf. The thick-walled terminal cell (***T***) of the hair is supported by a thin-walled absorbing cell (***A***). The wall between cell ***A*** and ***B*** (the basal cell) is thin, whereas the other epidermal cells have thick outer radial walls (***E***). (Redrawn from Haberlandt 1914)

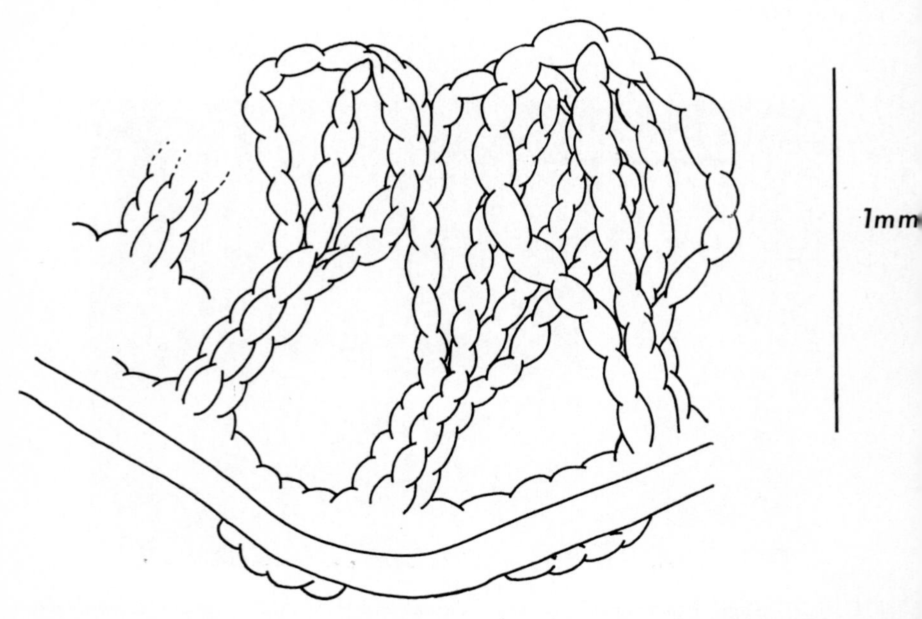

FIG. 4.12 Drawing of the trichomes on the upper surface of *Salvinia natans* leaf. (Courtesy P. R. Williams)

Leaf hairs or trichomes

All the unicellular and multicellular projections of the epidermis are known as trichomes (Greek, a growth of hair). Trichomes are sometimes distinguished from the so-called emergences, e.g. prickles and thorns, on the grounds that emergences are formed from sub-epidermal as well as epidermal tissues. However, a precise distinction between the two is not possible. Trichomes can be found on all parts of the plant; some persist throughout the life of an organ and some are ephemeral. Some retain their living cytoplasm and some persist, but in a dry state.

Metcalfe and Chalk (1950) list all the types of trichomes found on plants and indicate their taxonomic value. Amongst the many types that they recognise are stinging hairs, laticiferous hairs, salt glands, chalk glands, mucilage hairs and calcified and silicified hairs. Uphof (1962) also discusses their physical properties, growth and development, physiology and ecology. Some secretory and water-repelling types of trichome will be mentioned in Chapter 7. Often two or more different types of trichome may occur on a single leaf (see Fig. 4.9). Trichomes can conveniently be divided into non-glandular and glandular types. The former range from the simple unicellular or multicellular hairs such as are found on many leaves, petioles, and stems (Fig. 4.11) and also include the cotton (*Gossypium*) fibres, a seed hair up to 6 cm long. This group includes the papillate hairs, hairs like bladders, such as are found in *Atriplex*, where the vesicle dries out at maturity so that the salt content remains on the surface of the cuticle. It also includes the looped hydrophobic hairs of *Salvinia* (Fig. 4.12) and the straight lobed hairs of *Pistia* (Fig. 4.13), both of which are hydrophobic and hold water droplets away from the surface and prevent a wetting of the cuticle itself. Many of this type are of considerable importance in modifying the insolation, the wettability of, or water loss from, a leaf.

Glandular types of trichomes may be unicellular, multicellular or scale-like. Digestive glands, which are found on the surfaces of the leaves of *Nepenthes*, *Drosera*, *Drosophyllum*, *Sarracenia*, *Byblis* and *Pinguicula*, secrete digestive enzymes. Mucilaginous glands are formed not only on the adhesive surface of insectivorous plants such as *Pinguicula* and *Drosera*, but also on the surfaces of leaves, often with no obvious function (Uphof, 1962). Mucilage hairs are also found on seedcoats (Netolitsky, 1926). These mucilaginous glands are extremely diverse in structure, but generally they are covered by a thinner cuticle than that covering the normal epidermal cells. Occasionally mucilage is secreted through pores in the cuticle, but more often it accumulates until the cuticle is ruptured (Schnepf, 1966b). The glandular trichomes of *Urtica* are stinging hairs; the upper part of the trichome wall is impregnated with silica and calcium. The brittle stem breaks, penetrates the skin and injects its irritating contents of histamine and acetylcholine (Fig. 4.14). Both unicellular and multicellular glands exist from which nectar is secreted (Schnepf, 1964a). Some of these have no cuticle and the secretion of sugar takes place directly by diffusion. Some, with a cuticle, accumulate the secretion between the wall and the cuticle and finally

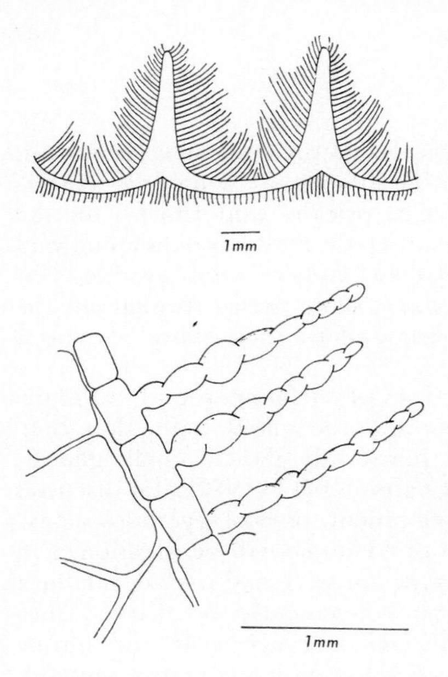

FIG. 4.13 Drawing of the trichomes on the upper surface of *Pistia stratiotes* leaf. (Courtesy P. R. Williams)

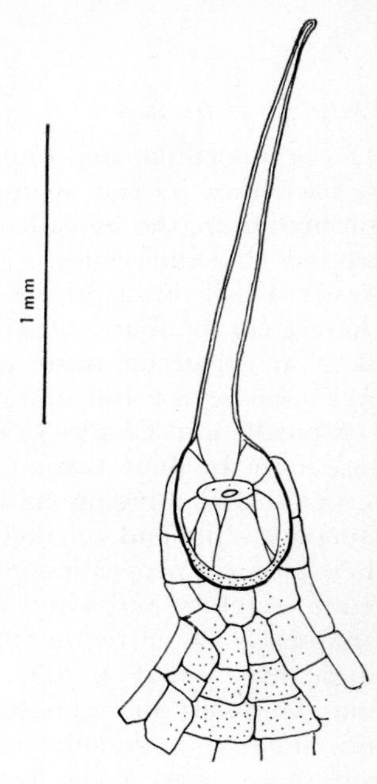

FIG. 4.14 Drawing of a trichome from *Urtica dioica*.

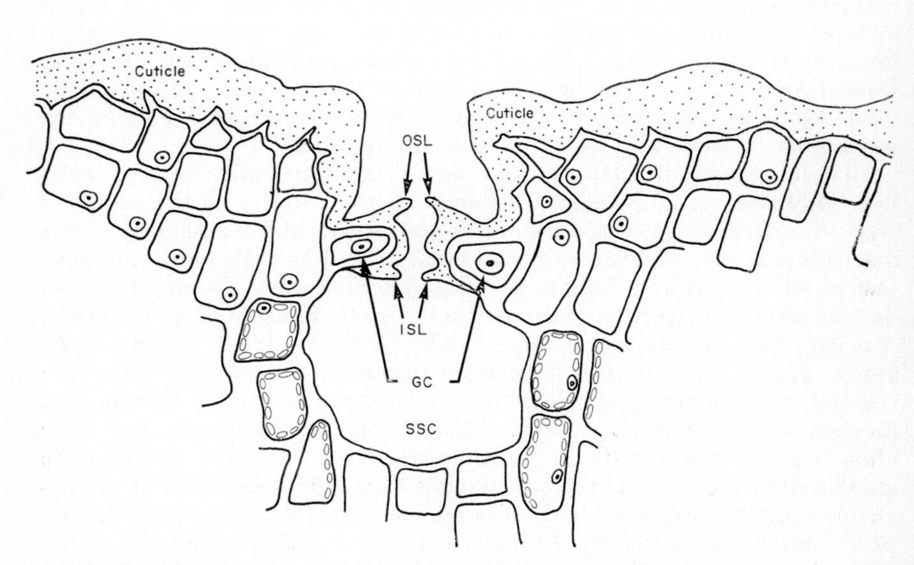

FIG. 4.15 Drawing of a stoma on the abaxial surface of a *Euonymus* leaf. The guard cells (GC) are surrounded by the cuticle which develops prominent inner (ISL) and outer (OSL) stomatal lips. Below the guard cells is the space known as the sub-stomatal cavity (SSC).

rupture the cuticle to release the secretion. A few glands are modified as hydathodes and secrete water through pores in the cuticle (Schnepf, 1965). The silvery bloom shown by the young leaf of fat-hen (*Chenopodium album*) is due to small globules containing silica attached to the leaf by capillary stalks (Brian and Cattlin, 1968).

Stomata

Stomata are apertures in the epidermis, each bounded by two guard cells which are themselves modified epidermal cells (Fig. 4.15). Stomata are such important features both of the structure and the physiology of the leaf that their own development and anatomy will be described in some detail. Their various roles will be discussed in Chapter 7. By changes in the shape of the guard cells the shape of the stomatal aperture is altered. This aperture leads into the substomatal cavity, which is continuous with the intercellular spaces of mesophyll (see Figs. 1.1 and 1.2). The distortions undergone by the plant's guard cells to achieve different degrees of opening are very varied indeed. However, mechanisms of opening ultimately all depend upon different patterns of thickening and occasionally lignification of the walls of the guard cells and it is this combination of rigid and flexible parts of their walls and changes in turgor that bring about the significant changes in shape. The mechanism which causes the changes in turgor, which in their turn distort the cells, is not yet understood, nor has the study of the fine structure of these cells thrown any light on the problem. The electron microscope has, however, revealed a number of interesting features of these specialised cells, one of which is the existence of large cytoplasmic continuities between guard cells both immature and mature. These continuities were discovered in the Gramineae by Brown and Johnson (1962). They have been confirmed in *Triticum* by Pickett-Heaps and Northcote (1966), but have not so far been confirmed outside the Gramineae.

In most plants the cells adjacent to the guard cells can be distinguished morphologically from other cells of the epidermis. They are thought to be associated with the functioning of the stomata and are called subsidiary or accessory cells. Stomata may occur on any of the aerial parts of a plant, but are especially abundant on leaves, ordinary stems and rhizomes. Some of the more extraordinary positions in which stomata are found are on petals, carpels and seeds of some plants and the staminal filaments of *Colchicum* (Fahn, 1967). Most of these stomata are not functional. Stomata are generally absent from plants lacking in chlorophyll, e.g. the parasites *Monotropa* and *Neottia*. In the parasitic chlorophyll-lacking *Orobanche* they are, however, found on the stem. Stomata are most frequently or exclusively present on the abaxial surface of dorsi-ventral leaves; this is clearly indicated in Table 4.1. However, in certain water plants such as *Nymphaea*, stomata are present only on the adaxial surface.

Only rarely does the adaxial surface have a higher stomatal frequency than the abaxial surface and the shade leaf usually has an overall lower total, often with no adaxial stomata.

In leaves with reticulate venation, e.g. most dicotyledons, the stomata are randomly distributed. In those with parallel venation, e.g. most monocotyledons, and the needles of conifers the stomata are arranged in parallel rows. Guard cells may occur below, at the same level, or above the surrounding

Table 4.1 Frequency of stomata on the adaxial and abaxial surfaces of leaves

Species	Counts of stomata per mm²		
	Adaxial	Abaxial	Together
Olea europaea	0	545	545
Stachys recta	77	270	347
Leontodon incanus	104	210	314
Impatiens noli-tangere			
Sun	31	251	282
Shade	0	100	100
Coronilla varia	137	123	260
Cynanchum vincetoxicum			
Sun	0	178	178
Shade	0	152	152
Buphthalmum salicifolium			
Sun	65	82	147
Shade	29	47	76
Sedum maximum	36	72	108
Pulmonaria officinalis			
Sun	0	104	104
Shade	0	57	57
Mercurialis perennis	0	63	63
Oxalis acetosella	0	37	37

Data from Pisek and Cartellieri (1932) and Fahn (1967)

epidermal cells (see Fig. 4.15). In *Ficus glandifera*, for example, they lie five cells below the surface (Corner, 1965). In some plants stomata are restricted to depressions in the epidermis called the stomatal crypts. Often these stomatal crypts are lined with epidermal hairs although the rest of the epidermis may be hairless as in *Nerium oleander*. Occasionally, as on the peduncle of *Cucurbita pepo*, each stoma lies at the tip of a conical or cylindrical papilla, raised at least five cells above the surrounding epidermis (Haberlandt, 1914). The stomatal chamber is joined to the leaf by a cylindrical passage. The ecological advantage of such exposed stomata is obscure. According to Metcalfe and Chalk (1950) and Stebbins and Khush (1961) the stomata of the dicotyledons and of the monocotyledons can be divided into eight groups, a classification based on the presence or absence of subsidiary cells and their relationship to the surrounding epidermal cells.

Stomata develop from protoderm, i.e. that region of the shoot meristem from which the epidermis, the protective tissue, develops. The sequence of development of a stoma in a grass leaf is shown by Pickett-Heaps and Northcote (1966). The complete stoma is developed from two successive asymmetrical divisions followed by one symmetrical, but incomplete division. The first asymmetrical division of a protoderm cell forms guard mother cells. The partially vacuolate cells surrounding the guard mother cells divide asymmetrically to form the subsidiary cells and the guard cell finally divides symmetrically to form the stomatal cavity. This final division is, at least in the grasses, incomplete and the two guard cells remain in contact at their bulbous tips. A band of microtubules appears in these dividing cells just prior to mitosis and these show where the cell plate will join with the mother cell wall. This band is most prominent before prophase, becomes less prominent as the chromosomes start to condense and is almost invisible at the end of prophase. In neither of these asymmetrical divisions is there any evidence of the polarity of any of the cytoplasmic particles other than the microtubules.

The last and symmetrical division of the guard mother cell which gives rise to the stomatal aperture itself is also characterised by a pre-prophase band of microtubules, this time in the strictly equatorial position. But there are fewer microtubules in the previous asymmetrical division. Again they disappear before the beginning of prophase. After the cell wall is complete several holes are left at each end. Pickett-Heaps and Northcote (1966) favour the idea that the wall is never completed rather than that it subsequently breaks down. They noticed that elements of the endoplasmic reticulum frequently passed through these gaps, but there is no reason to think that the endoplasmic reticulum is directly connected with the formation of these holes.

The way in which the guard mother cell influences the subsidiary cells, where present, to divide in their asymmetrical fashion is the subject of much speculation. Pickett-Heaps and Northcote suggest that this influence by the guard mother cells acts by causing the band of microtubules to form in a specific position in the subsidiary mother cells. It is not impossible that the whole pattern of epidermal cell shape is determined in this way. The influence of one cell upon another in determining its planes of division is extremely important in the understanding of patterns of differentiation.

The construction of the guard cells both in grasses and other groups is complicated. The nucleus, at least in grasses, is in two halves, linked by a thin thread through a central tunnel in the cell only 1·5 μm–2·5 μm in diameter. In some guard cells the two bulbous ends of the nucleus may finally separate entirely from one another. The central tunnel may contain numerous mitochondria as well as the nuclear thread. Plastids, which possess only a few lamellae and are often filled with starch, are found along with many mitochondria only in the ends of the guard cells. There is no evidence that chlorophyll develops in these plastids although they can obviously synthesise starch. Neither the tunnels nor the ends of the guard cells possess any vacuoles, nor, interestingly enough, are there any plasmodesmata between the guard cells and the subsidiary cells although these occur, as already mentioned, between normal epidermal cells. However, the walls between the guard cells and the subsidiary cells are so thin, only between 100–150 nm thick, that diffusion could perhaps occur just as rapidly over this distance in the absence of plasmodesmata. Guard cells differ from the surrounding cells in many other ways. Frequently the outer wall of the guard cell is very much thicker either than its other walls or the walls of normal epidermal cells. Often this wall material is carried up to form two ridges both above and below the stomatal aperture (see Fig. 1.2, **OSL**). Frequently, there are differences in the thickness of the cuticle and epicuticular wax over the surface of the guard cells in comparison with the surface of the normal epidermal cells. For example, in *Citrus*, which has a thick cuticle over the rest of the leaf, the epidermis is said to have no cuticle at all on the wall of the guard cell facing the stomatal aperture (Turrell, 1947).

According to Fahn (1967) in some desert plants the extent of cutinisation of the guard cells may vary during the season. During the drought period the deposition of cutin may proceed so far that the lumen of the guard cells is almost completely obliterated. It has been suggested that in this way guard cells may keep the stomata permanently closed for a while. However, these observations require confirmation.

Salt glands

As explained in the section on trichomes, *Atriplex* species may be able to lose salt from their leaves by passing it into the vesiculate salt trichomes on their leaf surfaces. These trichomes then die and the salt remains on the outside surface of the leaf. Sometimes the trichomes are shed (auto-amputation) with their salt load (Uphof, 1962). There are, however, in a number of plants more complex tissues for the excretion of salt. Some of these, in *Tamarix* species, have now been studied under the electron microscope (Ziegler and Lüttge, 1966, 1967; Thomson and Liu, 1967). These glands usually comprise two groups of cells (Fig. 4.16). There are usually two inner vacuolate 'collecting' cells and over these usually six or more outer secretory cells which have very prominent nuclei and dense cytoplasm. The most striking feature of these glands in *Tamarix* is that not only does a fenestrated cuticle cover the top of the gland, but also the cuticle extends down the side of the gland, between the collecting cells and the secretory cells of the leaf. This is therefore one of the few instances where a cuticle is developed not at a cell/atmosphere interface, but between adjacent cells and often at some distance removed from the atmosphere. On the surface of the gland four pores penetrate the outer cuticle, one usually to each secretory cell (Fig. 4.17). At the base of the gland, between the collecting cells and the secretory cells there

---

FIG. 4.16 Drawing of section through a salt gland on the adaxial surface of a leaf of *Statice* (*Limonium*) *gmelinii*. ***Cu***, the cuticle which also surrounds the gland, ***Pa***, the palisade cells, ***E***, the epidermal cells, ***S***, the secretory cells, ***P*** the pores through the cuticle and **C** the collecting cells. The pores in surface view, in a related species, can be seen in Fig. 4.17. Redrawn from Ruhland (1915)

FIG. 4.17 Scanning electron micrograph of a polystyrene replica of the adaxial surface of a leaf of *Limonium vulgare*. ***P*** indicates the four pores through which the salt is secreted. Ms/Sc. (Courtesy B. Landau-Schachar)

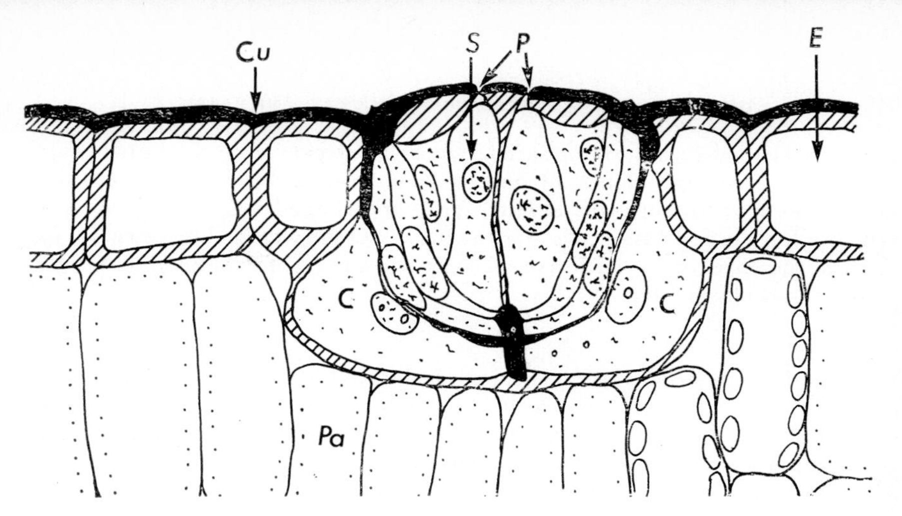

FIG. 4.16

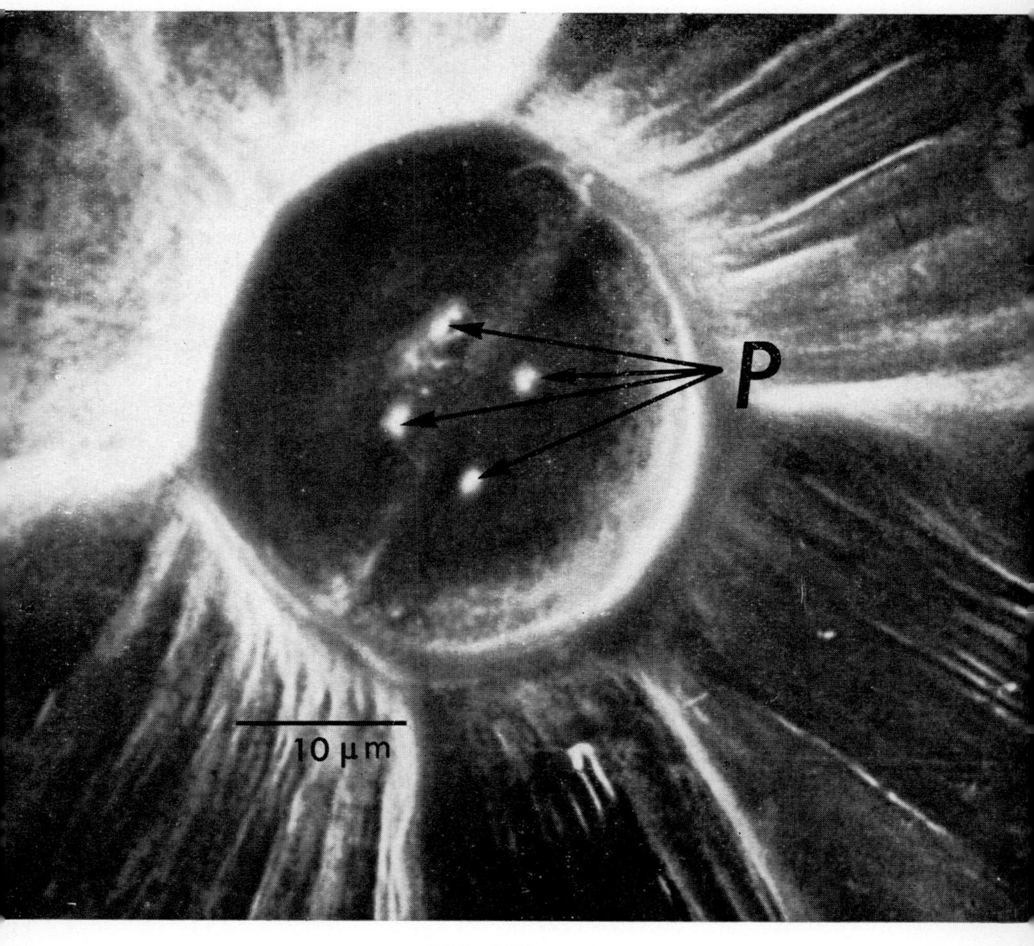

FIG. 4.17

is a cross-shaped formation of cuticle whose arms extend to the cuticle cap (Fig. 4.16). Thus the gland is completely surrounded by cuticle except for the few pores at the surface and four irregularly-shaped areas on the boundary between the collecting cells and the basal secretory cells. The secretion of salt and the effect of the release of salt onto the cuticle of secreting plants will be discussed in Chapter 7.

Hydathodes

Hydathodes discharge water from the interior of a leaf to its surface. This is achieved either through structures like stomata or through specialised trichomes. The process of water loss from uninjured leaves is usually called guttation and along with the water, other substances such as salts, sugars and amino acids, may flow in solution. A full discussion of the phenomenon is given by Stocking (1956a, b).

In the commonest type of hydathode in the angiosperms the terminal tracheids of the vascular system lead into a region of thin-walled parenchyma just below the epidermis known as the ***epithem***. The cells of the epithem lack chloroplasts and have numerous intercellular spaces. The water from the ends of the tracheids moves through these spaces to the epidermis. In the epidermis over the epithem are a number of openings very similar to, although usually somewhat smaller than, stomata. The equivalent of guard cells do not, however, appear to be functional. Like the guard cells of a normal stoma the cells surrounding the aperture of the hydathode appear to be covered with a cuticle. Below the pore lies a space between the cells of the epithem equivalent to a substomatal cavity, but it is not known to what extent this sub-poral cavity is lined with cuticle. Since these pores are continually open and the hydathode only exudes water at night or under conditions when the water balance of the plant is extremely favourable, these pores could theoretically also act as sites of ingress of solutions.

The water-absorbing scales of the Bromeliaceae

The Bromeliaceae are a group of epiphytes and xerophytes widely distributed in tropical America and the West Indies, possessing vestigial roots and almost completely dependent upon rainfall upon their leaves to maintain their water balance. Although they have a thick cuticle, it is penetrated in places by a number of pores and each pore is surmounted by a complex cellular structure, the whole unit forming an effective water-absorbing area and non-return valve (Fig. 4.18). This inverted cone of cells is usually called the scale and when dry the large number of scales on the leaf surface causes the whole leaf to appear white and, when wet, green. The cells of the upper part of the cone are dead and are thought to be completely uncutinised (Haberlandt, 1914). The base of the cone comprises three or four flat, thin-walled, living cells, called by Dolzman (1964, 1965) the 'Kuppelzellen'. These provide the continuous water-absorbing path between the dead cells above and the epidermal and palisade tissue below. When dry, the covering dead cells of the scale flatten and form a more or less impermeable lid to the valve (Fig. 4.18 **B**). When wetted, they rapidly absorb water, swell, raise the lid of the scale and allow water to flow by capillarity over the cuticle (Fig. 4.18 **A**). The living cells at the base in *Vriesia psittacina* have been studied under the electron microscope (Dolzman, 1964, 1965). These highly hydrophilic cells have an extraordinary fine structure which may be connected

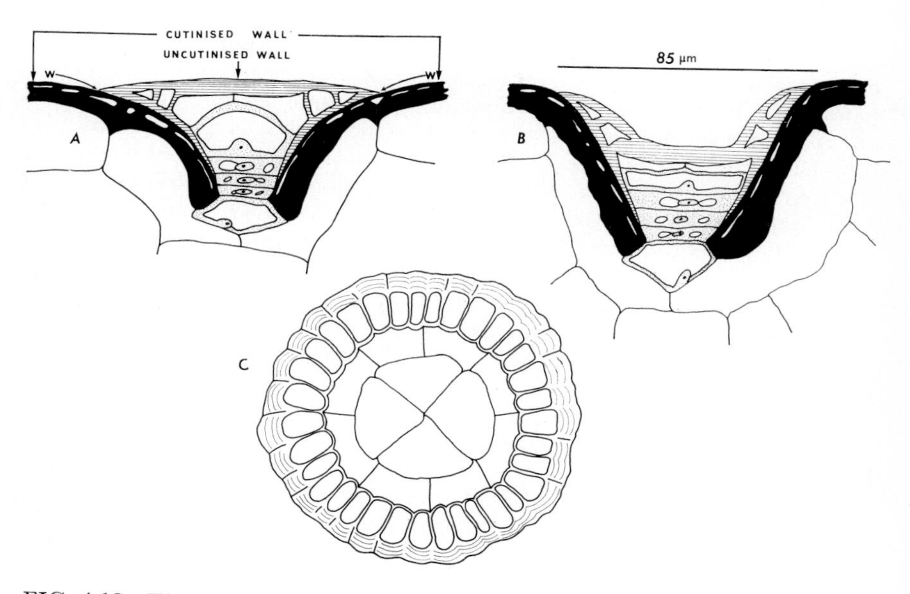

FIG. 4.18 Three drawings of the water-absorbing scales on the adaxial leaf surface of *Vriesia psittacina*. ***A***, the scale when wet; water enters over the surface of the cuticle (**W**). ***B***, the scale when dry. ***C***, the scale in surface view. (Redrawn from Haberlandt, 1914)

with their absorbing powers and are also connected to each other by large numbers of plasmodesmata, a possible route for the passage of water.

These modifications of the epidermis show that it cannot be considered a homogeneous unit. Virtually no leaf surface fails to possess one or more of these specialised structures and, as indicated in the section on stomata, the numbers of these modifications per unit area may be very large. Therefore, whenever calculations are being made on the absorptive capacity of a leaf, or the rate at which water or other substances might be lost through a leaf, these modified epidermal cells and their special contributions must be taken into account.

The cuticle

Not all plants possess a cuticle, although probably all terrestrial angiosperms do. According to Lee and Priestley (1924) the aquatic algae have no obvious cuticle. Little is known about the fungi, but electron microscopy suggests that when a cuticle is present it is vestigial. Roelofsen (1950) believes a very thin cuticle occurs in *Phycomyces*; this layer readily detaches itself from the cell wall beneath and bubbles may form between it and the wall below. A layer, which stains with Sudan III, can just be detected on the surfaces of thalloid liverworts and mosses. Amongst the Pteridophyta there seems to be a vestigial cuticle in the Selaginellaceae and virtually nothing in the Equisetaceae and Ophioglossaceae. In the Marattiaceae and in the Leptosporangiate ferns, although there is said to be some accumulation of fat in the epidermis or endodermis, the cuticle remains thin or in some cases incomplete (Lee and Priestley 1924). The cuticle of *Phyllitis scolopendrium* is reported to be 0·6 μm thick (Gäumann and Jaag, 1935). This absence or indifferent development of a cuticle does not, however, seem to prevent certain Leptosporangiate ferns, e.g. *Pteridium aquilinum*, from colonising relatively arid habitats. According to Arber (1920), all aquatic angiosperms possess a cuticle, although invariably a thin one.

Over all the aerial surfaces of the cells of most angiosperms and gymnosperms is formed a cuticle and parts of the other walls of epidermal cells may also be cutinised. Some workers believe that some of the epidermal cells of the root may be cutinised too. In stems and roots with secondary growth, the epidermis is normally replaced by the periderm and the outer cells of the ***periderm*** are commonly suberised. Cutin and suberin are closely related chemical compounds, both are polymers with a high proportion of fatty acids and have sometimes been mistaken for one another. The chemical relationships between cutin and suberin are discussed in Chapter 5.

Cutin is not entirely restricted to outer epidermal walls. It is known to occur within the substomatal cavity (see Fig. 1.2). Huber, Kinder, Obermüller and Ziegenspeck (1956) report that the cuticle lining the substomatal cavity of *Helleborus niger* is about 0·15 μm thick. Frey-Wyssling and Mühlethaler (1959) state that the cuticle of the substomatal chamber of *Passiflora edulis* is just under 1 μm thick. A cuticle is thought to occur within the spaces between the mesophyll cells of a leaf (Artz, 1933; Häusermann, 1944).

FIG. 4.19 (*top, opposite*) Electron micrograph of a section through the adaxial surface of a leaf of *Eucalyptus cinerea*. The epicuticular wax was stabilised with gold/palladium deposited onto it (***GP***) before the leaf was fixed and sectioned. Compare this section through the wax layer with the replica of a similar wax layer shown in Fig. 4.26. A two-layered cuticle (**C**) can be seen and below it the epidermal cell wall (**ECW**). Glut/Osm/Pb/ Ur. (Courtesy N. D. Hallam)

FIG. 4.20 (*bottom, opposite*) Electron micrograph of a section through the adaxial surface of a leaf of *Eucalyptus cinerea*. The epicuticular wax (***GP***) was also metal-stabilised before fixation. The leaf was fixed *in situ* by attaching a small chamber directly to the leaf surface and introducing the fixative and stains into the living tissue. The fixed material was then embedded in methacrylate rather than the more usual epoxide resins. After sectioning the cuticle (**C**) can be seen to be prominently banded; these bands do not appear after usual fixation and embedding methods. Epidermal cell wall ***ECW*** Glut/Osm/Pb/Ur. (Courtesy N. D. Hallam)

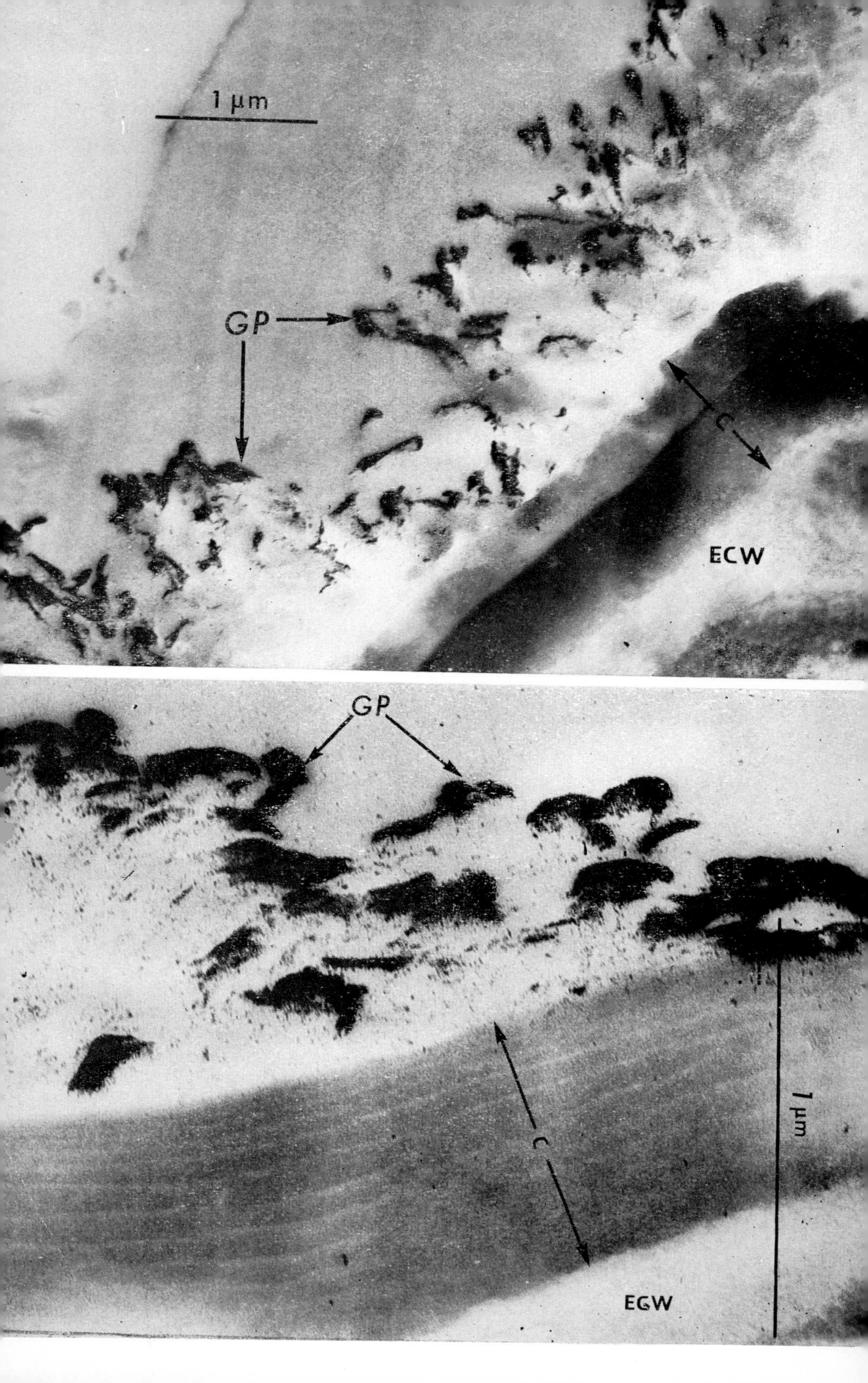
1 μm
GP
C
ECW
GP
C
ECW
1 μm

Goodman and Addy (1962b) found an inner cuticle in the apple leaf. A cuticle is also known to occur between the cells of a leaf and the glands of salt-secreting plants (see Fig. 4.16): as a lining to stylar canals: below the water-absorbing scales of the Bromeliaceae and as inner cuticles during the development of integuments into seedcoats in some seeds. Therefore, although the formation of cuticle has been likened to the drying of varnish in air, it is clear that the development of a cuticle can often take place *between* cells.

There is not, either in the chemical sense or in the anatomical sense, an abrupt boundary between the cutinised layers and the cuticle proper. In most plants they merge imperceptibly into one another and in some cannot be distinguished at all. There is enormous variation in the development of these layers in different plants, a development that cannot always be obviously correlated with the habitats in which the particular plants live. For example, according to Stace (1965), the epidermal walls of *Lumnitzera racemosa* and *Macropteranthes kekwickii*, two xeromorphs, are comparatively thick. However, in the former, the cuticular membrane is also thick with deep stout cuticular flanges, whereas in the latter, the cuticular membrane is thin and has hardly any flanges. In the latter plant the non-cutinised epidermal wall accounts for a greater part of thickness of the outer compound wall. Table 4.2 shows the thickness of the cutinised and cuticularised layers in contrast to the cellulose lamellae of the outer epidermal walls of leaves of several trees and shrubs.

Table 4.2 Thickness of cuticular and cellulose layers of the outer epidermal walls of leaves

	Thickness of the cuticular layers in microns	Thickness of the cellulose layers in microns
Prunus laurocerasus	0·7	6·0
Rhododendron catabiense	0·9	8·1
Rhamnus glandulosus	2·6	6·5
Rhamnus alaternus	3·3	5·2
Acer syriacum	3·5	4·0
Quercus coccifera	8·0	0·8
Quercus ilex	9·5	0·9
Olea europaea	10·0	1·0
Nerium oleander	13·2	1·3
Olea lancea	13·5	1·2

Data from Kamp (1930) and Gäumann and Jaag (1935)

The cuticle is, however, usually thicker in plants growing in dry habitats and under high light intensities and may vary on leaves from different parts

FIG. 4.21 Electron micrographs of carbon replicas of the adaxial and abaxial surfaces of the same leaf of pea (*Pisum sativum* var. *Alaska*). Note the difference in the fine structure of the epicuticular wax. Cr/Au/Pd

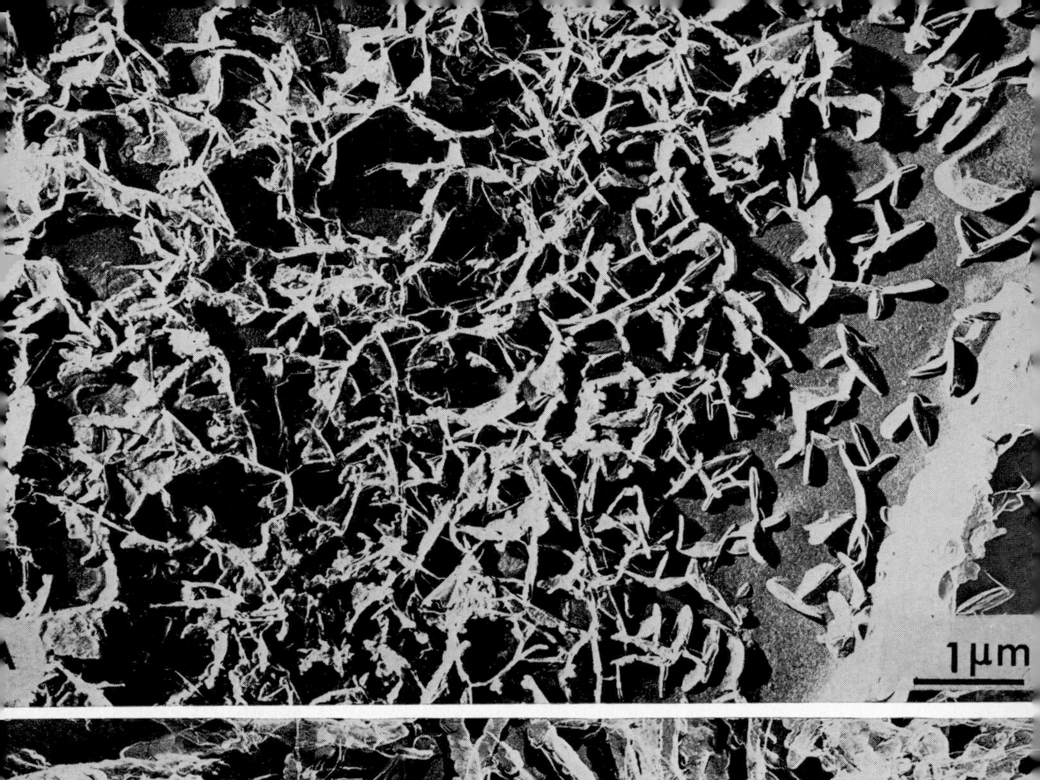

FIG. 4.21

of the same plant. The surface of the cuticle may be smooth, rough, ridged, furrowed or pimpled (Martens, 1934). Frequently it is covered by a layer of epicuticular wax whose fine structure may be incredibly diverse.

Priestley (1943) was convinced that a close analogy could be drawn between the drying of a varnish film and the formation of cuticle. He said 'As certainly as a vegetable oil spread over a picture oxidises and "dries" to varnish, so do these oil surfaces over any air-water surface on the growing shoot. This means that the apex is covered by a continuous film of "drying" fat in which oxyacids and waxes are present in varying proportions'. There is no doubt that the apex is covered from its inception and even at its most rapid rate of growth with a cuticle, but the varnish analogy cannot now be upheld since there are a number of situations already listed where a substantial cuticle is formed within the plant between the cells.

The cuticles of apices, primordia and leaves

Even the shoot apex, the terminal part of the shoot immediately above the uppermost leaf primordium has a thin, but detectable cuticle. Loomis and Schieferstein (1959) have shown that a cuticle can be separated from young leaves as soon as the bud opens. However, when chemical methods are used to remove these young cuticles they separate into cell-sized flakes rather than detaching as sheets as older cuticles usually do. Loomis and Schieferstein believe that this indicates that the cuticle, at least in very young leaves, is incomplete. They believe also that the outer wall of the epidermal cells grows at the margins, creating zones of weakness, or actual breaks, over the surfaces of the anticlinal walls.

The cuticle may vary greatly in thickness over the surface of a single leaf. For example, in *Nerium oleander* the adaxial cuticle is very much thicker than the abaxial (Fahn, 1967). According to von Lange and Schulze (1966), the thickness of cuticle varies over the surface of a spruce (*Picea*) needle; it is thicker on the corners than on the sides. The development of the cuticle over leaf trichomes or emergences is frequently different from that over the surface of the normal epidermal cells of the leaf. For example, the vesiculate salt trichomes on the leaf surface of *Atriplex portulacoides* have thin cell walls and a negligible cuticle whereas the normal cell walls of the epidermis are thick and have a substantial cuticle. The cuticle is thinner over the oil glands on the surface of the leaves of *Eucalyptus* spp. (Hallam, 1967). On the other hand, the trichomes on the surface of the leaves of mesquite (*Prosopis juliflora*) are more heavily cutinised than the epidermal cells themselves (Hull, 1958). Thick projecting lips of cuticle, the stomatal ledges, can develop both on the outside and on the inside of stomata, often arching over the stomatal cavity (see Figs. 1.2 and 4.15).

As a general rule the adaxial cuticle is more substantial than the abaxial cuticle (Lee and Priestley, 1924). From a consideration of the effects of the environment on the thickness and properties of the cuticle observed by Lee and Priestley, it is highly likely that sun and shade leaves, immature and mature leaves, cotyledon and adult leaves and leaves growing on different parts of the same plant will all have different thicknesses of cuticle. Very little comparative work seems to have been done on this problem. Von Lange and Schulze (1966) have shown that the sun and shade needles of spruce can clearly be distinguished from one another by the thickness of their cuticles. Hülsbruch (1966) has shown that the thickness of the cuticle on stems and leaves of *Ilex integra* increases markedly with age. The wax on the surface of the cuticle is very variable too. Its distribution, form and density may vary within a single plant. For example, the adaxial and abaxial leaf surfaces of the pea (*Pisum sativum*) and the oak (*Quercus robur*) differ strikingly from one another, although both carry considerable quantities of wax (Fig. 4.21). The hybrid tea rose var. *Ena Harkness* and *Rubus cockburnianus* on the other hand have no crystalline wax on their adaxial leaf surfaces, but their abaxial surfaces have very well-developed waxy layers. The prominent spathe and peduncle of the flowers of *Strelitzia reginae* has a most striking waxy bloom whereas the rest of the plant, leaves, petioles and stems, is almost completely non-waxy.

In the extensive taxonomic survey of the genus *Eucalyptus*, Hallam (1967) came to the conclusion that virtually all *Eucalyptus* were similar in the structure of the wax on the two surfaces of the mature leaves except for members of the series *Corymbosae*. The adaxial surfaces of such species as *E. gummifera* and *E. ficifolia* of the Corymbosae have only a few isolated small plates whereas the lower surface is densely covered. *Nelumbo speciosa* is prominently waxy on the adaxial surface of its leaves, but bears little wax on the abaxial or petiolar surfaces.

Apart from differences between adaxial and abaxial surfaces, the cotyledon, juvenile and adult forms of leaves, where these exist, often show differences in glaucousness. Barber (1955) has recorded some of these variations also in the genus *Eucalyptus*. In *E. risdoni*, *perriniana* and *cordata*, amongst others, the glaucousness persists throughout growth. In *E. rubida*, *urnigera and gunnii* the juvenile foliage is heavily glaucous, but in the adult the glaucousness is either absent or confined to the young leaves, stems and flower buds, rapidly fading as the leaves mature. In *E. gigantea* glaucousness is restricted to the young leaves and flower buds. The most complex of all is *E. coccifera* where the juvenile stems and the juvenile leaves lack any wax. The intermediate foliage and stems are usually heavily waxed. In the adult tree the waxiness may persist, particularly on the stems, but often it is transient as in *E. rubida* or is restricted to the flower buds and young fruit as in *E. globulus*.

Cotyledonary leaves are usually strikingly different from mature leaves both in thickness of the cuticle they develop and the amount and type of superficial wax they form. Usually the cotyledonary leaves have thin cuticles with no epicuticular wax, even when this is well developed in the mature leaf. A good example of this type of plant is kale (*Brassica oleracea* var. *acephala*). However, one striking exception to this rule is *Chrysanthemum segetum* in which both surfaces of the mature leaves, their petioles, the stems and both surfaces of the cotyledonary leaves are covered with a dense mat of

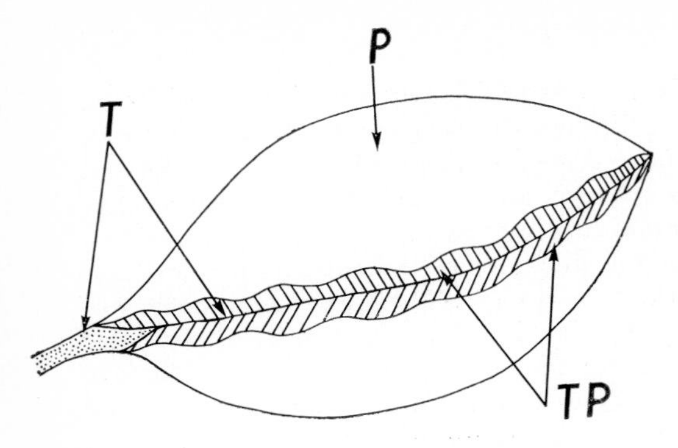

Fig. 4.22 Diagram of the distribution of wax on the adaxial leaf surface of *Eucalyptus polyanthemos*. ***P*** is the plate-type of wax found over most of the surface of the leaf; the tube-type (***T***) occurs on the petiole and in a narrow band along the midrib and the mixture of plate and tube types ***TP*** lies on either side of the midrib. (Redrawn from Hallam, 1967)

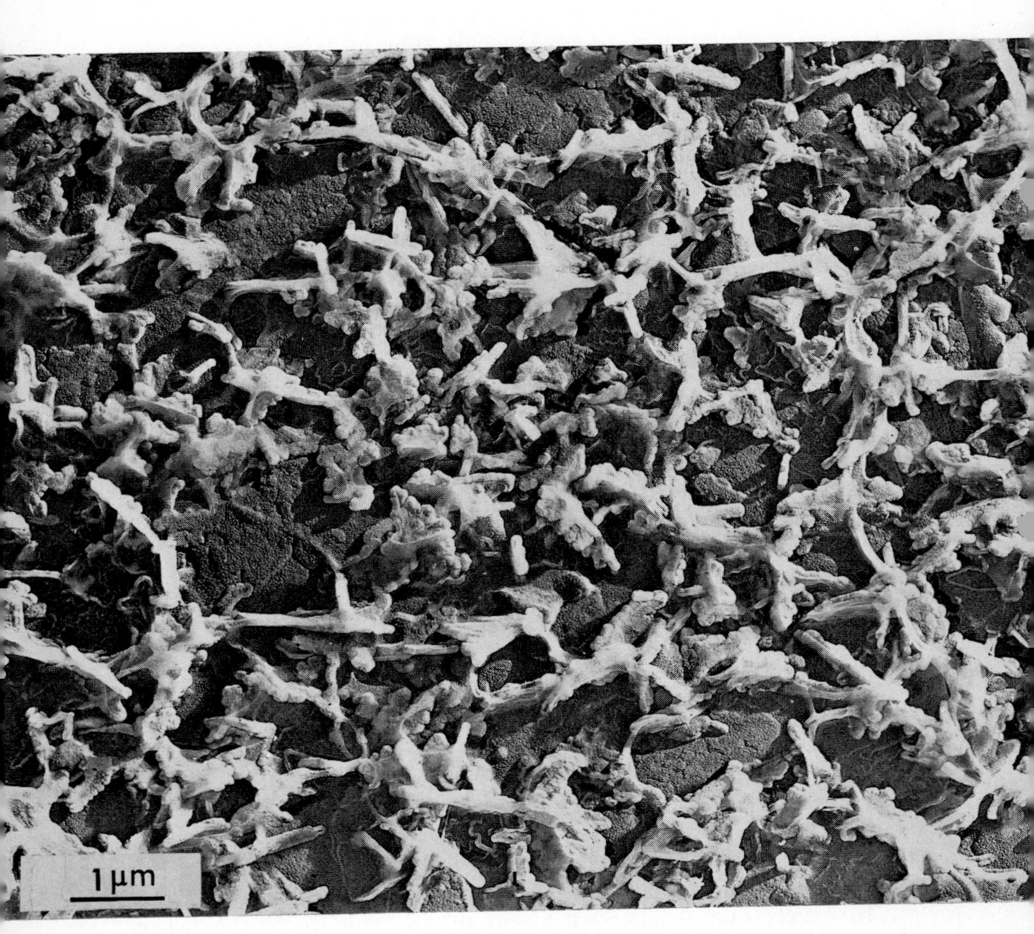

FIG. 4.23 Electron micrograph of a carbon replica of the adaxial surface of a leaf of *Zea mays*. Cr/Au/Pd

crystalline wax (see Fig. 2.10). In the normal maize (*Zea*) only the first five or six leaves are glaucous, the rest are glossy, i.e. non-waxy (Bianchi and Marchesi, 1960). The leaves of *Eucalyptus polyanthemos* have a most interesting variation of surface wax. Over the midrib is a tubular type of wax, close to the midrib a mixture of tubular and plate-like wax and over the main part of the lamina an exclusively plate type wax (Fig. 4.22). The petioles and stems have a tubular type of wax similar to that found over the midrib (Hallam, 1967). The leaves of the mutant of maize (*Zea*) known as *Corn grass* (*Cg*) have similar glossy and glaucous areas. The glaucous areas have the normal ultrastructure of a maize surface (Fig. 4.23) whereas the glossy areas have no wax projections at all (Bianchi and Marchesi, 1961).

The epicuticular waxes are frequently modified in their density or form around trichomes, emergences, stomata or glands. This irregular distribution was noticed by de Bary in 1871. Very commonly, as in the pea, the cabbage (*Brassica oleracea* var. *capitata*) and many grasses the number and size of epicuticular projections are markedly reduced or eliminated completely over the subsidiary cells and guard cells (see Fig. 2.7). In the genus *Eucalyptus* a few species show this exclusion or partial exclusion of wax from the guard cells and the cells immediately surrounding them. Some, however, such as *E. fastigiata* and *E. obliqua*, which have very little, if any, wax on their leaves, are conspicuous in having virtually all their particulate wax on their guard cells. Some, such as *E. panda* and *E. radiata*, have a high density of wax particles over the normal epidermal cells, but an even higher concentration over the guard cells (Hallam, 1967). In *Euphorbia tirucalli* and some species of *Strelitzia* each stoma is surrounded by a ring of wax, thus forming an external air chamber (Haberlandt, 1914).

In pitcher plants (*Nepenthes* spp.) the pitcher is formed from a modified leaf. Part of the trap mechanism is a layer of secreted wax on the inside (the adaxial surface) of the pitcher. Full details of this mechanism will be given in Chapter 9. Secretion of this wax takes place over only a very limited area of what is morphologically the adaxial surface of the leaf and no bloom of crystalline wax is formed at all either on the abaxial leaf surface or on any other part of the plant. Where trichomes of any type are present no instances have so far been found where the epicuticular wax is developed over their surfaces as well as those of the normal epidermal cells. However, hairy leaves are difficult to prepare for examination and it would be premature to generalise that trichomes do not develop wax.

The cuticles of petioles, stems and culms

Petioles and stems frequently possess both a similar cuticular development and epicuticular wax formation and density to that possessed by the leaves. However, stems such as those of some varieties of the raspberry (*Rubus idaeus* vars. *Viking*, *Latham*, *Chief*, *Newman* and *Ontario*), of the black raspberry (*Rubus occidentalis*), *Rubus giraldianus*, the dewberry (*Rubus caesius*), *Rubus biflorus*, the palm *Ceroxylon andicola*, *Salix daphnoides aglaea*, and *Berberis dictophylla* are heavily coated with a wax bloom, although the leaves are non-waxy or have little wax. The segmented stems of the succulent composite *Kleinia articulata* have the same prominent 'bloom' as the leaves, although this bloom is not formed by a wax but a polysaccharide (Fig. 4.24). The stems (and the abaxial surfaces of the leaves) of *Rubus cockburnianus* are prominently waxy although a waxy bloom is absent from the adaxial leaf surfaces.

Peat (1928) found in *Ricinus communis* that three genes determine the distribution of bloom. The first gene, if dominant, results in bloom on the stem petioles and capsules. The second gene intensifies the bloom already present and adds to it a bloom on the abaxial surface of the leaves. A third gene intensifies the effects of the other two. These and other examples of green mutants and the genetics of blooming will be discussed in Chapter 6.

In his study of the genus *Eucalyptus*, Hallam (1967) noticed that, although both surfaces of the lamina of the species *E. pauciflora* possess a plate-like wax, there is a tube-like wax on the surface of the young stems. *Eucalyptus polyanthemos* has a tube-like wax on the petioles and stem and a mixture of tube or plate waxes on the lamina. The young culms, but not the leaves, of many bamboos and particularly those of *Lingnania chungii*, have a prominent grey-white bloom which completely covers the green surface of the internodes (McClure, 1966).

The cuticles of fruits and seeds

The development of the mature fruit or seed is highly complex and a wide range of accessory tissues may be involved in their production (Esau, 1965). However, all of these tissues, wherever they may be derived, if they come to form the epidermal cells of the mature fruit or seed, are capable of forming a recognisable cuticle. The cuticle may not necessarily be on the outside of the seed or fruit. The cuticle of the pericarp of *Asparagus*, according to Roelofsen (1959), may consist of up to 65 layers. Sometimes as in the Leguminosae the cuticle of the seeds is very thick and provides a considerable barrier to the passage of water and gases. In many fruits a detectable cuticle may develop between the pericarp and the nucellus or endosperm. Examples of this are seen in *Triticum* and *Lactuca*. In seeds a cuticle can develop between the testa and the remains of the nucellus or endosperm, as in *Plantago lanceolata*, *Malva sylvestris*, *Lythrum salicaria*, *Lycopersicum esculentum* and *Asparagus officinalis*. These cuticles appear to originate from the ovule. Although they are present at an early stage in the development of the fruit or seed, cuticle-free channels to the embryo sac always exist at the chalazal region.

Most fruits have a very well-developed cuticle and in many of them, e.g. the apple (*Malus*), the cuticle develops and thickens as the fruit matures. In some apples, even after harvesting, there is a continuous accumulation of waxy or greasy materials on the surface of the cuticle (Fig. 4.25). Most fruits with a prominent waxy bloom such as *Benincasa cerifera*, the fig (*Ficus*

FIG. 4.24 Electron micrographs of carbon replicas of the adaxial surfaces of leaves of *Kleinia articulata*. ***A***, the intact dry leaf; ***B***, a leaf after a 1-sec dip in distilled water; (***C***) after a 10-sec dip in distilled water. Cr/Au/Pd

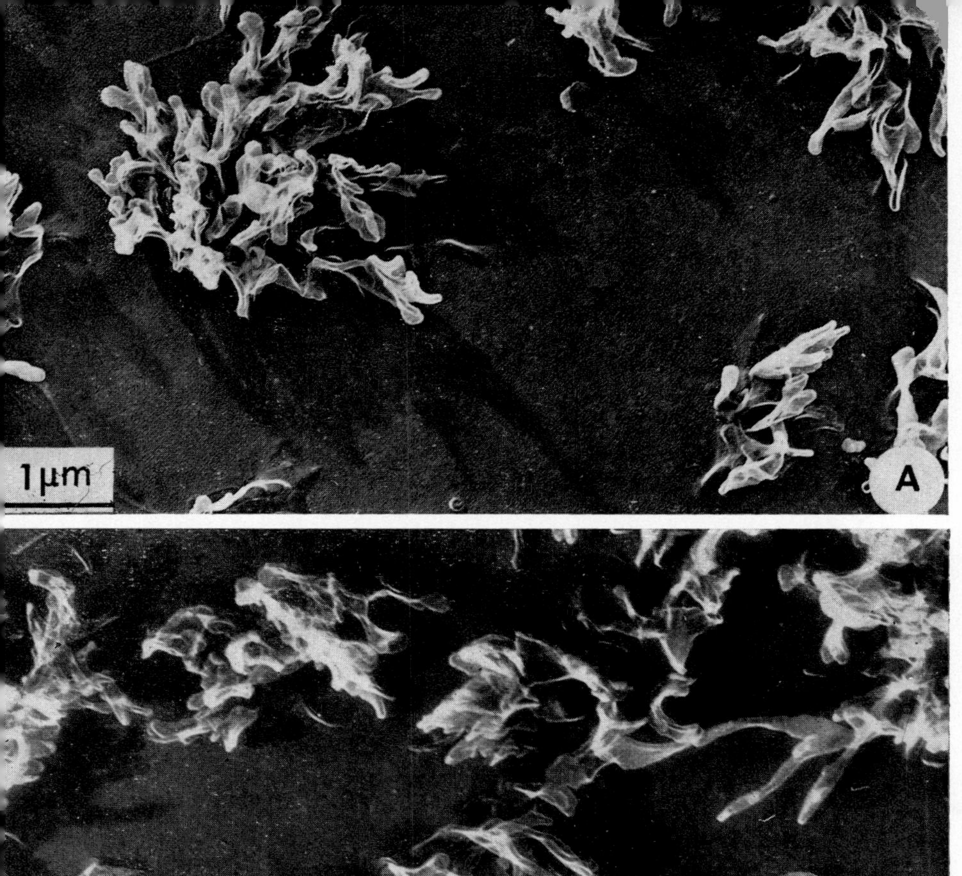
1 µm
A

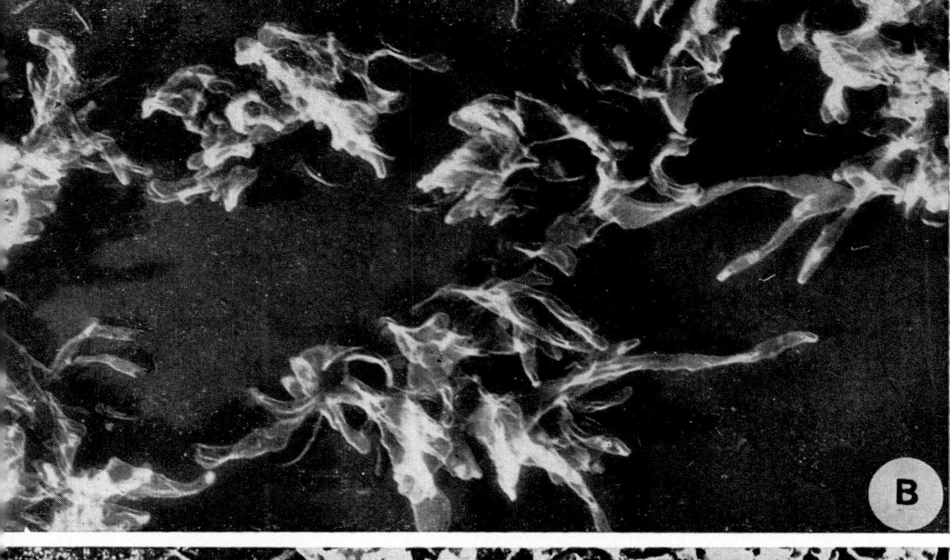
B

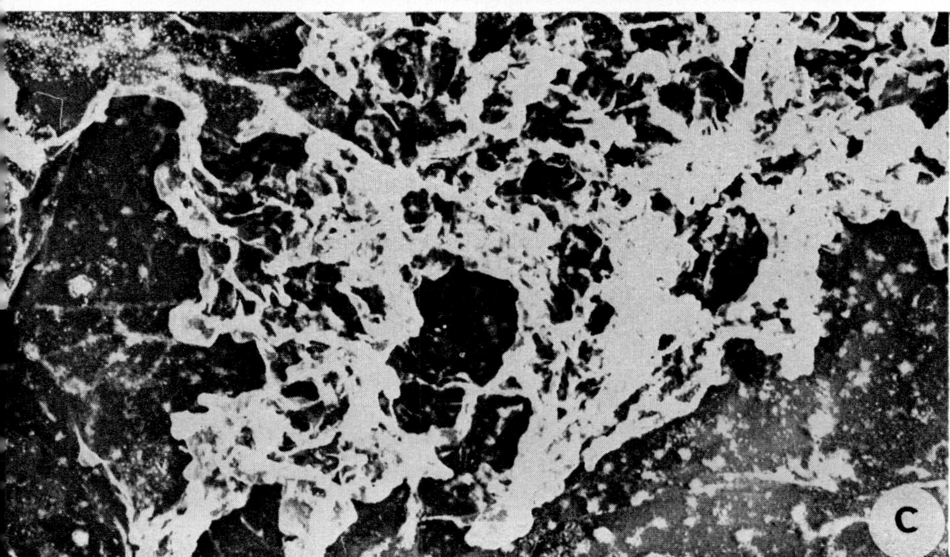
C

FIG. 4.25 Electron micrograph of a carbon replica of the skin of a Cox's Orange Pippin apple fruit. Cr/Au/Pd (Courtesy D. S. Skene)

carica), *Mahonia aquifolium* and *M. bealei*, the grape (*Vitis vinifera*), some varieties of the plum (*Prunus domestica*), *Myrica cerifera* and *Juniperus communis* do not develop the bloom until the fruit has reached or is approaching maturity.

The pericarp surfaces of some hard fruits, e.g. acorns (*Quercus* spp.), develop a prominent waxy coating. *Quercus nuttallii* and *Q. palustris* of North America have thick waxy coatings whereas other species such as *Q. falcata pagodaefolia* and *Q. rubra* have only a thin wax layer (Bonner, 1968).

The cuticles of sepals and petals

Both sepals and petals are essentially leaf-like in form; like leaves they have a vascular system and epidermal cells covered by a cuticle and penetrated by stomata. Some of these stomata are not fully differentiated and some are modified as nectaries for the secretion of sugary materials. In other nectaries excretion takes place by the bursting of the cuticle, in yet others there exist regions of the epidermis with thin walls over which no cuticle is formed and through which excretion takes place, unimpeded, by diffusion.

Martens (1934) has shown that the cuticles of the petals of many species, which do not possess a waxy bloom, have a very finely rippled or pleated surface. Examples of such petals are those of *Centaurea cyanus*, *Tradescantia virginiana*, *Pelargonium zonale* and *Viola tricolor*. An example of this surface detail is given in Fig. 58/2 in Fahn (1967). Martens believes that this pleating of the cuticle is a safety device to avoid damage when extension and contraction takes place. The petals and occasionally sepals of very many species such as *Azalea*, *Rosa*, *Primula*, *Camellia* and *Cyclamen* develop a fine waxy bloom, but have no corresponding bloom on the surface of the leaves, petioles or stems. This bloom contributes both to the iridescent quality of many flowers and also to their resistance to rain damage by shedding water from their surfaces. A few flowers such as *Hoya carnosa* and *Lapageria rosea* are prominently waxy, but the wax of the petals is in the form of a smooth layer and not crystalline.

The cuticles of other floral parts

Between the tissue of the stigma and the ovary there is a specialised area through which the germinating pollen tube grows; this is commonly known as the transmitting tissue. This tissue is also believed to supply nutrients to the growing pollen tube. Sometimes one or more canals develop in the style and, bordering these canals prior to pollination, the transmitting tissue will be covered with a cuticle. According to Fahn (1967) the cuticle on the transmitting tissue disappears before pollination and the walls of this tissue soften and swell.

There is no evidence for the existence of a cuticle on the surface of pollen tubes as seen under the electron microscope either growing *in vitro* or down through the tissue of the style (Rosen, Gawlik, Dashek and Siegesmund, 1964; Dashek and Rosen, 1966).

The cuticles of roots

Scott, Hamner, Baker and Bowler (1958), Scott (1963, 1966) believe that a cuticle may be present on young roots with or without root hairs. They also believe that a cuticle exists on root hairs of *Vicia* from the youngest, about 10 μm long to the oldest, about 60 μm long. These conclusions are not, however, supported by recent work on other plants using section techniques for the electron microscope on primary roots and root hairs. No cuticle is visible on the surface of the primary roots of maize and barley studied by Juniper and Clowes (1965) and Juniper and Roberts (1966) nor on the surfaces of root hairs studied by Newcomb and Bonnett (1965) and Bonnett and Newcomb (1966). Lee and Priestley (1924) grew a seedling of *Pisum sativum* and placed 4 in. (10 cm) of its stem underground. Under these conditions epidermal cells broke through the existing cuticle and formed 'root hairs', but did not develop any cuticle while underground.

The surface layer on *Elodea* water roots grown in the light gives a positive stain with Sudan III, sometimes used to detect cutin, and is regarded by some workers as a cuticle. Chloroplasts are present in the epidermal cells, and

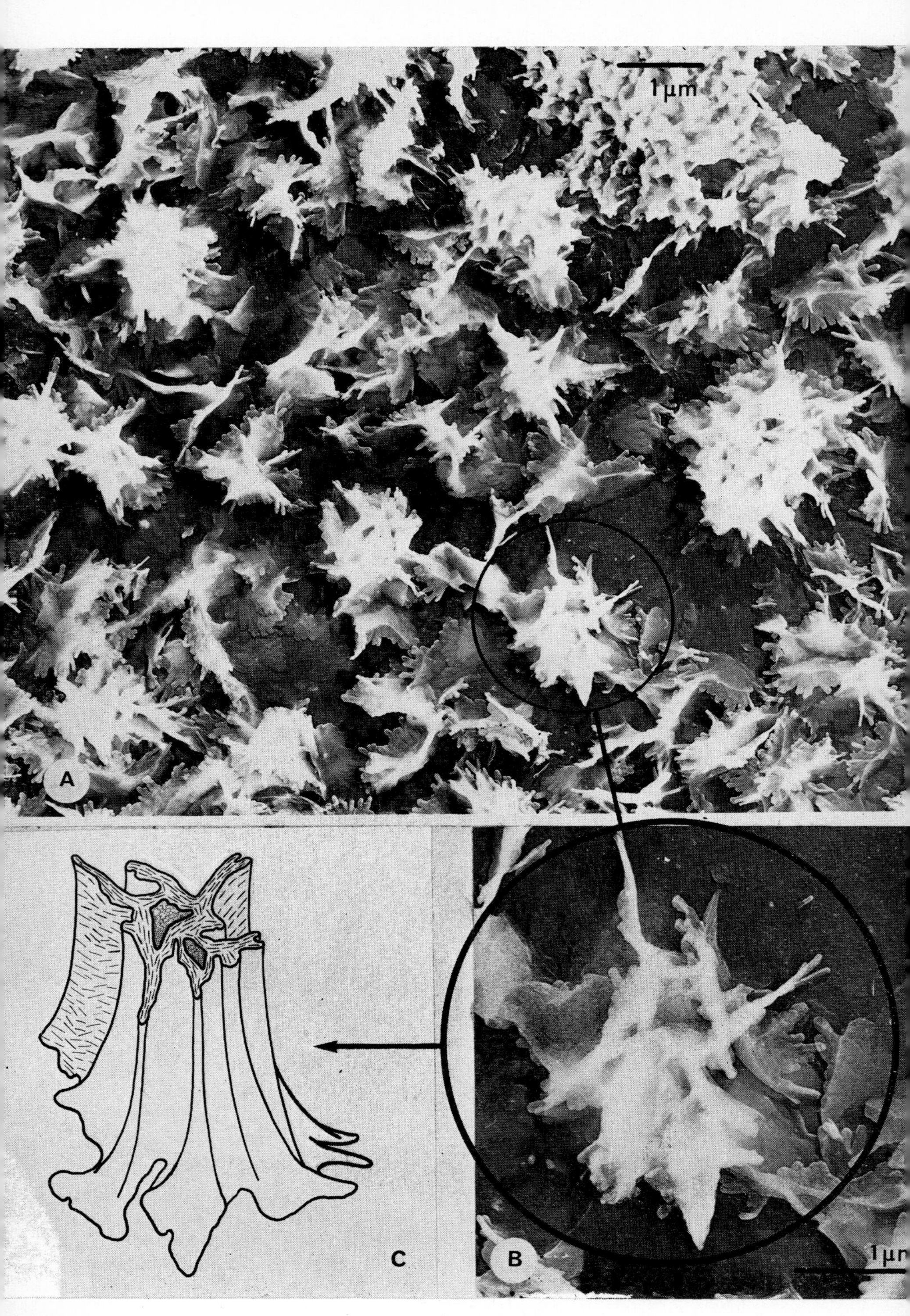

FIG. 4.26

it was thought that this cuticle might be formed by the oxidation of fats caused by the oxygen liberated in the photosynthesis (Cormack, 1937). Dale (1951) suggests that the surface layer of the *Elodea* root is not a true cuticle of oxidised fats, but a film of unsaturated acids.

The surface of the cuticle and the structure of the epicuticular wax

The surface of the cuticle may be sculptured and develop surface detail in a way which does not reflect the shape of the epidermal cells underneath, but is a phenomenon of the cuticle itself (Rao, 1963). Fahn (1967) shows a micrograph of the abaxial surface of a petal of *Pelargonium zonale* and Rao (1963) a figure of the surface of the leaf epidermis of *Hevea brasiliensis*. The cuticle has prominent striations which are most marked in the centre of the epidermal cells. The cuticle of the adaxial surface of *Aesculus hippocastanum* is not smooth under the electron microscope, but is traversed by a pattern of sinuous cuticular ridges which appear to bear no relation to the shape of the epidermal cells underneath (see Fig. 2.9). The cuticle of *Atropa belladonna* has striations which follow the wavy anticlinal walls of the epidermis. Most of the surface detail of leaves on the microscopic or submicroscopic scale is provided not by the cuticle, but by the fats, waxes, terpenoids, oils, carbohydrates and other substances that are secreted, often in a crystalline form, onto the surface of the cuticle.

Very occasionally, as in sugar cane (*Saccharum officinarum*), the carnauba or wax palm (*Copernicia cerifera*), and *Ceroxylon andicola* where the wax layer is 5 mm thick, the structure of the wax is visible either to the naked eye or with a hand lens. A few wax patterns can clearly be seen under the light microscope. The projections from the leaf surfaces of *Strelitzia ovata* and *Heliconia farinosa* are between 10 and 20 μm long (de Bary, 1871). The majority of species with glaucous leaves possess a superficial structure which is resolvable at best at the limit of the resolution of a light microscope or only under the electron microscope. Whenever a plant surface possesses a visible bloom the electron microscope has revealed, without exception, a pattern of acicular projections from the surface (Fig. 4.26). For a bloom to be formed, i.e. for the incident light to be scattered in all directions from the surface, it seems that one or more of the dimensions of the projections must be of the same order of size or only slightly larger than the wavelength of light.

De Bary in 1871 proposed a classification for types of wax coating and his groups were (1) single layers of granules; (2) small rodlets perpendicular to the cuticle; (3) several layers of very small needles or granules and (4) membrane-like layers or incrustations. To group 1, according to de Bary,

FIG. 4.26 Electron micrograph of a carbon replica of the adaxial surface of a *Eucalyptus cloeziana* leaf. ***B***, a higher magnification micrograph of a single projection from ***A***. ***C***, an interpretation of its shape, truncated to show the possible distribution of the wax molecules within it. Cr/Au/Pd

belonged such genera as *Allium*, and *Saccharum*; to group 2, *Musa* and *Canna*; to group 3 *Eucalyptus* and *Secale* and to group 4, *Sempervivum* and *Euphorbia* species. Amelunxen, Morgenroth and Picksak (1967) have re-examined the multiple forms of the excretions from leaves, shoots and fruits. They classify the waxy coverings into six types. These are (1) granular coatings, consisting of a single layer of granular or spherical masses (e.g. the waxes of syncarps of *Rosa canina* and of leaves of *Sedum cauticolum*); (2) rodlets and threads which may be straight, curved or coiled and which project from the epidermis (e.g. the waxes of leaves and fruits of *Tulipa gesneriana*); (3) platelets and scales which lie flat on the cuticle or may be elevated and closely packed (e.g. the waxes of *Dasylirion serratifolium*, *Polygonatum multiflorum* and *Iris pseudacorus*); (4) layers and crusts which may be smooth or show rod-like or other projections (e.g. the waxes of *Brassica oleracea* var. *capitata*, *Vitis vinifera* and *Papaver orientale*); (5) aggregate coatings which consist of superimposed secretions of multiple structure (e.g. the wax of *Eucalyptus globulus*) and (6) liquid or soft wax coatings, consisting of droplets or irregular flat cakes (e.g. the wax of the apple fruit). Several of the types may occur in a single species or even on the same organ. The authors suggest that the chemical composition of the wax plays a part in determining the form of crystallisation.

It no longer seems possible either to maintain or to confirm any classification which has any value. The crystalline structures under the scanning or transmission electron microscope may take the form of tubes or rods; sometimes simple, sometimes branched; they may be grains, either heaped or as single layers; they may occur as plates or most commonly of all as tufts either isolated or joined together (Figs. 4.23–4.27). Usually one pattern of crystalline structure occupies a single leaf surface. Occasionally, as in the surfaces of plum and damson (*Prunus insititia*) fruits and the surface of the needles of *Pinus sylvestris*, two or more different crystalline structures occur side by side with one another. Very rarely, as in *Eucalyptus polyanthemos*, a crystalline pattern of one type over part of a leaf surface is wholly replaced by another type over a different part of the same surface. All of them are susceptible to changes in the environment.

Individual acicular forms are often very complex and difficult to interpret. Figure 4.26 shows an electron micrograph of a single extrusion from the surface of a leaf of *Eucalyptus cloeziana* and the drawing shows a three-dimensional reconstruction of its probable shape. The projection has been truncated to show its probable internal structure and orientation of wax rodlets making up the whole mass. Often within a genus or family there may be general resemblances in wax structure, but frequently the resemblances

---

FIG. 4.27 (*top, opposite*) Electron micrograph of a carbon replica of the surface of a needle of *Pinus sylvestris*. Cr/Au/Pd (Courtesy L. Leyton)

FIG. 4.28 (*bottom, opposite*) Electron micrograph of a carbon replica of the adaxial surface of a cabbage (*Brassica oleracea* var. *capitata*) leaf after deliberate mechanical damage. Cr/Au/Pd

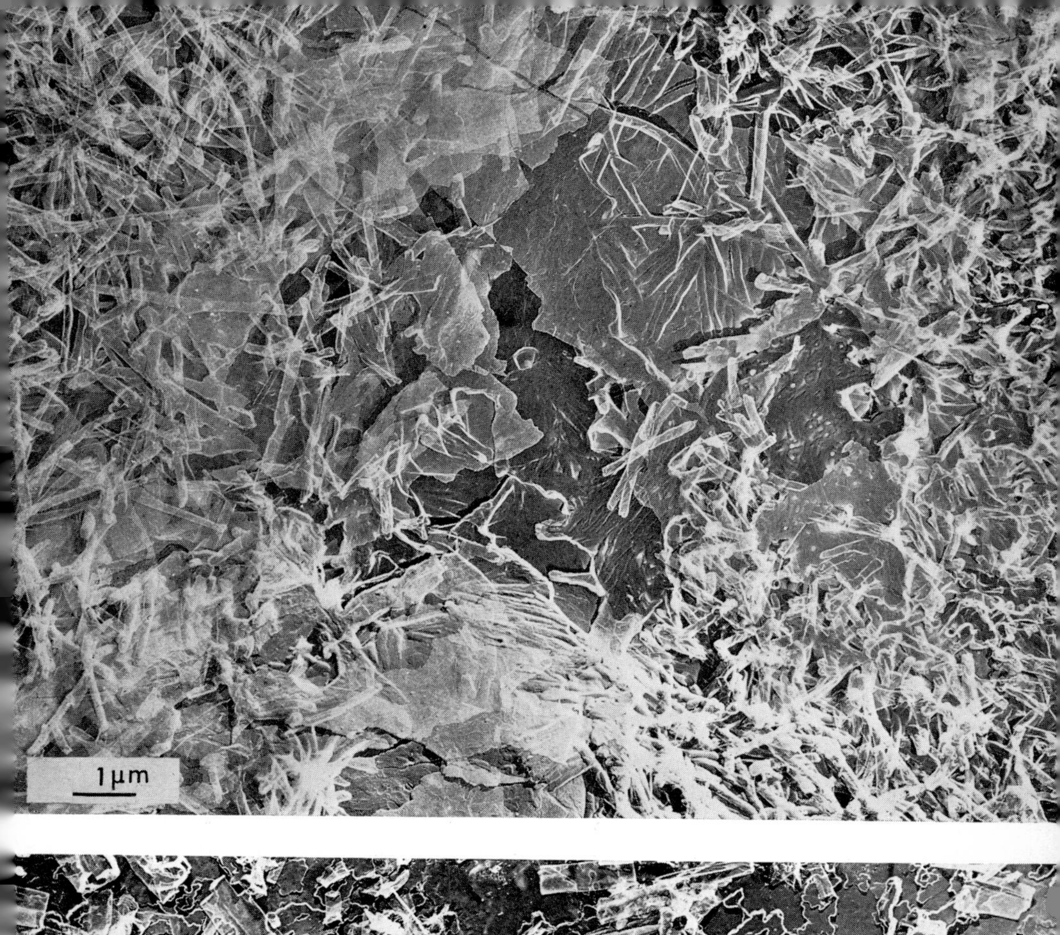
1 µm

10 µm

infringe taxonomic boundaries. Within the group of plants having tufted crystals on their surfaces are the Papilionaceae. Within this family the wax of the garden pea (*Pisum sativum*) closely resembles that of the white lupin (*Lupinus albus*), but so also do the waxes of the totally unrelated snowdrop (*Galanthus nivalis*) and *Oxalis corniculata*. Most of the Gramineae waxes are extremely similar to barley (*Hordeum sativum*) (see Fig. 2.4) and those of most of the glaucous Crassulaceae resemble *Bryophyllum tubiflorum* and *Echeveria glauca*. However, the coverings on the leaves of some *Eucalyptus* species resemble those on the stems of *Berberis dictophylla* and on the leaves of the grass *Poa colensoi*. All wax forms of *Brassica oleracea*, so far examined, look alike, but their curious annulate ridged tubes seem to be restricted to this species (Fig. 4.28). No surface structure resembling that of *Chrysanthemum segetum* with its curious coiled tubules, or that of *Hyacinthus orientalis* (Figs. 2.10 and 4.29) has been seen in any other species of the genus or outside the genus. This, however, may be due to the so far limited use of the electron microscope for the study of wax structure and not to restriction on the types of epicuticular wax.

Cuticular canals

Interest has been taken for many years in the possible presence, in cuticles, of canals through which wax may migrate to the surface. Hall (1967a) has studied the structure of the cuticle with the freeze-etching technique for electron microscopy (see Fig. 2.12). He claims that the inability to see canals in conventionally sectioned material is because each canal is not bounded by a membrane and that the usual fixatives and stains do not give sufficient contrast between the wax in the pathways and in the cell wall and cuticle. He floated pieces of *Trifolium repens* leaf on 40% glycerol overnight, to prevent the formation of ice crystals in the tissues during subsequent freezing. The leaf was then frozen in liquid Freon to −150 °C, transferred to a standard freeze-etching unit, fractured, etched and replicated in the conventional way. Figure 2.12 is a micrograph of a cuticle prepared with this technique. The top edge shows the leaf surface with wax deposits very similar to those observed in micrographs of standard surface replicas of *Trifolium* sp. Below this wax layer is the cuticle and under the cuticle can be seen the cellulose fibrils of the cell wall. The canals that Hall believes transport the wax are indicated (MC). They have a central core of about 6–10 nm and an overall diameter of about 40 nm. These dimensions are similar to the dimensions of the canals seen by Hall with surface replica methods. Although they have not been seen to pass uninterrupted from the plasmalemma of the epidermal cell to the cuticle, their distribution and orientation suggest that they travel the full distance. Hall (1967b) substantiates the existence of wax-exuding pores in the cuticles of cauliflower (*Brassica oleracea* var. *botrytis*), *Trifolium repens* and *Eucalyptus urnigera* leaves and of apple fruits by the use of an improved technique of preparing replicas. Pores were observed in the cuticles of other plants, including pea and wheat; they occur in greatest numbers in those heavily coated with wax. Hall suggests that the wax exudes almost certainly in liquid form and that stresses in the cuticular membrane influence the orientation of the pores and affect the shape of the wax deposits.

However, neither Bolliger (1959) nor any of the subsequent workers looking at the cuticle by thin section methods have seen any sign of distinct

FIG. 4.29 Electron micrograph of a carbon replica of the adaxial surface of a leaf of *Hyacinthus orientalis*. Cr/Au/Pd

canals. The epidermal walls and cuticle appear to be remarkably uniform. The only other suggestion of canals in the walls comes from the work of Gaff, Chambers and Markus (1964) who postulated the existence of water canals about 30 nm in diameter along which small particles such as ferritin could be moved.

Some interesting observations by Hallam (1964) provide support for the view that wax lamellae occur within the cuticle itself. The wax bloom is always destroyed by conventional fixation and embedding methods. Hallam stabilised the surface wax on *Eucalyptus cinerea* by evaporating gold/

palladium onto the leaf surface (see Figs. 4.19 and 4.20); the leaf was then fixed in osmium tetroxide and embedded in the usual way, but the methacrylate was polymerised by ultra-violet light and not by heating. When this was done the cuticle showed lamellations of a lighter staining material, decreasing in thickness towards the external surface. Material fixed and embedded without the protecting layer of gold/palladium or polymerised by heat treatment did not show these lamellations. These differences suggest, according to Hallam, that the lamellae are wax, or its precursors, and could be indicative of pathways through the cuticle. Pathways, he thinks, that are anastomosing channels rather than a large number of pores directly connecting the epidermal cell wall with the exterior.

Shellhorn and Hull (1961) have also delineated distinct layers in the cuticle of the mesquite leaf by the use of incandescent illumination and special staining techniques. An outermost purple-coloured layer merged through shades of magenta and yellow to the green-coloured cellulose wall.

Cuticular structure and plant taxonomy

Four different approaches have been used to incorporate information from the study of cuticles into the taxonomic framework. Firstly there is the classic morphological approach of, for example, Edwards (1935), Stace (1965) and Sharma (1967). These authors use, amongst other characters, stomatal index, trichome frequency, stomatal size, epidermal cell shape, epidermal cell size and cuticular thickness and surface striations as contributory factors in the understanding of general taxonomic problems. Sax and Sax (1937) and Sax (1938) have used stomatal counts as a guide in studying chromosome numbers. Uphof (1962) has reviewed the limited literature on the taxonomic value of trichomes. Secondly there is the chemical taxonomic approach using the chemistry of selected constituents of the epicuticular wax. In general, the most valuable chemical substances taxonomically are not those which are involved in primary metabolic processes, but those which are end-products of metabolism. The wax hydrocarbon fraction meets this desideratum and provides a potentially valuable taxonomic fingerprint. This approach has been adopted by Eglinton and Hamilton (1967); Borges del Castillo, Brooks, Cambie, Eglinton, Hamilton and Pellitt (1967); Eglinton, Gonzalez, Hamilton and Raphael (1962) and Purdy and Truter (1961) amongst others and will be dealt with more fully in Chapter 5. Thirdly there is the approach adopted by Hallam (1967) in his study of the genus *Eucalyptus*; he used the fine structure of the crystalline waxes themselves as a contributory taxonomic character. Fourthly there is the 'spodogram' technique as developed by Molisch (1920). 'Spodograms' (Greek *spodos*, ash or dross; *gramma*, something drawn) are the siliceous skeletons of cell walls or hairs, cystoliths, calcium oxalate crystals and wall inclusions that are found in the residue of leaves slowly reduced to ash.

Regardless of the technique adopted, features of the epidermis come into the category of characters that, although not used as a basis for classification, are often safe criteria for the identification of individual species. Epidermal characters, as studied by Stace and Sharma, often in single instances provide the greatest differences between two taxa that are being studied (*taxon*, any taxonomic group). However, conclusions based on cuticular studies on one group of plants may not be applicable to another. The cuticle and the epidermis are also, as Stace and Sharma point out, as variable within the taxon as any other superficial morphological factor. Epidermal characters are, however, particularly useful in the study of sterile and fossil material. In summary Stace (1965) states that cuticular characters, although useful, are not of any outstanding fundamental or all-important significance, as has been claimed for some other features. They should be regarded as characters contributing to systematic evidence, and at times, may be of greater value in identification than any other characters.

Sax (1938) concluded that although stomatal counts were not an exact guide to the degree of ploidy of the plant, they were often a suggestive indication of variations in chromosome number. They could moreover be used, as Sax and Sax (1937) point out, on herbarium material and possibly, one supposes, on fossil material.

In developing their chemical taxonomic methods Eglinton and his colleagues examined the leaf waxes of a compact group of closely related genera of the sub-family Sempervivoideae of the family Crassulaceae. These genera are endemic to the Canary Islands and it was thought that one of the genera, *Aeonium*, would serve as an evolutionary model, similar to that of Darwin's finches of the Galapagos. The conclusions that Eglinton *et al.* were able to draw proved somewhat inconclusive. They showed that the alkane carbon-number patterns in *Aeonium* species confirmed relationships between closely related species, but that the differences between related genera were often insufficiently discriminating. Moreover even similar species, on conventional taxonomic criteria, sometimes had widely differing patterns and there was in general only a rough parallelism of hydrocarbon pattern and conventional botanical classification (see Chapter 5).

Hallam (1967) studied the wax types of the leaf surfaces of 316 species of *Eucalyptus* and divided them into three morphological types; those with tube-like waxes alone; those with plate-like waxes alone and those with both tube and plate waxes. Sub-groups of these were differentiated, for example, by the presence of simple or branched tube waxes. The pattern of wax morphology confirms the homogeneity of the *blood wood* group, and also the affinities of the Miniatae and Clavigerae with some *Angophora* species. The series Exsertae and subexsertae of the section Macrantherae with their sinuate-edged plates and the species of the series Globulares of the section Macrantherae (Normales) with compoundly-branched tubes each form natural groupings as assessed by the structure of their leaf surface wax. Three species, *Eucalyptus preissiana*, *E. megacarpa* and *E. coronata*, are thought on many characters to be out of place in the Globulares; this is confirmed by their possessing a different wax type from all the others in the series. The arid zone species (mostly from Western Australia) are separated by some conventional taxonomic keys, but it is suggested on cotyledon shape and supported by wax

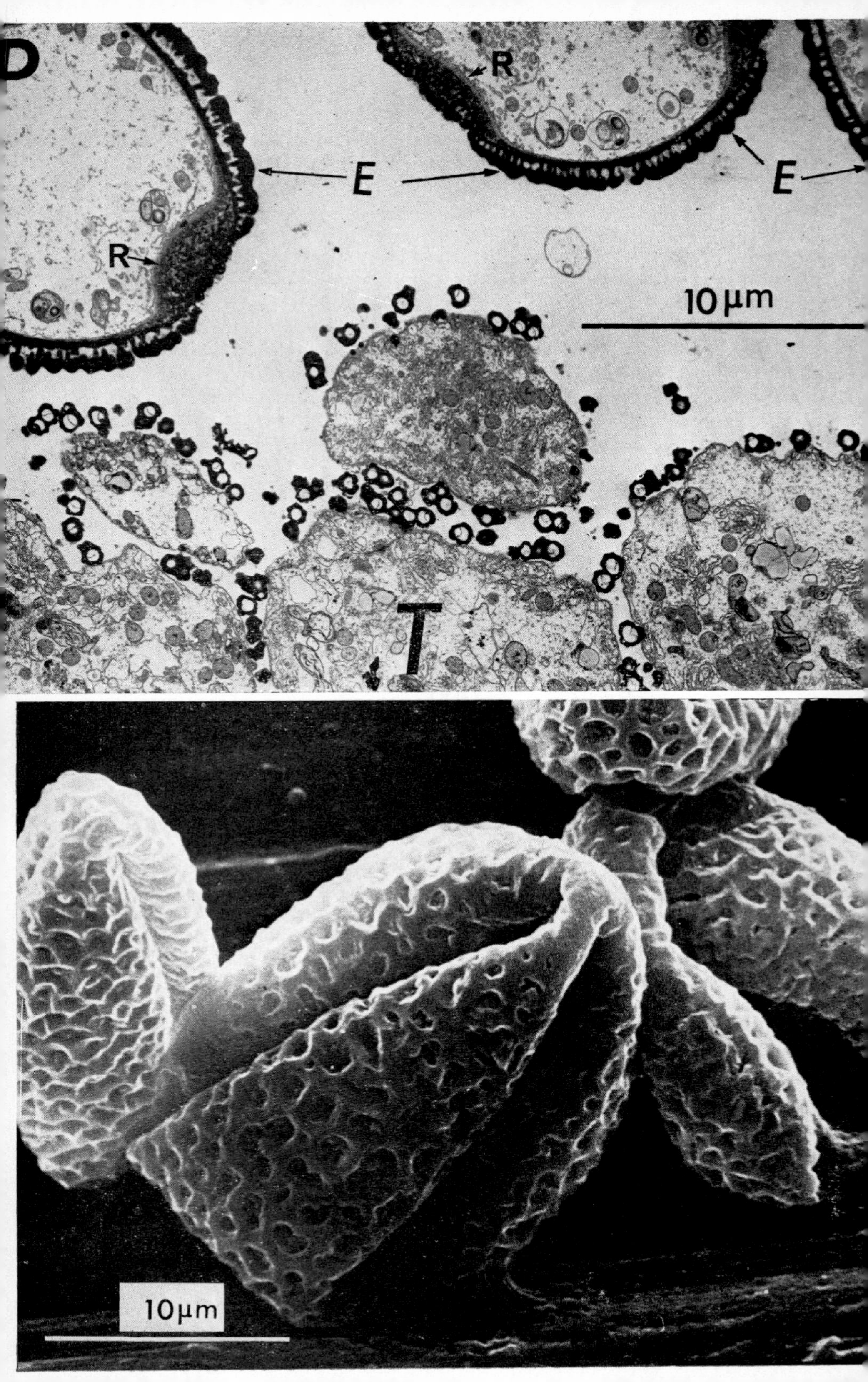
D
R
R
E
E
10 µm
T
10µm

morphology that they form a homogenous group. Wax morphology does not, however, assist in determining the boundaries of the existing *Eucalyptus* genus and its distinction from other closely related genera such as *Symphyomyrtus*. In general then the electron microscopic morphology of leaf waxes is useful not only in confirming some existing taxa, but sometimes also in providing additional information to establish other distinctions. This study also suggests evolutionary trends within the genus, not so far made apparent by the study of other features.

The spodogram technique is obviously limited to the few families which accumulate silica in significant quantities. The Gramineae have large amounts of silica in the leaves and a few other families, e.g. the Rubiaceae and Urticaceae have silicified hairs. Molisch (1920) indicated that spodograms could be useful in the taxonomy of the Gramineae, Werner (1928) used the method to evaluate the relationship of certain grasses and Ohki (1932) to separate members of the Bambusaceae. Neither silicified hairs nor calcium carbonate crystals appear to have been used for taxonomic purposes (Uphof, 1962). Recently a modification of the technique has been used in the identification of ancient buried plant remains. Under unfavourable conditions for preservation, e.g. alkaline or aerobic soils, the soft tissue disintegrates, but leaves behind the spodogram of inert silica. A study of the spodograms, and of pollen ***exines*** many of which are almost as resistant to decay as the spodograms, supplemented by the known associations of plants, many of whose members do not form spodograms, can often give valuable information about the flora of an area at a given time (Dimbleby, 1967).

Apart from the conventional use in taxonomy of epidermal and cuticular characters, Stewart (1965) and others have used such features to identify the grasses eaten by wild animals. In East Africa it is almost impossible to identify directly the food plants being grazed by herbivorous wild animals. But it is possible to identify the species selected both from stomach contents and the fragments remaining undigested in the faeces.

BARK OR PERIDERM

The term bark usually includes the phloem of stems and roots and all the tissues outside it. Since bark is a non-technical term and embraces tissues with which we are not concerned, it will be more useful to use the term periderm

FIG. 4.30 Electron micrograph of a section through the pollen grains and tapetal cells of an anther of *Saintpaulia ionantha*. The pollen grains (***P***) are already covered by their exine (***E***). Poral regions (***R***) in the pollen grains show where the exine is interrupted. The tapetal cells (***T***) are still surrounded by the dark-staining 'Ubisch' bodies. Glut/Osm/Pb/Ur. (Courtesy M. C. Ledbetter)

FIG. 4.31 Scanning electron micrograph of *Salix* pollen grains. Ms/Sc. (Courtesy P. Echlin)

which is the protective tissue that replaces the epidermis when the epidermis is killed or sloughed away. The process of formation of a periderm is common in stems and roots of dicotyledons and gymnosperms that increase in thickness by secondary growth. Monocotyledons rarely develop such a protective growth. The periderm consists of three parts; a ***meristem*** or cork ***cambium,*** the cork produced by this meristem to the outside, and a parenchymatous tissue produced by it on the inside. As the periderm develops certain areas may be separated off by non-living layers of cork cells. These layers subsequently die and are seen as the sloughing skins of the trunks of such trees as London plane (*Platanus acerifolia*).

Periderm develops at different times in different woody plants. It normally appears after primary elongation of the stem or on surfaces exposed after abscission of plant parts such as leaves or branches and beneath wounds. The cork cells produced by the meristem of the periderm are arranged in radial rows and owe some of their protective ability to the presence of suberin in their walls. Apart from the more obvious constituents of cells and cell walls, cork cells may contain considerable quantities of lignin, fats, tannins and terpenoids. Cork cells lose their protoplasm after differentiation and then are either filled with air or may remain filled with coloured resinous or tannin-rich compounds. Bottle cork consists of thin-walled, suberin-rich, air-filled cells. It is compressible, light, buoyant and highly impervious to water and decay.

Lenticels

Lenticels, passages through the periderm, are regions of loose cells and incomplete suberisation. Seen from the surface they look like a tuft of dry cells pushing up through a hole in the periderm. They vary in size and shape ranging from about 250 μm to 1 cm across. Some lenticels increase in size as the circumference of the stem increases. Lenticels are usually initiated beneath a stoma. They can be detected before the stem has ceased its primary growth and before the rest of the periderm has developed. A localised meristem develops just below the original stoma. A tuft of colourless loose tissue is developed which finally ruptures the epidermis. The meristem of the lenticel usually divides faster than that of the periderm surrounding it so that the lenticel comes to protrude through the surface.

Tetley (1931) showed that the stomata of apple fruits are usually converted into lenticels at an early stage in the development of the fruits. The walls of the cells lining the substomatal cavity become 'suberised' and the cells are killed; cork is formed and isolates the dead cells from the flesh below. Edney (1956) differentiates between 'open' and 'closed' lenticels of apple fruits according to the degrees of completeness of suberisation of the substomatal tissues. The outermost cells of the lenticel, as they protrude through the periderm, die and are continually sloughed away. The cells of the lenticel are thinner walled than those of the periderm and contain little or no suberin. They readily absorb water and, when wetted, swell up. Having substantial air spaces between the cells a lenticel can act as a pathway for gas exchange.

Although they are not as important as stomata in this respect Haberlandt (1914) has demonstrated by artificially plugging lenticels that they can substantially affect both gas exchange and water loss.

POLLEN GRAIN SURFACES

Pollen grains, although exposed to the atmosphere for at least part of their existence, are not covered by cutin, but by a complex material called sporopollenin. The chemistry of this substance, in so far as it is known, will be discussed in Chapter 5. The whole of the pollen grain wall is called the sporoderm and this is divided into an inner part, the ***intine*** and an outer the exine. In other texts much more complicated subdivisions are made. Only the exine contains sporopollenin; the intine is largely composed of carbohydrates including cellulose and possibly pectin. The inner and outer layers of the exine are joined by systems of hollow funnels and pillars of great complexity (Figs. 4.30 and 4.31). As in the cuticle of leaves the exine comprises two or more different layers probably having different chemical properties. Whether the various layers are derived from within the pollen grain itself or from the nursing tapetal cells has been much debated. The evidence so far suggests that there may be contributions from both sides of the young microspore membrane. Part of the shape and surface characteristics of the pollen grain may be determined by the fact that the four products of meiosis (at the tetrad stage) fit together very closely (Echlin, 1968). The general shape of the grain is apparently closely related to the contact geometry of the microspores during their association at the tetrad stage.

When the pollen grain begins to develop the layer of sporopollenin on its surface, small bodies known as spheroids or Ubisch bodies, also apparently containing sporopollenin, appear within the tapetal cells (Fig. 4.30) (U). These Ubisch bodies are later discharged through the plasmalemma of the tapetal cells and accumulate in the space between the tapetal cells and the now well-developed pollen grains. Evidence is lacking to suggest that the Ubisch bodies themselves form the exine, but it is interesting that, at the same time, both the pollen grains and the tapetal cells have the synthetic ability to form sporopollenin. There is more evidence to suggest that most of the formation of the exine is directed from within the pollen grain. Heslop-Harrison (1963a, b) believes that lengths of endoplasmic reticulum, which align themselves against the wall of the developing pollen grain, organise in some way the deposition of the exine. Heavy deposition of the layers of the exine takes place between, but not opposite, the regions where the endoplasmic reticulum lies against the plasmalemma. Thus the endoplasmic reticulum is opposite what becomes the thinnest regions of the wall. These thin regions may become the furrow or furrows of the pollen grain. There are other reports of sporopollenin accumulating around what appears to be stacks of lamellae of unit membrane dimensions (Rowley and Southworth, 1967). The lamellae become embedded in the exine and remain visible until the maturity of the grain.

Thus some of the factors controlling the general shape of the pollen grain are environmental, but the evidence obtained so far suggests that, at least after the tetrad stage, the pollen grain cytoplasm determines its own surface structure and chemistry with little or no contribution either from the tapetal cells or anything in the anther cavity.

5

The Chemistry of Cuticles and Barks

CUTICLES

Cuticle waxes

Up to 1955, evidence from the analysis of plant waxes largely supported the view of Chibnall and his co-workers that they contain *n*-paraffins (*n*-alkanes) only of odd carbon numbers, and *n*-primary alcohols only of even numbers (see Chapter 3). Prophète (1926) reported that rose petal wax contains both odd and even carbon-numbered paraffins in the overall range C_{16}–C_{30}. Wanless, King and Ritter (1955) produced further evidence of the occurrence of even carbon-numbered paraffins in waxes by showing that those of pyrethrum and bean (*Phaseolus aureus*) contain all the *n*-paraffins, both odd and even carbon-numbered, in the approximate range C_{24} to C_{36}. The paraffins were obtained from the acetone-insoluble fractions of the waxes by elution from activated alumina, fractionated by elution from activated charcoal and examined with the mass spectrometer. Carruthers and Johnstone (1959) detected both odd and even carbon-numbered paraffins in the wax from tobacco leaves, and Kranz, Lamberton, Murray and Redcliffe (1960) showed that sugar-cane wax contains both odd and even carbon-numbered homologues of paraffins and other compounds.

In 1961, Waldron, Gowers, Chibnall and Piper re-examined, with the mass spectrometer, the paraffin and primary alcohol fractions of waxes characterised in former years. All the *n*-paraffins from C_{28} to C_{34} were found in the waxes of apple and Brussels' sprout; even carbon-numbered paraffins from C_{22} to C_{30} occurred as minor constituents in rose petal wax. The primary alcohol fractions were not composed exclusively of even carbon-numbered members; odd carbon-numbered compounds were present in some, also as minor constituents. An interesting new development was the detection, in rose petal and tobacco leaf waxes, of branched chain mono-methyl paraffins, the *iso*-paraffins. Chibnall and his colleagues concluded that the older methods of fractionation were not sufficiently selective to resolve mixtures of neighbouring homologues and that the composition of such mixtures could not be predicted with certainty from the analytical data obtained earlier. The recognition that even carbon-numbered paraffins occur alongside odd, and odd carbon-numbered alcohols alongside even, led to new views on the biosynthesis of these compounds. Wanless, King and Ritter (1955), for example, proposed possible mechanisms of formation of paraffin and fatty acid homologues (see Chapter 6).

Recent work has revealed even greater complexity in the constituents of plant waxes. Many new types have been examined and reappraisals made of some that had received attention in the past. The methods of isolation of the waxes have varied greatly and difficulty arises in assessing the extent to which a particular material may be regarded as being of cuticular origin. Some are clearly derived from the cuticle; the wax may be scraped from the plant surface, removed by washing it rapidly with cold solvent, or extracted from a surface layer containing a minimum of cellular tissue. At the other extreme, the treatment of dried powdered plant material with a hot lipid solvent removes cuticular wax and cellular lipids, although the subsequent fractionation of the extract may concentrate the components derived from the cuticle. Considerable evidence is now available that nearly all of the hydrocarbon fraction found in plant extracts, however made, is derived from the cuticle. Eglinton, Hamilton and Martin-Smith (1962) found similar proportions of alkane constituents in the wax of *Arundo conspicua* obtained by extracting the powdered leaf or washing the leaf surface, and Hill and Mattick (1966) showed that about 90% of the total hydrocarbon fraction of the cabbage leaf is located on the surface. Some of the other reported components of waxes obtained by the extraction of dry plant tissue with hot solvent are clearly of cellular origin. In the following discussion the waxes that, from a consideration of the method of isolation, are beyond reasonable doubt derived solely from the cuticle are designated by (C).

Apple fruit wax (C)

This was further examined by Huelin and Gallop (1951a, b). Apple peel was treated with a solution of ammonium oxalate and oxalic acid of pH 4 at

37 °C and then at 50 °C. The acid solution dissolved a pectic layer between the epidermal and underlying cells, and so liberated a skin which consisted of the cuticle and epidermis. The skin was washed, dried and extracted with petroleum ether, and hard and soft wax fractions were obtained by the acetone precipitation procedure. Subsequent extraction of the skin with ether yielded the ursolic acid fraction. The analyses were made quantitatively; the lipid (petroleum ether-soluble) fraction of the skin amounted to 0·5 mg/cm^2 or to about half of the total lipid content of the whole fruit. The soft wax, or oil, fraction was found to contain unsaturated esters, which on saponification yielded fatty and some hydroxy-fatty acids. The oil fraction especially increased during storage of the fruits; non-volatile esters were produced most rapidly during the early period of storage and volatile esters, derived from the lower acids and alcohols with maximum chain length of 6 carbon atoms, during the latter. Some evidence was obtained of an association between the lipid material and cellulose in the skin.

Davenport (1956, 1960) examined the acids obtained by saponification of the soft wax fraction of the apple cuticle. The saturated fatty acids comprised 10% of the cuticle oil, and consisted chiefly of stearic and arachidic acids with smaller amounts of palmitic, behenic and acids of larger molecular weight than that of behenic; the unsaturated acids were linoleic and oleic, with a smaller amount of linolenic. Hydroxy-fatty acids comprised 7% of the cuticle oil. The unsaturated acids predominated among the acids obtained from the oil, the approximate percentage contents being linoleic 39, oleic 22, stearic 7, arachidic 6, linolenic 4, palmitic 1, behenic 1, and lauric and myristic each less than 1. Evidence was obtained of the presence of small amounts of caproic (C_6), enanthic (C_7) and caprylic (C_8) acids. A wide range of saturated acids ($C_6 - > C_{22}$) thus occurs in the cuticle oil, probably largely in combination as esters.

Mazliak (1960a, 1961a) identified in the wax straight-chain saturated *n*-alkanes from heptadecane to tritriacontane and *n*-primary alcohols from hexadecanol to tetratriacontanol. The wax was saponified, and the hydrocarbons and alcohols were successively eluted from a column of alumina with petroleum ether and a petroleum ether-methanol mixture. Most of the hydrocarbon fraction consisted of nonacosane. Fatty acids found after saponification of the wax were lauric, myristic, palmitic, stearic, oleic, linoleic, arachidic, behenic, lignoceric, cerotic, montanic and melissic; the most abundant acid was palmitic (Mazliak, 1960b). In 1962, Mazliak showed that the wax contains even and odd carbon-numbered α,ω–diols from C_{20}–C_{28}. The compounds were obtained by elution of the unsaponifiable fraction from alumina with a benzene-ethanol mixture and examined by gas chromatography of their diacetates. In addition to the fatty acids, the wax was found to contain ω-hydroxy-acids within the range C_{10}–C_{23}; the hydroxy-acids as methyl esters were separated from the normal acids by elution from a column of magnesium silicate and examined by gas chromatography of their methyl acetoxy esters (Mazliak, 1963a). The principal constituents of the hard and soft fractions of apple wax, as summarised by Mazliak and Pommier-Miard (1963), are shown in Table 5.1.

Table 5.1 Principal constituents of apple wax

	Hard wax	Soft wax
Hydrocarbons	C_{27} C_{29} C_{31} C_{33}	C_{15} C_{16} C_{17} C_{18} C_{19} C_{20} C_{21} C_{22} C_{23} C_{24} C_{25} C_{26} C_{27} C_{28} C_{29}
Saturated acids	C_{16} C_{18} C_{20} C_{22} C_{24} C_{26} C_{28} C_{30}	C_{10} C_{12} C_{14} C_{16} C_{18} C_{20} C_{22} C_{24}
Unsaturated acids	oleic linoleic	palmitoleic oleic, linoleic linolenic
Primary alcohols	C_{20} C_{22} C_{24} C_{26} C_{28} C_{30} C_{32}	C_{16} C_{18} C_{20} C_{22} C_{24} C_{26} C_{28} C_{30}
Secondary alcohol	nonacosan-10-ol	
Diols	C_{22} C_{23} C_{24} C_{25} C_{26} C_{27} C_{28} C_{29} C_{30}	
Hydroxy acids	C_{14} C_{15} C_{16} C_{17} C_{18} C_{19} C_{20} C_{21} C_{22} C_{23}	

Mazliak's investigations revealed the presence of over 50 compounds in apple wax. Mazliak and Pommier-Miard point out that since the analyses were made after saponification, the constituents listed do not necessarily reflect the true composition of the wax; any mono-esters derived from the primary alcohols and fatty acids may contain about 50 carbon atoms. Mazliak (1963b, 1964) has reviewed the morphological, biochemical and physiological investigations on apple wax.

Brieskorn and Schneider (1961) identified 16-hydroxypalmitic (juniperic) acid and 14-hydroxymyristic acid in the wax of apple peel. Meigh (1964) examined the fatty acids and hydrocarbons of apple skin in connection with the condition known as scald which the fruits develop during storage. Odd and even carbon-numbered fatty acids from C_9 to C_{22} were found, the C_{18} group of acids predominating followed by C_{16} and C_{20}; the principal hydrocarbon was C_{29} followed by C_{27}, C_{25} and C_{28}, and odd and even carbon-numbered members between C_{10} and C_{31} were present. Murray, Huelin and Davenport (1964) and Huelin and Murray (1966) have also investigated the constituents of apple peel in relation to scald. The presence of a volatile sesquiterpene hydrocarbon, α-farnesene, was demonstrated and evidence was obtained that it plays a part in the development of the condition. Huelin and Murray (1966) suggest that a farnosyl or nerolidyl intermediate may be involved in the biosynthesis of triterpene acids (see Chapter 6).

The detection of α,ω-diols extended the range of alcohols known to be present in apple wax; diols had been found earlier by Murray and Schoenfeld (1955a) in carnauba wax. The hydrocarbons, alcohols, fatty and hydroxy-fatty acids discussed above represent possibly not more than 60% of apple wax. Comparatively little attention has been paid to the triterpenoid fraction which consists chiefly of ursolic and oleanolic acids (Brieskorn and Klinger,

1963). It may amount to about 40% of the wax. Two 20-hydroxylated ursanes were isolated by Lawrie, McLean and Younes (1966, 1967) from apple skin.

The hydrocarbons of the cuticles of apple and pear were examined by Wollrab (1967). The fruit cuticles contain C_{16}–C_{31} *n*-alkanes, two series of C_{19}–C_{33} branched alkanes (chiefly *iso*- and *anteiso*-alkanes) and a small proportion of C_{20}–C_{34} alkenes (chiefly unbranched). The leaf cuticles contain C_{17}–C_{33} *n*-alkanes, two homologous series of C_{17}–C_{31} branched alkanes and C_{16}–C_{33} alkenes with a terminal vinyl group.

Carnauba wax (C)

Koonce and Brown (1944, 1945) isolated tetracosanoic acid, octacosan-l-ol, triacontan-l-ol and dotriacontan-l-ol from carnauba wax, the C_{32} compound being the chief constituent of the unsaponifiable fraction. The products of hydrolysis of the wax were examined by Murray and Schoenfeld (1951, 1953, 1955a, b). Straight-chain even carbon-numbered alcohols in the range C_{24}–C_{34} were present, with dotriacontan-l-ol and tetratriacontan-l-ol predominating. Almost 40% of the *n*-fatty acids were even carbon-numbered compounds from C_{18}–C_{30}. Seven hydroxy-fatty acids were isolated from the wax and identified as the ω-hydroxy derivatives of the C_{18}, C_{20}, C_{22}, C_{24}, C_{26}, C_{28}, and C_{30} acids; the C_{24} acid derivative was the most abundant, followed by C_{26} and C_{28}. Diols comprised 6–7% of the wax; they were isolated from the unsaponifiable fraction and identified from melting points, melting points of their diacetates and long crystal spacings as *n*-docosane-1,22-diol, *n*-tetracosane-1,24-diol, *n*-hexacosane-1,26-diol and *n*-octacosane-1,28-diol. The chief diol was the C_{24} compound, followed by C_{28} and C_{22}. Savidan (1956a) reported the presence of pentacosane, heptacosane, nonacosane and lupeol in the wax.

Downing, Kranz and Murray (1961) quantitatively analysed hydrolysed carnauba wax by gas chromatography. The unsaponifiable fraction (56% of the wax) was separated on an alumina column into *n*-alcohols (83%) *n*-α, ω-diols (12%) and smaller amounts of hydrocarbons, sterols and a fraction of high molecular weight. The acidic fraction was resolved by chromatography of the methyl esters into *n*-fatty acids (62%) and ω-hydroxy-fatty acids (33%). The naturally occurring hydrocarbons consisted of both odd and even carbon-numbered compounds with the odd predominating; the hydrocarbons prepared from the alcohols, fatty and hydroxy-fatty acids were mostly even carbon-numbered compounds. The sterols gave a positive reaction with acetic anhydride and sulphuric acid (Liebermann-Burchard test); analysis indicated a molecular formula of $C_{27}H_{46}O_2$ or $C_{28}H_{48}O_2$, containing two hydroxyl groups. Barnes, Galbraith, Ritchie and Taylor (1965) report the isolation of carnauba-diol $C_{31}H_{54}O_2$ from the hydrolysed wax.

Mazliak (1961b, 1962, 1963a) studied carnauba wax with methods similar to those used for apple wax. The paraffin fraction contained odd and even carbon-numbered members from C_{19}–C_{35}, the chief ones being *n*-heptacosane, *n*-octacosane, *n*-nonacosane and *n*-triacontane. The primary monoalcohols were chiefly *n*-C_{28}–C_{36}; 70% of the fraction consisted of *n*-dotriacontan-1-ol. All even carbon-numbered α,ω-diols, and some odd, from C_{20}–C_{32}, were found. Fatty acids from C_{16}–C_{30}, with *n*-tetracosanoic, *n*-octacosanoic and *n*-docosanoic acids most prominant, and odd and even carbon-numbered ω-hydroxy-fatty acids from C_{10}–C_{30} were also identified.

Carnauba wax contains a trace only of alkanes and 14% of alkyl esters. The chief esterified primary alcohol is *n*-dotriacontan-1-ol and the chief ester probably dotriacontanyl tetracosanoate. No triterpenoids or sterols are detectable by thin-layer chromatography in carnauba, candelilla, esparto (*Stipa tenacissima*) or ouricury waxes (Holloway, 1967). Vandenburg and Wilder (1967) have isolated *p*-hydroxy- and *p*-methoxy-cinnamic acids from a fraction of the unsaponifiable material of carnauba wax. These aromatic acids, previously unreported, occur as part of a polymerised diester. Only trace amounts occur in the free state in the wax.

Carnauba wax is truly cuticular; it occurs abundantly on the plant surface. Its constituent groups of alcohols and acids resemble those of apple wax, but it differs from apple wax in its much lower contents of hydrocarbons and, in particular, cyclic compounds.

Sugar-cane wax (C)

Hexacosanol, triacontanol, saturated and unsaturated fatty acids and sterols occur among the components of sugar-cane (*Saccharum officinarum*) wax (Mitsui and Matsuda, 1942). A hard wax fraction comprises 70% of the wax; *n*-hexadecanoic and *n*-eicosanoic acids predominate among the lower fatty acid constituents, but a considerable proportion of the acid fraction consists of members with chain lengths greater than C_{22} (Kapil and Mukherjee, 1954, 1963). Kreger (1948) showed by X-ray studies that the chief alcohol of the wax is *n*-octacosan-1-ol, and this was confirmed by Horn and Matic (1957) and Kranz, Lamberton, Murray and Redcliffe (1960). Horn and Matic extracted the wax from sugar cane with *iso*heptane; the unsaponifiable fraction contained a large proportion of compounds of about 30 carbon atoms including hydrocarbons in the range C_{27}–C_{31}, alcohols, saturated ketones and α,β-unsaturated ketones. Compounds with three conjugated double bonds comprised 1% of the wax. Octacosan-1-ol, the chief alcohol, amounted to about 7% of the wax or 65% of the alcohol fraction. Wiedenhof (1959) concluded from chromatographic and X-ray diffraction studies that paraffins, esters, alcohols and acids are the main constituents of sugar cane and some other waxes. Bose and Gupta (1961) found that sugar-cane wax contains β-sitosterol (13% of the unsaponifiable fraction) and stigmasterol (6%). Sugar-cane wax contains 24% of free primary alcohols C_{19}–C_{32}, chiefly octacosan-1-ol and hexacosan-1-ol, and 10% of free fatty acids C_{14}–C_{34} chiefly octacosanoic and triacontanoic (Holloway, 1967). Warth (1956) reports 14% of free fatty acids in the wax and Kranz, Lamberton, Murray and Redcliffe (1960) 8%.

That sugar-cane wax has an unusual composition was first reported by Lamberton and Redcliffe (1959, 1960). The wax was obtained by scraping sugar-cane stalks with wire brushes and extracted with petroleum ether. Distillation of the wax under reduced pressure gave a high yield of long-chain aldehydes, which were considered to occur in the wax in polymeric form. Hydrocarbons, free alcohols and free acids were also present, but no evidence was obtained of the occurrence of esters. Kranz, Lamberton, Murray and Redcliffe (1960) examined sugar-cane cuticle wax by gas chromatography. Aldehydes, alcohols and acids, all straight-chain compounds principally of even carbon numbers with the C_{28} members predominating, were found but appreciable amounts of odd carbon-numbered homologues were also present. The occurrence of odd carbon-numbered fatty acids in a plant wax was thus reported for the first time. The hydrocarbons consisted of a wide range of odd and even carbon-numbered compounds and included *n*-heptacosane as the major constituent. The composition of the cuticle wax was reported to be aldehydes *c.* 50%, alcohols 25–27%, acids 7–8% and hydrocarbons 8–9%. Free aldehydes were present in solutions of the cuticle wax in chloroform, but some doubt remained as to the form in which the aldehydes occurred in the wax. The position was re-examined by Lamberton (1965) who found good evidence from nuclear magnetic resonance studies that the aldehyde occurs in the free state, or more probably in the form of an aldehyde hemihydrate R—CH(OH)—O—CH(OH)—R ($R = n-C_{27}H_{55}$ or homologues), which dissociates readily in solution. The possibility of a more highly polymeric structure of the hemihydrate was not excluded.

The hydrocarbons of sugar-cane wax were further investigated by Šorm, Wollrab, Jarolimek and Streibl (1964). The *n*-alkanes are accompanied by mono-methyl (*iso* and *anteiso*) alkanes C_{10}–C_{30}, by dimethyl alkanes C_{22}–C_{31} and by three homologous alkene series. The chief component of the alkene fraction is an homologous series C_{15}–C_{33} with the C_{31} and C_{33} members predominating; the *cis* double bond occurs at position 10. The second homologous series consists of alk-1-enes, with partial isomerisation of the double bond to position 2 (*trans*). The third homologous series is believed to contain aliphatic dienes. Streibl and Stránský (1968) describe the chromatographic and other methods used in the isolation and identification of alkenes in sugar-cane and other natural waxes.

The work on sugar-cane wax further reveals the complexity of cuticular waxes. Not only was the wax found to be devoid of esters, but alkenes and dimethyl alkanes were detected among the hydrocarbons, odd carbon-numbered members among the fatty acids, and aldehydes were found for the first time.

Eucalyptus wax (C)

Examination of *Eucalyptus* leaf wax, assessed from the method of isolation as being of cuticular origin, revealed for the first time the occurrence in waxes of long-chain β-diketones (Horn and Lamberton, 1962). The compounds also exist in the waxes of acacia, Gramineae and carnation (*Dianthus*) spp. (Table 5.2, p. 130). The chief β-diketone in the waxes of *E. globulus*, *E. behriana*, *E. cinerea*, *E. pulverulenta*, *E. gamophylla* and *E. viminalis* is *n*-tritriacontan-16,18-dione; the waxes of *E. risdoni* and *E. coccifera* yield a mixture of β-diketones consisting chiefly of *n*-nonacosan-12,14-dione and *n*-hentriacontan-14,16-dione. Hydrocarbons in the wax are low in amount (up to 3% of the whole wax) and the most important members are in the range C_{23} to C_{31}. The esterified *n*-alcohols of the waxes of *E. globulus* and *E. risdoni* belong to two series: long-chain alkan-1-ols of predominantly even carbon numbers and alkan-2-ols of medium chain length (C_9–C_{15}) of predominantly odd carbon numbers. A flavone named eucalyptin (5-hydroxy-7,4′-dimethoxy-6,8-dimethylflavone) is present in the waxes of *E. globulus*, *E. cinerea*, *E. risdoni*, *E. deglupta*, *E. viminalis* and *E. regnans*, and another, 5-hydroxy-7,4′-dimethoxy-6-methylflavone, along with eucalyptin in the waxes of *E. urnigera* and *E. torelliana*. Eucalyptin also occurs in the leaf waxes of *Angophora* spp. (Horn and Lamberton, 1963; Horn, Kranz and Lamberton, 1964; Lamberton, 1964). Waxes from *E. globulus*, *E. cinerea*, *E. risdoni* and *E. urnigera* contain small amounts of 11,12-dehydroursolic lactone acetate and other triterpenoids (Horn and Lamberton, 1964).

Grape wax (C)

Radler and Horn (1965) isolated wax from grapes by extraction with hot chloroform. The wax, which was probably largely cuticular, was extracted with petroleum ether to yield a soft, low-melting fraction (*c*.30%) and a crude triterpene acid residue (*c*.70%). High vacuum sublimation of the original wax gave a volatile fraction which consisted of alcohols, aldehydes and hydrocarbons and a residue of esters, free acids, oleanolic acid, products of polymerisation and compounds of high molecular weight. Separation into fractions was also achieved by column chromatography on silicic acid or alumina. The findings are of interest in showing again the occurrence of aldehydes in a plant wax. Radler and Horn point out that the use of alumina in chromatographic separations destroys the aldehydes; because of the frequent use of this chromatographic agent aldehydes in other waxes may have escaped detection and may be much more prevalent than the analytical results suggest.

An interesting comparison was made of the cuticular waxes of berries and leaves of several varieties of *Vitis vinifera* and *Vitis* hybrids (Radler, 1965a, b). Hydrocarbons comprise 1–5% of the petroleum ether-soluble, soft wax fractions, free acids 4–13% and primary alcohols 40–67%. The hydrocarbons are odd and even carbon-numbered *n*-paraffins from C_{18}–C_{35}; C_{25}, C_{27}, C_{29} and C_{31} predominate. Free acids occur between *n*-C_{14} and *n*-C_{34}, and the pattern of acids in the leaf wax differs from that in the stem, young grape and mature grape waxes. The alcohols are *n*-C_{20} to *n*-C_{34}; the C_{26} compound is the predominant alcohol of the berry wax and the C_{28} and C_{30} compounds the chief alcohols of the leaf wax. Ester-aldehyde fractions are second in amount to the alcohol fractions of the berry and mature leaf waxes. The chief aldehydes of the wax of the mature leaf are C_{28} and C_{30} and of the fresh berry C_{24}, C_{26} and C_{28}; the wax of the young leaf contains no aldehydes but, instead, a series of esters of long-chain acids with an alcohol, either *n*-nonanol or *n*-decanol. The petroleum ether-insoluble hard wax fraction of the berry wax is chiefly oleanolic acid, but this occurs only in a small amount in the hard wax fraction of the leaf wax. Differences in the composition of the waxes of comparable parts of varieties and species are usually small.

Brassica wax (C)

Purdy and Truter (1963a, b, c) resolved cabbage (*Brassica oleracea* var. *capitata*) leaf surface wax by thin-layer chromatography into nine fractions, including long-chain hydrocarbons, esters, ketones, ketols, primary and secondary alcohols, and acids. The hydrocarbons contain ten *n*-paraffins, odd and even carbon-numbered, with *n*-nonacosane predominant. Seven free alcohols, five free acids, and six alcohols and five acids from the esters, all even carbon-numbered compounds, were identified. The primary alcohols range from *n*-dodecanol to *n*-octacosanol, and the acids from *n*-dodecanoic to *n*-tetracosanoic. The secondary alcohol fraction is a mixture of *n*-nonacosan-10-ol (about 80%) and *n*-nonacosan-15-ol, the ketone fraction a mixture of *n*-nonacosan-15-one (about 70%) and *n*-nonacosan-10-one and the ketol fraction a mixture of *n*-nonacosan-10-ol-15-one (about 80%) and its 10-one-15-ol isomer. The composition of the wax was reported to be: hydrocarbons 36%, esters 13%, ketones 14%, primary alcohols 9%, secondary alcohols 11%, acids 9% and ketols 1%. Seoane (1961) obtained *n*-nonacosan-10-ol from the lipids of leaves of the greater celandine (*Chelidonium majus*); he proposes the name celidoniol for the alcohol and celidonione for the corresponding ketone *n*-nonacosan-10-one.

The lipid fraction of the cabbage leaf was examined by Laseter, Weber and Oró (1968) by combined gas chromatography-mass spectrometry. The saturated hydrocarbons ranged predominantly from C_{27} to C_{31} with *n*-nonacosane the chief constituent. Free unsaturated and saturated fatty acids were identified; the major components were linolenic, linoleic and palmitic acids and these constituted more than 50% of the total fatty acids present. The ketonic fraction contained a component in high concentration which was identified from the mass spectrum as *n*-nonacosan-15-one.

Cabbage wax contains 32% of *n*-alkanes in the range C_{20}–C_{33} chiefly nonacosane, 16% of alkyl ketones including *n*-nonacosan-15-one, 14% of secondary alcohols chiefly C_{29} and 10% of alkyl esters (Holloway, 1967). These values are in general agreement with those reported for alkanes by Purdy and Truter (1963c), Horn, Kranz and Lamberton (1964), Baker, Batt, Silva Fernandes and Martin (1964) and Hall, Matus, Lamberton and Barber (1965); for alkyl ketones by Purdy and Truter (1963c), Horn, Kranz and Lamberton (1964) and Hall, Matus, Lamberton and Barber (1965); for secondary alcohols by Purdy and Truter (1963c) and for alkyl esters by Purdy and Truter (1963c) and Horn, Kranz and Lamberton (1964).

Tomato wax (C)

The wax of the tomato (*Lycopersicum esculentum*) fruit was examined by Brieskorn and Reinartz (1967a). Its alkane fraction contains 60% of *n*-hentriacontane, 17% of *n*-dotriacontane, 8% each of *n*-nonacosane and *n*-tritriacontane and 1% of *n*-tetratriacontane. Palmitic, stearic, oleic, linoleic and linolenic acids, α- and β-amyrin and sterols are also present. *p*-Coumaric acid also occurs in the skin.

The waxes of non-glaucous mutants

Hall, Matus, Lamberton and Barber (1965) have provided interesting information on the composition of the surface waxes from the normal waxy, or glaucous, and the green, or non-glaucous, variants in cauliflower (*Brassica oleracea* var. *botrytis*), *Eucalyptus urnigera* and New Zealand blue tussock grass (*Poa colensoi*). The non-glaucous, sometimes called non-waxy, variants always possess wax deposits; on cauliflower the deposit amounts to 60% of that on the glaucous plant. The wax of glaucous cauliflower leaves contains about three times more ketone (chiefly C_{29} ketone, probably *n*-nonacosan-15-one) than that of the green leaves, with a corresponding reduction in the amount of the combined alcohol and ester fraction. The wax of glaucous *E. urnigera* leaves is much richer in β-diketones (chiefly *n*-tritriacontan-16,18-dione) and much poorer in hydrocarbons than that of the non-glaucous. The wax of glaucous *Poa colensoi* also contains an appreciable proportion of β-diketones; that of the non-glaucous variant contains none, but is richer in hydrocarbons. The physical forms of the waxes on the bloomed and green leaves differ, and this is reflected in differences in the ease with which the surfaces can be wetted (see p. 188).

Other plant waxes

Many other plant waxes have been examined chemically in recent years; some examples to demonstrate the range of constituents are shown in Table 5.2. Mazliak (1963c, 1968) has tabulated the principal constituents (hydrocarbons, alcohols, saturated and unsaturated acids) of the cuticular waxes of the leaves, flowers and fruits of many plants of tropical, subtropical or temperate regions.

Holloway (1967) examined the unsaponified surface waxes of the leaves of 66 plants by thin-layer chromatography. Compounds corresponding in R_F

Table 5.2 Constituents of plant waxes

Source of wax	Constituents and references
Acacia spp. (C) Wattle (leaf and stem)	*n*-alkanes C_{21}–C_{29}; β-diketones, chiefly *n*-tritriacontan-16,18-dione; lupeyl acetate Horn and Lamberton (1962, 1964); Horn, Kranz and Lamberton (1964)
Aeonium spp. (C) (leaf)	*n*-alkanes, chiefly C_{31}, C_{33}; *iso*-alkanes, including 2-methyldotriacontane; diterpene diol: labdane-8α,15-diol (lindleyol) Eglinton, Gonzales, Hamilton and Raphael (1962); Baker, Eglinton, Gonzalez, Hamilton and Raphael (1962); Eglinton, Hamilton, Kelly and Reed (1966)

Source of wax	Constituents and references
Agave sisalana (C) Sisal hemp (leaf)	hentriacontane, tritriacontane; alcohols C_{15} and C_{28}–C_{31}; diols; acids C_{18}, C_{24}, C_{32}, C_{34}; hydroxy-acids Razafindrazaka and Metzger (1963)
Aichryson spp. (C)	*n*-alkanes chiefly C_{31}, C_{33}; *iso*-alkanes Eglinton, Gonzalez, Hamilton and Raphael (1962)
Allium porrum (C) Leek	*n*-alkanes C_{18}–C_{33}, chiefly C_{31}, C_{29} and C_{27}; ketone C_{31}; *n*-alkyl esters Holloway (1967)
Arbutus sp. Strawberry tree (leaf and stem)	hentriacontane; triacontanol, dotriacontanol, nonacosanol; ursolic acid Sosa (1950)
Avena sativa (C) Oat	alkanes, alcohols, aldehydes, ketones, esters, fatty acids, flavonols Tribe, Gaunt and Wynn Parry (1968)
Bulnesia retamo	*n*-alkanes, chiefly hentriacontane and nonacosane; *n*-primary alcohols C_{16}–C_{34}; diols; acids Brenner and Fiora (1960)
Centaurea aspera (leaf)	esters of C_{20}–C_{31} alcohols and C_{20}–C_{30} acids; acids C_{27}, C_{31} Viguera Lobo, Sánchez Parareda and Sánchez Parareda (1964)
Chamaecyparis obtusa Hinoki cypress (leaf)	hydrocarbons C_{15}–C_{20}, nonacosane, pentatriacontane; dodecane-1,12-diol; triacontan-1-ol; nonacosan-10-ol; nonacosan-10-one; acids lauric, sabinic, juniperic, thapsic, dodecanedioic, triacontanoic Kariyone, Ageta and Isoi (1959); Fukui and Ariyoshi (1963)
Chamaenerion angustifolium Rose-bay willow herb (leaf)	2α-hydroxyursolic acid Glen, Lawrie, McLean and Younes (1965)
Clarkia elegans (C)	*n*-alkanes C_{23}–C_{33}, chiefly C_{29}; ketones C_{27}–C_{31}, chiefly C_{29} Holloway (1967)
Cortaderia toe-toe	β-amyrin methyl ether Eglinton, Hamilton, Martin-Smith, Smith and Subramanian (1964)
Cycas revoluta (leaf)	nonacosane; octacosan-1-ol, nonacosan-10-ol; hexadecane-1,16-diol; nonacosan-10-one; acids sabinic, juniperic Kariyone, Ageta and Tanaka (1959)

Source of wax	Constituents and references
Cymbopogon citratus (C) Lemon-grass	Alcohols in range C_{30}–C_{34}; alkyl esters Crawford and Menezes (1963)
Dianthus caryophyllus (C) Carnation (leaf and stem)	β-diketones, chiefly *n*-hentriacontan-14,16-dione Horn and Lamberton (1962)
Diplopterygium glaucum (leaf)	nonacosane, pentatriacontane; hexacosan-1-ol, triacontan-1-ol, tritriacontan-1-ol, nonacosan-10-ol; nonacosan-10-one; acids hexacontanoic, triacontanoic Kariyone and Ageta (1959)
Dracaena draco (C) Dragon-tree	*n*-alkanes, chiefly C_{29}, C_{31} Eglinton, Gonzalez, Hamilton and Raphael (1962)
Echeveria secunda (C) (leaf)	*n*-alkanes C_{29}–C_{33}; β-amyrin acetate Horn and Lamberton (1964); Horn, Kranz and Lamberton (1964)
Ephedra gerardiana (leaf)	nonacosane; nonacosan-10-ol, triacontan-1-ol Ageta (1959)
Euphorbia cerifera (C) Candelilla	hentriacontane, nonacosane, tritriacontane; alcohols, acids, alkyl esters, hydroxy-esters; *n*-alkanes C_{28}–C_{35}, chiefly C_{31}; fatty acids C_{18}–C_{36}, chiefly C_{30}, C_{32} Schuette and Baldinus (1949); Findley and Brown (1953); Holloway (1967)
Euphorbia spp. (C)	*n*-alkanes, chiefly C_{27}, C_{29}, C_{31}; *iso*-alkanes Eglinton, Gonzalez, Hamilton and Raphael (1962)
Festuca ovina var. *glauca* (C) Blue fescue (leaf)	β-diketones, chiefly *n*-tritriacontan-12,14-dione Horn and Lamberton (1962)
Ficus elastica Rubber plant (leaf)	hydrocarbons C_{21}–C_{34}; alcohols C_{25}–C_{31}; ketones C_{21}–C_{37}; aldehydes C_{28}–C_{34}; acids Nevenzel and Rodegker (1962)
Ginkgo biloba Maidenhair tree (leaf)	nonacosane; octacosan-1-ol, nonacosan-10-ol; nonacosan-10-one Ageta (1959)
Gossypium hirsutum Cotton (leaf and calyx)	triacontane, dotriacontane, hexatriacontane, octacosane; octacosanol, triacontanol, dotriacontanol Sadykov, Isaev and Ismailov (1963)
Greenovia spp. (C)	*n*-alkanes, chiefly C_{33}; *iso*-alkanes Eglinton, Gonzalez, Hamilton and Raphael (1962)
Hordeum vulgare (C) Barley	alkanes, alcohols, aldehydes, ketones, esters, fatty acids, flavonols Tribe, Gaunt and Wynn Parry (1968)

Source of wax	Constituents and references
Humulus lupulus Hop	*n*-alkanes C_{12}–C_{33}; monomethyl alkanes C_{13}–C_{33}; dimethyl alkanes Jarolimek, Wollrab, Streibl and Šorm (1964); Wollrab, Streibl and Šorm (1965b)
Leptochloa digitata (C) (stem)	*n*-alkanes C_{23}–C_{33}, *n*-dohexacontane (C_{62}); alcohols; ketones; acids C_{19}–C_{32}; hydroxy acids Kranz, Lamberton, Murray and Redcliffe (1961)
Lolium multiflorum (C) Italian rye-grass (leaf)	*n*-alkanes, chiefly C_{29}, C_{31} Eglinton, Gonzalez, Hamilton and Raphael (1962)
Marrubium vulgare White horehound	*n*-alkanes, mono- and di-methylalkanes Brieskorn and Feilner (1968)
Medicago arabica	*n*-alkanes, chiefly C_{29}, C_{31} and C_{33} Oró, Nooner and Wikström (1965)
Monanthes spp. (C)	*n*-alkanes, chiefly C_{33}; *iso*-alkanes Eglinton, Gonzalez, Hamilton and Raphael (1962)
Musa sapientum Banana (leaf)	*n*-paraffins C_{19}–C_{30}, aromatic hydrocarbons Nagy, Modzeleski and Murphy (1965)
Nepeta ruderalis (flowers and stems)	hentriacontane; ester; β-sitosterol; triterpenoid 'nepetol' Siddiqui and Ahsan (1967)
Nicotiana tabacum Tobacco (leaf)	*n*-alkanes C_{27}–C_{31}, *iso* and *anteiso* alkanes, cyclic hydrocarbons Gladding and Wright (1959); Carruthers and Johnstone (1959); Bendoraitis, Rusaniwsky, Stedman and Swain (1960); Mold, Stevens, Means and Ruth (1963)
Papaver spp.	*n*-alkanes C_{12}–C_{37}, *iso*- and *anteiso*-alkanes, alkenes C_{15}–C_{33} Stránský and Streibl (1969)
Papaver somniferum (C)	*n*-primary alcohols C_{16}–C_{32}, chiefly C_{26}; *n*-secondary alcohols, chiefly *n*-nonacosan-15-ol Holloway (1967)
Persea americana (C) Avocado	*n*-alkanes: C_{29} also C_{23}, C_{25}, C_{27}, C_{31}; alcohols, C_{14}, C_{16}, C_{18} also C_{15}, C_{20}, C_{24}; acids palmitic, oleic, linoleic, lignoceric, cerotic Mazliak (1965a, b)
Phragmites communis (C) (leaf)	saturated and unsaturated acids Simionescu, Diaconescu and Feldman (1960)
Pinus thunbergii (leaf)	triacontan-1-ol; dodecane-1,12-diol, hexadecane-1, 16-diol; acids palmitic, sabinic, juniperic Kariyone and Isoi (1956); Kariyone, Isoi and Yoshikura (1959)

Source of wax	Constituents and references
Pisum sativum (C) Pea (leaf)	*n*-alkanes C_{27}–C_{35}, chiefly C_{31}; *n*-primary alcohols C_{16}–C_{32}, chiefly C_{26}, C_{28} Holloway (1967)
Populus nigra Black poplar (leaf)	dimethylalkanes Jarolimek, Wollrab, Streibl and Šorm (1964)
Rhus succedanea (C) (fruit)	dibasic acids chiefly C_{20}, C_{22}, also C_{16}, C_{18}, C_{24} and C_{26} Tsujimoto (1935); Ueno and Tsuchikawa (1942); Shiina (1946); Lamberton (1961)
Rhus typhina (C) (fruit)	hydroxy-fatty acids; saturated and unsaturated fatty acids Tischer (1960)
Rosa spp.	*n*-alkanes C_{17}–C_{33}, heptacosane, nonacosane, hentriacontane, tritriacontane; *iso, anteiso* alkanes C_{13}–C_{34}; dimethylalkanes C_{14}–C_{34}; *cis*-alk-3-enes C_{17}–C_{33}; *trans*-alkenes C_{19}–C_{33} Šorm, Wollrab, Jarolimek and Streibl (1964); Wollrab, Streibl and Šorm (1965a)
Ruta montana (leaf)	hydrocarbons C_{25}–C_{33}; alkyl ester C_{54}; acids C_{26}–C_{30} Viguera Lobo and Sánchez Parareda (1963)
Ruta pinnata (leaf)	outer layer *n*-alkanes chiefly C_{31}–C_{33}; *iso*-alkanes; inner layer *n*-alkanes chiefly C_{20}–C_{23} Bermejo-Barrera, Estevez-Reyes and Gonzalez-Gonzalez (1964)
Sedum anglicum (C) (leaf)	*n*-alkanes, chiefly C_{31} Eglinton, Gonzalez, Hamilton and Raphael (1962)
Stipa tenacissima (C) Esparto	*n*-alkanes C_{18}–C_{35}, chiefly C_{31}, C_{33} and C_{29}; esters; lupeol Savidan (1956b, 1959); Holloway (1967)
Syagrus (*Cocos*) *coronata* (C) Ouricury	hydroxy acids, free and combined as estolides; hydrocarbons C_{24}–C_{36}, chiefly C_{31}; triacontanol; hexacosanoic acid; hydrocarbons; alcohols; acids; esters; hydroxy mono- and di-esters; di-esters; polyesters Findley and Brown (1953); Schuette and Khan (1953); Cole (1956); Cole and Brown (1960)
Triticum vulgare (C) Wheat	*n*-alkanes C_{29}, C_{31}; alkyl esters; primary alcohols; β-diketones; hydroxy-β-diketones Tulloch and Weenink (1966)
Ulex europaeus Gorse	triacontane; β-amyrin acetate, lupeol McLean and Thomson (1963)

value to alkanes were found in all the waxes. Those of *Narcissus pseudo-narcissus*, *Hordeum vulgare*, *Triticum aestivum*, *Iris pseudacorus*, *Yucca filamentosa*, *Hedera helix*, *Atriplex patula*, *Aster novi-belgii*, *Senecio squalidus*, *Fuchsia* sp., *Ligustrum vulgare*, *Laburnum anagyroides*, *Phaseolus vulgaris*, *Pisum sativum*, *Plantago lanceolata*, *Rumex obtusifolius*, *Ficaria verna*, *Atropa belladonna* and *Datura innoxia* were particularly rich in alkanes. Alkyl esters were the next most frequently encountered compounds. Fifteen of the waxes showed carbonyl compounds of different types; these were prominent in *Hordeum vulgare*, *Triticum aestivum*, *Saccharum officinarum*, *Allium porrum*, *Clarkia elegans*, *Eucalyptus* and *Brassica* waxes. Secondary alcohols were major constituents (*c.* 40%) of the waxes of *Chelidonium majus*, *Papaver lateritium* and *P. somniferum*, all members of the Papaveraceae; this provided the only example of a taxonomic relationship among the waxes studied. Cyclic compounds (triterpenoids or steroids) were detected in 37 waxes, and appeared as major constituents in *Ilex aquifolium*, *Viburnum lantana*, *Senecio laxifolius*, *Rhododendron ponticum*, *Chaemaenerion angustifolium*, *Epilobium montanum*, *Syringa vulgaris*, *Tilia europaea*, *Arundinaria*, *Prunus*, *Malus sylvestris* (*Pyrus Malus*) and *Eucalyptus* waxes. Ursolic acid, lupeol and β-amyrin were found to be of common occurrence.

The waxes in Table 5.2 that may be regarded as truly cuticular (C) show the familiar constitutional pattern of hydrocarbons, alcohols (including diols) ketones (including diones) acids (including hydroxy acids) and esters. Of special interest is the occurrence of a straight-chain hydrocarbon containing 62 carbon atoms in *Leptochloa* wax, dibasic acids in *Rhus* wax, and hydroxy-β-diketones in wheat wax. Some *Aeonium* waxes and tobacco (*Nicotiana tabacum*) wax are especially rich in branched alkanes; these may amount to approximately 50% of the total alkane content. The constituents of the waxes which cannot be regarded as being exclusively of cuticular origin closely resemble those of waxes known to be derived solely from cuticles.

Wollrab, Streibl and Šorm (1967) studied further the branched chain hydrocarbons in natural waxes. The odd carbon-numbered homologues are chiefly *iso*-alkanes, and the even carbon-numbered *anteiso*. The C_{21}–C_{34} series of branched alkanes from lilac (*Syringa vulgaris*) blossom contain three lower homologues, 2-methyleicosane, 3-methylheneicosane and 2-methyldocosane. Stránský, Streibl and Herout (1967) examined the distribution of wax hydrocarbons in plants at different evolutionary levels. Alkane fractions of over 40 plants were isolated and their components identified by gas chromatography. The distribution of the alkanes varied according to the evolutionary level of the plant, the season, the locality, the form, the morphologically distinct parts and the dorsal or ventral parts of leaves. Some plants at a lower level of evolution are characterised by a relatively high occurrence of the lower *n*-alkanes (up to C_{23}), with approximately equal ratios of odd and even carbon-numbered members. The waxes of the more highly organised plants contain the higher homologues, with the odd carbon-numbered members predominating. Wollrab (1968, 1969) examined in detail the hydrocarbons and secondary alcohols of the waxes of the leaves, stems, flowers and fruits of members of the Rosaceae. *n*-Alkanes range from C_{10}–C_{33} and *n*-alkenes from C_{17}–C_{33}. The alkenes occur in the *cis* and *trans* forms and include *cis*-5-alkenes, *cis*-7-alkenes and *cis*-9-alkenes. 1-Alkenes are present including probably 2-ethylnonacos-1-ene in one species. The wax of rose hips contains 24% of hydrocarbons of which 40% are alkenes. The occurrence of secondary alcohols in the overall range C_{18}–C_{33} was found to be unexpectedly specific; they were detected only in the waxes of rose flowers, the leaves and fruits of hawthorn (*Crataegus oxyacantha*) and the apple. Rose flower wax contains nonacosan-7-ol, nonacosan-10-ol, hentriacontan-9-ol, *cis*-nonacos-5-ene and *cis*-hentriacont-7-ene.

Pollen wax has received attention from Scott and Strohl (1962). Wax that was probably part of the outer spore coat comprised 1·6% of the pollen of *Pinus taeda* and consisted chiefly of octacosan-1-ol, hexacosan-1-ol and smaller amounts of esters and free fatty acids. The wax of *P. montana* pollen contains tetracosan-1-ol, hexacosan-1-ol and octacosan-1-ol; that of *Zea mays* pollen *n*-pentacosane, *n*-heptacosane and saturated and unsaturated fatty acids; and that of the pollen of *Cedrus spp.* pentacosan-1-ol and heptacosan-1-ol. Spada, Coppini and Monzani (1958) obtained heptacosan-1-ol and a compound $C_{20}H_{40}$ from the wax of *Cedrus deodara*, and pentacosan-1-ol and a compound $C_{34}H_{68}$ from that of *C. atlantica*.

Interest has also been taken in the wax on the surface of the spores of fungi. Oró, Laseter and Weber (1966) steeped the spores of *Ustilago maydis*, *U. nuda* and *Sphacelotheca reilana* successively in benzene-methanol and *n*-heptane at 50°; the waxes obtained, although not necessarily cuticular, contained *n*-alkanes from C_{14}–C_{37} with C_{27}, C_{29} or C_{35} predominating. The wax from each type of spore has a characteristic alkane pattern. The wax of *U. maydis* also contains free fatty acids from C_{12} to C_{20} and the methyl esters of both saturated and unsaturated fatty acids predominately in the range C_{16}–C_{20}. Naturally occurring methyl esters of fatty acids in fungal spores are reported for the first time (Laseter, Weete and Weber, 1968).

The red-tinted galls or 'oak apples' produced by the gall wasp *Biorhiza pallida* yield a white wax which contains all the *n*-alkanes from C_{18}–C_{31} of which the chief constituents are the C_{25}, C_{27}, C_{23} and C_{29} compounds (Calam, 1968).

Cyclic constituents of leaf and fruit waxes

Much is known of the nature of the aliphatic constituents of cuticular waxes, but as yet comparatively little information is available on the aromatic or cyclic constituents. Triterpenoid acids such as ursolic acid occur in many cuticular waxes, e.g. in the peel of the pomegranate (*Punica granatum*) (Brieskorn and Mustafa Keskin, 1954), the skin of the American cranberry (*Vaccinium macrocarpon*) (Arnold and Hsia, 1957) and the cuticles of many Japanese plants (Kariyone and Hashimoto, 1953). Oleanolic acid is present in the cuticle wax of the fruit of *Fagraea borneensis* (van Die and Soemarsono, 1957). As already reported, ursolic acid is a major constituent of apple leaf and fruit wax; oleanolic acid is present in tomato fruit wax and accounts for 70% of the grape berry wax. Friedelin is a constituent of the surface wax of leaves of *Euonymus japonicus* (Baker, Batt, Silva Fernandes and Martin, 1964) and has been isolated, with related compounds, from *Rhododendron*, *Prunus*, *Salix* and other plants. Young culms of *Lingnania chungii* and some other bamboos possess a bloom which can completely cover the green surface of the internodes. The bloom consists of a triterpenoid identical with or closely resembling friedelin (McClure, 1966). The range of cyclic compounds known to occur in cuticular waxes is extended by reports of lupeyl acetate in *Acacia* wax, lindleyol in *Aeonium* wax, β-amyrin acetate in *Echeveria* wax (see Table 5.2) and 11,12-dehydroursolic lactone acetate in *Eucalyptus* waxes. Flavones occur in some waxes; primetin is a constituent of *Primula* wax (p. 138) and eucalyptin of *Eucalyptus* and *Angophora* waxes.

Cyclic compounds have also been found in plant extracts not exclusively of cuticular origin, e.g. 2α-hydroxyursolic acid in rose-bay willow-herb (*Chamaenerion angustifolium*) extract, β-amyrin acetate and lupeol in gorse (*Ulex* sp.) extract and β-amyrin methyl ether in *Cortaderia* wax (Table 5.2). The diterpene hydrocarbon phyllocladene has been obtained from the powdered leaves of *Podocarpus nivalis* (Aplin, Cambie and Rutledge, 1963). Aromatic hydrocarbons have been found in banana wax (Table 5.2). Sterols, e.g. β-sitosterol and stigmasterol, have been reported as constituents of a number of waxes.

Estolides in plant waxes

There is little doubt that the ω-hydroxy acids of carnauba wax occur largely esterified together in the form of estolides, many of high molecular weight; it is to these that the wax owes its outstanding properties (Murray and Schoenfeld, 1955b; Hatt and Lamberton, 1956). The hydroxy acids of ouricury wax probably also occur as estolides (Cole, 1956; Cole and Brown, 1960). The wax of needles of *Pinus thunbergii* contains fatty and hydroxy-fatty acids, combined with alcohols or in the form of estolides (Sakurai, 1933). The neutral wax of Colorado spruce (*Picea pungens*) consists of 12-hydroxydodecanoic (sabinic), 14-hydroxytetradecanoic, and 16-hydroxyhexadecanoic (juniperic) acids with only small amounts of *n*-fatty acids and non-saponifiable material; the wax is composed chiefly of a cyclic polymer or estolide formed from about four molecules of the hydroxy acids esterified with each other (von Rudloff, 1959). Fujita and Yoshikawa (1951) isolated 24-hydroxytetracosanoic acid from the saponified wax of *Juniperus rigida*; the acid polymerised into a gummy estolide assumed to be composed of seven units.

Kariyone and Isoi (1955) identified palmitic, juniperic and thapsic acids in the ester fraction of hinoki (*Chamaecyparis obtusa*) leaf wax. Isoi (1958) found that the leaf waxes of conifers could be divided into estolide and non-estolide types. Those of the Pinaceae and Cupressaceae were estolide in character and those of the Taxaceae, Podocarpaceae and Taxodiaceae non-estolide. Information on the constituents of the estolides was obtained by Kariyone and his associates in an examination of the leaf waxes of *Cycas revoluta* and *Pinus thunbergii* (see Table 5.2). The wax of *Cycas revoluta* contains an acid estolide composed of condensed juniperic acid (4 to 6 molecules) and sabinic acid, and a neutral estolide formed from hexadecane-1,16-diol and juniperic and sabinic acids. *Pinus thunbergii* wax contains neutral estolides composed of hexadecane-1,16-diol, sabinic acid (4 to 5 molecules) and juniperic acid; dodecane-1,12-diol, palmitic and sabinic acids; and dodecane-1,12-diol and sabinic acid. Kariyone, Takahashi, Isoi and Yoshikura (1959) investigated the effect of grafting in the Taxodiaceae; the characteristic estolide feature of the stock was not evident in the leaf component of the scion. It is noteworthy that our knowledge of the occurrence of estolides comes largely from the study of conifer waxes. The presence of estolides in carnauba wax may indicate that they occur in the waxes of plants more frequently than has been supposed.

Glyceride plant waxes

From time to time reports have been made of the occurrence of glycerides in plant waxes. Tsujimoto (1935) estimated 10–14% of glycerol, combined with acids, in Japan wax from *Rhus* sp. The chief components of the wax from the fruit coat of *Myrica cerifera* are trimyristin and tripalmitin, with trimyristin the chief glyceride (McKay, 1948). Wax from the berries of *M. arguta* is believed to be of similar composition. The wax of *M. cordifolia* is also said to contain glycerides, with a greater proportion of tripalmitin than in that of *M. cerifera*. Another wax believed to be of the glyceride type is that of the common reed, *Phragmites communis* (Simionescu, Diaconescu and Feldman, 1960). The Cochin-China wax, from the cay-cay tree (*Irvingia oliveri*) of the Far East, is said to consist largely of glycerides of myristic acid and its homologues.

Other cuticular secretions

It has long been known that inorganic and organic substances are lost from leaves by the leaching action of rain. Among the substances that may be lost are potassium, carbohydrates (including free sugars and polysaccharides), amino acids and organic acids (Tukey and Morgan, 1964). It is probable that the substances migrate from the cellular tissues and intermingle with the surface waxes. Richmond and Martin (1959) showed that phenolic compounds occur in the surface wax of apple leaves, and in rain drippings from the leaves. A white incrustation found on the leaves of *Euonymus* sp. during a period of drought and abnormally high temperatures was dulcitol (Baker, 1949) and a water-soluble deposit on the surface of leaves of *Kleinia articulata* also is of a carbohydrate nature. An unusual apparently seleniferous wax was obtained by McColloch, Hamilton and Brown (1963) from the surface of leaves of *Stanleya bipinnata* plants grown in pots watered at intervals with a soluble selenium compound. The selenium was intimately associated with the wax, but no conclusive proof was obtained of chemical combination. Blasdale (1945, 1947) reported that at least half of the 500 or more species or subspecies of *Primula* bear minute glandular hairs which secrete a white or yellow powder, commonly called 'farina'. The powder is insoluble in water but is too loosely adherent to afford protection; it is most evident on the flowers and more abundant on the lower leaf surface than the upper. Müller (1915) showed that the secretion of *Primula pulverulenta* is a flavone ($C_{15}H_{10}O_2$) mixed with wax, and Karrer and Schwab (1941) identified 5-hydroxyflavone in that of *P. imperialis.* Primetin (5,8-dihydroxyflavone) was obtained by Hattori and Nagai (1930) from the leaves of *P. modesta.* Blasdale (1945) examined the farina of 21 species of *Primula* and showed that it contains at least 75% of flavone with variable amounts of waxy substances. 5-Hydroxyflavone was isolated from the secretion of *P. verticillata* and a small amount of a dihydroxyflavone from that of *P. denticulata.* He suggests that the compounds are end-products of metabolism, secreted to the surface to avoid a possible detrimental effect upon the cells. Harms and Wurziger (1968) have shown that 5-hydroxytryptamides of eicosanoic, docosanoic and tetracosanoic acids occur in the superficial wax of coffee (*Coffea arabica*) beans.

Schnepf (1968) studied, by light and electron microscopy, the secretion of a mucopolysaccharide by the gland hairs on the ***ochrea*** in buds of *Rumex* and *Rheum.* The mucilage is secreted by the Golgi apparatus and accumulates in spaces between the cell wall and the cuticle.

New identifications of wax constituents

Work in recent years on cuticular waxes has been exceptionally rewarding in revealing constituents for the first time. The identification, in waxes, of even carbon-numbered *n*-alkanes and of *iso*- and *anteiso*-alkanes has already been discussed. *n*-Alkenes were first detected in sugar-cane and rose-petal waxes by Šorm, Wollrab, Jarolimek and Streibl (1964) and branched (dimethyl) alkanes in hop, rose and *Populus* waxes by Jarolimek, Wollrab, Streibl and Šorm (1964). An exceptionally long-chain *n*-alkane, dohexacontane (C_{62}) was newly identified in *Leptochloa* wax by Kranz, Lamberton, Murray and Redcliffe (1961), and aromatic hydrocarbons, in banana wax, were first reported by Nagy, Modzeleski and Murphy (1965). The range of ketonic constituents was extended by the identification of α,β-unsaturated ketones in sugar-cane wax (Horn and Matic, 1957), β-diketones in eucalyptus, acacia, grass and carnation waxes (Horn and Lamberton, 1962) and hydroxy-β-diketones in wheat wax (Tulloch and Weenink, 1966). Aldehydes, in sugar-cane wax, were first reported as wax constituents by Lamberton and Redcliffe (1959). Free hydroxy-fatty acids were identified, in apple cuticle oil, by Davenport (1956) and diols, in apple wax, by Mazliak (1962). The occurrence of odd carbon-numbered fatty acids, in sugar-cane wax, was established by Kranz, Lamberton, Murray and Redcliffe (1960) and 14-hydroxytetradecanoic was found as an hitherto unreported constituent of estolides by von Rudloff (1959). Dodecanedioic acid was first identified, as a constituent of the wax of *Chamaecyparis obtusa,* by Kariyone, Ageta and Isoi (1959).

Despite the large amount of information available on the aliphatic compounds present in cuticular waxes, it is possible that their true nature to some extent remains obscure. More evidence is needed on the extent to which the different classes of constituents, and in particular the alcohols, fatty acids and hydroxy-fatty acids, occur in combination with each other. For many years it was almost standard practice to saponify extracts before the isolation of individual components, and naturally occurring esters and estolides may have been missed. Thin-layer chromatographic and other modern analytical procedures which preclude disruption give clearer pictures of component fractions, and indicate that esters are important constituents of many waxes. Aldehydes, too, may figure more prominently now that it is realised that they may be degraded by some of the analytical procedures used.

Among new identifications of cyclic compounds in cuticular waxes are those of lindleyol in *Aeonium lindleyi* wax (Baker, Eglinton, Gonzales, Hamilton and Raphael, 1962), lupeyl acetate in *Acacia* wax, β-amyrin acetate in *Echeveria* wax and 11,12-dehydroursolic lactone acetate in *Eucalyptus* wax (Horn and Lamberton, 1964). Cyclic compounds will probably increasingly be recognised as important constituents of waxes.

Plant waxes and chemotaxonomy

Interest has been taken in the chemical composition of waxes as an aid to the botanical classification of plants. Purdy and Truter (1961) observed that the composition of the surface lipid appears to be characteristic of the species of plant. Eglinton, Gonzalez, Hamilton and Raphael (1962) and Eglinton, Hamilton, Raphael and Gonzalez (1962) examined the surface waxes of plants of the subfamily Sempervivoideae of the Crassulaceae. Members of the closely related genera *Aeonium*, *Aichryson*, *Greenovia* and *Monanthes* (see Table 5.2) were studied and, to simplify the approach, attention was given to the hydrocarbon fractions of the waxes which were amenable to exact and rapid analysis. Some uniformity of hydrocarbon pattern, i.e. the relative proportions of individual compounds, was found in the waxes of members within each genus which suggested that comparisons of the hydrocarbon patterns may serve to establish botanical relationships and to indicate as yet unrecognised departures from the existing botanical classification.

Eglinton, Hamilton and Martin-Smith (1962) examined the alkane constituents of New Zealand plants within the families Scrophulariaceae, Ericaceae, Liliaceae, Thymeleaceae and Rosaceae and concluded that the alkane distribution patterns of these may also be of taxonomic value. The hydrocarbons of waxes of Gymnosperms were studied by Borges del Castillo, Brooks, Cambie, Eglinton, Hamilton and Pellitt (1967). The *n*-alkane constituents of 32 Podocarpaceae and related species were closely similar, the major component being either C_{29}, C_{31} or C_{33}. Earlier Aplin, Cambie and Rutledge (1963) had concluded that the diterpene hydrocarbons (not necessarily cuticular) of the Podocarpaceae were of doubtful taxonomic value.

Many *Eucalyptus* waxes contain as major components long-chain β-diketones, of which the most commonly occurring is *n*-tritriacontan-16,18-dione; some contain shorter chain length β-diketones (*n*-hentriacontan-14,16-dione and *n*-nonacosan-12,14-dione) and others none. The compositions of the waxes have been considered in relation to botanical classification, but no conclusion reached because of the small number of species examined; the large number of components present in the waxes, however, suggests that detailed examination might provide useful taxonomic information supplementing that based on essential oil composition (Horn, Kranz and Lamberton, 1964). Mecklenburg (1966) examined the alkanes of the waxes of the flowers of 22 tuber-bearing species of *Solanum* also in relation to their possible taxonomic significance. All the hydrocarbon fractions contained n-C_{25} to C_{31} components and relatively large proportionate amounts of branched C_{25} to C_{32} compounds. He concluded that the alkane distribution in the waxes tends to confirm the relationships between the species thought to exist on the basis of morphological, cytogenetic and inter-fertility data and is not a reflection of the ecological conditions of the areas in which the species occur. Closely similar alkane distribution patterns were found in the waxes of varieties of tobacco (Mold, Stevens, Means and Ruth, 1963).

Purdy and Truter (1961) point out that the surface wax pattern may be of taxonomic value provided that the same pattern is shown by foliage from different parts of the plant and that the pattern does not change as the plant ages. These criteria were met in the Brussels' sprout, *Rhododendron*, *Ilex* and Gramineae plants they used. Eglinton and his associates in their work on the Sempervivoideae found that the alkane pattern of the leaf wax of a given species is substantially constant during the life of the plant. Tribe (1967) examined the surface waxes of leaves of representatives of 24 genera of Gramineae in relation to taxonomy. The plants were grown under controlled conditions. He found only slight quantitative, and no qualitative differences in the proportions of alkanes, alkanols and other components in the wax from the lamina and petiole of *Saccharum officinarum*, and similar slight differences in the waxes of laminae of barley (*Hordeum vulgare* var. *distichon*) plants of different ages. Long-chain alkanols predominate in waxes of the Gramineae, except under conditions of extreme habitat, when they are replaced by β-diketones. Complex hydrocarbon patterns occur in the waxes of cultivated genera of the Triticeae and Aveneae, and an unusual hydrocarbon is present in the wax of species of *Poa*. Long-chain aldehydes are restricted to members of the Andropogonoideae and the Bambusoideae. Tribe finds that the patterns of waxes of the plants examined are distinctive and concludes that they appear to be valid in assessing taxonomic relationship.

Hallam (1967) examined with the electron microscope the leaf waxes of over 300 species of *Eucalyptus* and found that the various forms of each wax type could be related to the taxonomic groupings of the genera as a whole. Wax morphology was closely linked with wax chemistry; the tube-like waxes when present as a dense cover had a high proportion of β-diketones and the plate-like waxes a zero or low proportion. Hallam's study proved useful not only in providing additional information supporting previous changes made in taxonomic groupings, but also in suggesting others (see p. 114).

Herbin and Robins (1968a) examined the distribution of alkanes (C_{21}–C_{33}) in the hydrocarbon fraction of the leaf and perianth surface waxes of 63 East African species of *Aloe*. The chief alkane is n-hentriacontane (C_{31}) but considerable proportions of n-nonacosane (C_{29}) and n-heptacosane (C_{27})

occur in some of the waxes. Sometimes large differences are met in the relative proportions of the C_{31} and C_{29} compounds in the leaf and perianth hydrocarbon fractions; the leaf fraction of *A. kedongensis*, for example, contains 70% of C_{31} and 14% of C_{29}, while the perianth fraction has 9% of C_{31} and 79% of C_{29}. A correlation between the chemical data and classification is revealed more clearly by the perianth wax alkanes than by the leaf. The conclusion is reached that the chemical evidence broadly agrees with the botanical classification, although the alkane patterns of the leaf and perianth waxes of a single species may not be sufficiently discriminating either to differentiate it from other related species or to place it unambiguously within the framework of the present classification.

This work was followed by a similar examination of the alkanes of the leaf waxes of 21 species of *Agave*, 20 members of the Crassulaceae (*Kalanchoe*, *Echeveria*, *Crassula* and *Sedum* spp.) and 19 species of *Eucalyptus*, also in relation to botanical classification (Herbin and Robins, 1968b). High contents of *n*-tritriacontane (C_{33}) occur in the hydrocarbon fractions of *A. americana* (67%) and *A. sisalana* (54%); the C_{31} compound is the major constituent of many of the others. As in the case of *Aloe*, a wide range of leaf alkane patterns in *Agave* does not permit a close correlation with classification. The C_{33} and C_{31} compounds are the chief alkanes of the leaf waxes of the Crassulaceae; a uniformity in alkane distribution pattern occurs in the waxes of the species of the four genera examined. The alkane contents of *Eucalyptus* cuticular waxes are low (up to 6%). The C_{29} and C_{27} compounds frequently predominate in the *Eucalyptus* alkane fractions, which thus differ clearly from those of the Crassulaceae (C_{33}–C_{31}) and of *Aloe* and *Agave* (C_{31}–C_{29}). The data on *Eucalyptus* wax, it is suggested, provide a strong indication that its composition is controlled genetically rather than environmentally. Herbin and Robins (1968b) also report additional data on the alkane distribution patterns in leaf waxes obtained to determine whether such criteria provide evidence in support of an early evolutionary dichotomy into herbaceous and woody plants. They conclude that no differentiation is apparent on the basis of leaf alkane pattern.

An examination was also made of the value, as chemotaxonomic criteria, of the leaf waxes of members of the Cupressaceae and Pinaceae (Herbin and Robins, 1968c). In this work, the distributions of both the alkanes and ω-hydroxy-alkanoic acids were determined. The authors conclude that in the Gymnosperms studied the ω-hydroxy acids, which are major constituents of the waxes, appear to be of value both in the separation of genera and in the identification of species and of groups of species within genera. On the other hand, the alkanes, which constitute only a minor proportion of the leaf waxes, do not afford a useful guide in certain genera of the Cupressaceae or in the genus *Pinus*. Dyson and Herbin (1968) used the distribution of *Cupressus* leaf wax alkanes to discriminate between cypresses grown in Kenya. They found that species may be diagnosed by such analyses and that hybrids show alkane distribution patterns intermediate between those of the parent species. The wax analyses provided additional support for the conclusion, reached on morphological characters, that a Kenyan cypress had arisen by selection from

C. lusitanica seed introductions and was not a product of hybridisation. Herbin and Sharma (1969) studied the ω-hydroxy-fatty acids present in the estolides of the leaf waxes of 39 *Pinus* spp. and found broad areas of agreement between the composition of the estolides and botanical classification.

Other plants have surface waxes which differ in composition on different parts or between cultivars. The cuticular waxes of the sultana vine (*Vitis vinifera* var. *sultana*) are not uniform in different parts of the vine; oleanolic acid is a major constituent of the berry wax, but is present only in a small amount in the leaf wax and the berry, stem and leaf waxes differ in the relative proportions of individual hydrocarbons, free alcohols and free fatty acids. The leaf wax, for example, has n-C_{29} as its predominant alkane, whereas the stem wax contains almost equal amounts of n-C_{25}, n-C_{27}, n-C_{29} and n-C_{31} (Radler, 1965b). A significant difference occurs in the relative proportion of the C_{29} and C_{31} n-alkanes in the surface waxes of the Winnigstadt and Copenhagen varieties of cabbage (Purdy and Truter, 1963c; Hill and Mattick, 1966).

A detailed taxonomic comparison was made by Martin-Smith, Subramanian and Connor (1967) of the alkanes, alkanols, fatty acids and triterpenoid compounds of the surface waxes of three indigenous New Zealand species of *Cortaderia* and of two S. American species naturalised in New Zealand. No chemotaxonomic differences were found between the species in fatty acids and alkanols. The alkane distribution patterns of the species indigenous to New Zealand were similar, but there appeared to be a difference between the New Zealand and S. American species in the relative proportions of the C_{29} and C_{31} alkanes. Only the species indigenous to New Zealand contained triterpene methyl ethers. Moreover, a marked divergence was apparent in the relative proportions of the alkane constituents of the surface wax of leaves of *C. toe-toe* collected at different times. From these and other considerations Martin-Smith, Subramanian and Connor conclude that 'considerable caution must be exercised in any attempted application of plant alkane analysis to taxonomy. A systematic investigation into the possible influence of season, climate, geographical distribution and the kind and age of organs on the composition of plant surface waxes is essential before the method can be accepted without qualification.'

Tribe, Gaunt and Wynn Parry (1968) found that the proportions of the surface components of oat (*Avena sativa*) and barley (*Hordeum vulgare*) grown under controlled environmental conditions differed only slightly with age or in different parts of the plants. Their results indicated a broad agreement with the present classification of the Gramineae. Stránský, Streibl and Herout (1967), however, found variation in the distribution of n-alkanes in the leaf waxes of *Betula verrucosa* and other deciduous trees analysed at different times of the year, in the wax of needles of *Picea excelsa* obtained from different localities, in the wax of different organs of *Robinia pseudacacia* and in the wax from the adaxial and abaxial surfaces of *Juglans regia*. Stránský and Streibl (1969) showed that the n-alkanes and n-alkenes differ substantially in individual parts of *Papaver rhoeas* and other plants; only the morphologically analogous parts of plants should be compared. These workers concluded that the distribution of *n*-alkanes is of limited value as a taxonomic criterion.

Cutin

Lee's observation that oxy-fatty acids predominated in the products of hydrolysis of cutin and Markley and Sando's suggestion that cutin is an ester-like substance assumed significance from the work of Huelin and Gallop (1951a). The cuticle of apple skin was isolated by the action of ammonium oxalate solution, and wax and a triterpenoid fraction were removed by solvent extraction. The residual membrane on saponification with ethanolic potash yielded about 50% of ether-soluble acids which were identified as complex hydroxy-fatty acids. The work was extended by Huelin (1959), who used a purer form of cutin obtained from apple cuticle isolated enzymically with snail gut extract. On saponification of the cutin, ether-soluble acids (80% yield) were obtained. Huelin concluded that cutin is a polyestolide of hydroxy acids and other substances. He regarded the formation of the polymer as primarily due to the condensation of the hydroxy acids through ester linkages, probably under the control of specific and localised enzymes. He cast doubt on the hypothesis of Lee and Priestley (1924), repeated by Priestley (1943), that cutin arises from the spontaneous oxidation of unsaturated acids on reaching the surface.

Meanwhile, Matic (1956) identified the major hydroxy acids of the cutin of the leaf of *Agave americana*. The cuticle was stripped off, freed from wax by solvent extraction, boiled in water and treated with cuprammonium solution to remove attached cellulose. The residue, containing cutin, was ground and saponified by autoclaving at 130°C with 5% aqueous potassium hydroxide solution for 17 hr. The liberated ether-soluble acids (60% yield) on reduction gave palmitic and stearic acids, indicating that the fraction consisted essentially of *n*-C_{16} and *n*-C_{18} hydroxylated acids. It was resolved by counter-current distribution and preparative scale reversed-phase partition chromatography. Four acids were isolated in a pure state and characterised; they were 9,10,18-trihydroxyoctadecanoic (phloionolic), 10,16-dihydroxyhexadecanoic, 10,18-dihydroxyoctadecanoic and 18-hydroxyoctadecanoic acids; a fifth, 18-hydroxydec-*cis*-9-enoic acid, was obtained as its dihydroxy derivative. They comprised more than 80% of the total ether-soluble acids obtained from the cutin. Phloionolic acid was the major component, amounting to about 25% of the cutin; other acids, probably the corresponding hexadecanoic homologues, were also present. The 10,16-dihydroxyhexadecanoic, 10,18-dihydroxyoctadecanoic and 18-hydroxyoctadecanoic acids were obtained for the first time from a natural product.

The term 'cutin acids' was used by Matic to describe the hydroxy-fatty acids isolated from cutin and the term is now generally accepted. Those obtained earlier by Fremy and Legg and Wheeler were clearly mixtures, with the possible exception of Legg and Wheeler's substance of m.p. 107–8 °C (p. 68) which was probably phloionolic acid. No unsaponifiable residue was detected by Matic in cutin. He therefore suggested that cutin is a natural cross-linked polyester, built up by the inter-esterification of the hydroxy acids. The infra-red spectrum of cutin showed the presence of some free hydroxyl groups such as would be expected from the preponderance of hydroxyl over carboxyl in the isolated acids.

The occurrence in apple fruit cutin of long-chain saturated and unsaturated hydroxy-fatty acids combined in ester form was confirmed by Schneider (1960), Brieskorn and Schneider (1961) and Mazliak and Pommier-Miard (1963). Brieskorn and Böss (1964,) exhaustively extracted apple cuticle and obtained a residue consisting of cutin and cellulose. Ester linkages were opened by treatment with alcoholic potash, and 16-hydroxyhexadecanoic acid (juniperic acid), 10,16-dihydroxyhexadecanoic acid, 9,10,18-trihydroxy-octadecanoic acid and 9,10-*epoxy*-18-hydroxyoctadecanoic acids were obtained. A dicarboxylic acid, thapsic acid (hexadecanedioic) was also isolated, for the first time. Whether the epoxy acid occurred in the cutin was open to doubt; it may have arisen during the process of obtaining the cutin acids. Brieskorn and Böss thought it probable that a linkage occurs between the hydroxyl groups of cutin and cellulose. This possibility arose because a residue of cellulose was found after saponification of the membrane, despite the fact that it had been treated, before saponification, with cuprammonium solution; the polyester, however, may have mechanically shielded the cellulose from the cellulose solvent during the first treatment. Not all of the carboxyl groups seem to participate in the ester linkages because the cutin was stained with basic dyes, which is possible only when free carboxyl groups are present.

An examination of the hydroxy-fatty acid constituents of the cutin of *Agave americana* was made by Crisp (1965). The leaf was washed with ether to remove surface wax and the cuticle, stripped from the leaf, was successively treated with *n*-hexane, alcohol and water to remove lipid and water-soluble components and with cuprammonium solution to remove attached cellulose. The residual cutin was systematically degraded; ester (—CO—O—) linkages were broken by treatment with ethanolic sodium hydroxide, peroxide (—O—O—) linkages with sodium iodide in acetic acid-*iso*propyl alcohol, and ether (—O—) linkages with hydriodic acid in acetic acid. The completely degraded cutin yielded hydroxy-fatty acids from C_{13} to C_{18}, which were fractionated as their acetoxymethyl ester derivatives by preparative gas-liquid chromatography. By reduction of the acids to the corresponding fatty acids, oxidation to the dicarboxylic acids, hydrogenation and gas chromatographic retention times, nineteen fractions were identified. The following individual acids occurred in amounts greater than 1% of the polymer: 9,10,18-trihydroxyoctadecanoic, 9,10,12,18-tetrahydroxyoctadecanoic, 10, 16-dihydroxyhexadecanoic 9,18-dihydroxyoctadecanoic, 16-hydroxy-9,10-*epoxy*-hexadecanoic, 10,18-dihydroxyoctadecanoic, 10,17-dihydroxyhepta-decanoic, 9,10,13,16-tetrahydroxyhexadecanoic and 18-hydroxyoctadeca-noic. Other minor acids were reported. 9,10,18-Trihydroxyoctadecanoic acid predominated, comprising 55% of the polymer, and the next important acids were 9,10,12,18-tetrahydroxyoctadecanoic (14%) and 10,16-dihydroxyhexa-decanoic acids (11%). Ester linkages predominated in the cutin, the ratio of ester: peroxide: ether bonds being 7: 2: 0.1. Baker, Batt and Martin (1964) found a small proportion (c. 15%) of fatty acids among the acids liberated from the cutin of apple fruit.

The constituent acids of apple cutin were studied in detail by Eglinton and Hunneman (1968). Peel of *Cox's Orange Pippin* apple fruits was extracted with hot chloroform to remove the waxes, digested successively with dilute sulphuric acid, ammonium oxalate-oxalic acid solution and zinc chloride-hydrochloric acid solution to remove cellular tissue, and further extracted with methanol to give the cuticular membrane. The infra-red spectrum of the membrane showed the presence of hydroxyl groups and ester linkages. The cutin acids were liberated by hydrolysis of the membrane with 3% potassium hydroxide in methanol. Analysis by combined gas chromatography and mass spectrometry showed that the mixture of acids contained (as determined in the form of their methyl esters) 38% of dihydroxymonobasic acids, 31% of monohydroxy-monobasic, 27% of trihydroxy-monobasic, 1% each of monobasic and dibasic and a small proportion of dihydroxy-dibasic. Acids present in amounts of the order of 5% or more of the mixture were 10,16-dihydroxyhexadecanoic acid (24%), *threo*-9,10,18-trihydroxyoctadecanoic (17%), *erythro*-9,10,18-trihydroxyoctadecanoic (7%), 16-hydroxyhexadecanoic (8%) and the unsaturated 18-hydroxyoctadec-9,12-dienoic (13%) and 18-hydroxyoctadec-9-enoic (5%). Small amounts of the unsaturated acids heptadec-9-ene-1,17-dioic, octadec-9-ene-1,18-dioic and octadec-9,12-diene-1,18-dioic were detected. Interestingly, a rather larger proportion (3%) of an unsaturated trihydroxy-monobasic C_{18} acid was present. The monobasic acids included palmitic, stearic, oleic and linoleic and also the saturated C_{20}, C_{22} and C_{24} compounds. No hydroxy-fatty acids of chain length greater than C_{18} were found. Brieskorn and Reinartz (1967b) have shown that the cutin of the tomato fruit cuticle is composed chiefly of 10,16-dihydroxyhexadecanoic acid, with small amounts of 16-hydroxyhexadecanoic, 9,10,16-trihydroxyhexadecanoic and 9,10,18-trihydroxyoctadecanoic acids also present in the cutin complex.

Information on the hydroxy-fatty acids of cutin was greatly enlarged by the recent work on *Agave*, tomato and apple. An interesting feature was the detection of tetrahydroxy acids among the degradation products of *Agave* cutin. 9,10,16-Trihydroxyhexadecanoic acid occurs in *Agave* and tomato cutin, but was not detected in apple cutin. Small amounts of unsaturated dioic acids are reported, in apple cutin, for the first time. The trihydroxy acids are extended by the new identifications of the unsaturated 9,10,18-trihydroxyoctadecenoic acid and a considerable proportion of *erythro*-9,10,18-trihydroxyoctadecanoic acid in apple cutin.

The basic chemistry of cutin is now known. Recent work supports the view of Roelofsen and Houwink (1951) that cutin is a polymolecular network of carboxylic and hydroxycarboxylic acids connected through ester and ether bridges except that, from Crisp's work, peroxide bridges may be more important than ether bridges. Most of the information on the acids of cutin has been obtained from a study of the cuticles of *Agave* and apple. 9,10,18-Trihydroxyoctadecanoic acid is prominent in both, and 10,16-dihydroxyhexadecanoic acid an important, but not so prominent a constituent. Baker and Martin

(1963) examined the cutin of leaves and fruits of different species to ascertain whether cutin is constant in composition. The cutin of apple and pear, of the Pomoideae, yielded both 9,10,18-trihydroxyoctadecanoic and 10,16-dihydroxyhexadecanoic acids while that of the gooseberry (*Ribes uva-crispa*) and blackcurrant (*Ribes nigrum*), of the Ribesoideae, gave largely 10,16-dihydroxyhexadecanoic acid. The possibility that the cutin acid make-up of cutin might be useful for taxonomic purposes was examined by comparing the cuticles of plants of the Saxifragaceae, Rosaceae and Leguminosae (Baker and Martin, 1967). The dominant acid obtained throughout was 10, 16-dihydroxyhexadecanoic acid; the 9,10,18-trihydroxyoctadecanoic acid was given in significant amount only by the cutin of members of the Rosaceae. The cutin of the legumes examined yielded largely 10,16-dihydroxyhexadecanoic acid. This acid clearly is a ubiquitous and important component of cutin.

Polyesters of large molecular weight can be prepared by the polymerisation of hydroxy-fatty acids. Compounds such as 10-hydroxydecanoic acid and 15-hydroxypentadecanoic acid on prolonged heating are transformed into stiff solids (Carothers and Hill, 1932; Carothers and van Natta, 1933). The hydroxy acids obtained from cutin tend to polymerise on standing.

Other components of cuticles

Pectin and cellulose occur along with cutin in the region where the cuticular membrane merges into the outer wall of the epidermal cell. In the cuticle of the *Euonymus japonicus* leaf, pectin is seen embedded within the membrane rather than as a cementing layer between the membrane and the wall of the epidermal cell. Cellulose likewise occurs within the membrane and is protected by the cutin from the action of cellulose solvents (Baker, Batt, Roberts and Martin, 1962).

Huelin (1959) obtained a water-soluble carbohydrate fraction on saponification of apple fruit cuticle. The fraction on hydrolysis yielded arabinose, galactose and galacturonic acid, clearly derived from the pectin component of the cuticle. Brieskorn (1959) and Schneider (1960) isolated carbohydrate fractions from apple cuticle, and characterised them as composed of arabinose, galactose, xylose and glucose. Huelin also obtained an insoluble fraction containing protein which, on hydrolysis, gave aspartic, glutamic and other amino acids; he thought that part of the protein fraction may have consisted of enzymes secreted into the cuticle and subsequently denatured. Schneider too, found evidence of protein and phenolic material in the apple fruit cuticle.

A wide range of phenolic compounds is known to occur in the apple fruit cuticle. Sando (1924, 1937) showed that it contains quercetin, quercetin-3-galactoside and cyanidin-3-galactoside (idaein); Duncan and Dustman (1936) confirmed the presence of idaein. Siegelman (1955) found 6 quercetin glycosides, with the sugar moiety in the 3-position, in the skin of the *Grimes Golden* variety of apple. Hulme and Edney (1960) detected at least 10 compounds, including quercetin and chlorogenic acid, in the fruit peel of *Cox's Orange Pippin* apple and Hulme and Wooltorton (1958) shikimic acid (and citramalic acid) in the peel of *Bramley Seedling*. Höster-Auer (1964) identified 3 quercetin glycosides and quercetin in the peel of *Golden Delicious*, and Fisher (1966) found 7 quercetin glycosides, chlorogenic acid, phloridzin and other phenolics in the skins of *Jonathan*, *Granny Smith* and *Democrat* fruits. Fisher estimates that the free phenolic compounds cannot exceed 1% of the fruit cuticle.

An important and often overlooked component of many cuticles is a material which appears to be a condensed tannin. The tannin may be removed from isolated membranes by refluxing them in dilute aqueous alkali (e.g. 1% potassium carbonate solution). The extract sometimes becomes deeply coloured, and on acidification deposits ***phlobaphene***-like material. The tannin amounts to 26% of the dry weight of the apple fruit cuticular membrane (Fisher, 1966) and 27–52% of the membranes of leaves of *Ribes* spp. (Baker and Martin, 1967). It protects the cutin from the hydrolytic action of alcoholic alkali; the membranes dissolve much more quickly after the tannin is removed by extraction with dilute aqueous alkali.

The chemistry of plant cuticles has been reviewed by Mazliak (1968).

BARKS

The term bark is given to the outer part of a stem or branch which surrounds the wood; its structure has been described on p. 117 *et seq*. The outer bark differs considerably in chemical composition from the inner. The outer bark contains up to 40% (on dry weight basis) of waxy, fatty and other material including triterpenoid compounds extractable by ether or other solvents, up to 40% of suberin and a small proportion of lignin; the inner bark contains little waxy and fatty material or suberin and about 20% of lignin. The outer bark comprises approximately 5 to 35% of the complete bark layer according to the species and age of the tree. Many classes of compounds have been isolated from barks by successive extraction with solvents. Petroleum ether, ether, benzene, alcohol and hot water extraction may be used in turn, but there is no generally accepted method of bark analysis. The residue obtained after exhaustive solvent extraction consists largely of suberin.

Bark waxes

The wax content of bark is usually low. The dry bark of Western yellow pine (*Pinus ponderosa*) contained 3·4% of a hexane-soluble lignocerate wax; subsequent extraction with benzene and ether yielded an additional 1·3% of wax and a pentahydroxyflavone which had not previously been isolated. The bark of sugar pine (*P. lambertiana*) contained 2·2–3·1% of wax (Kurth and Hubbard, 1951). Red fir (*Abies magnifica*) bark gave 2·5% of wax in which lignoceryl alcohol, behenic acid and phytosterol were found (Becker and Kurth, 1958). The bark of white spruce (*Picea abies*) yielded a wax from which lignoceryl alcohol, behenic acid and lignoceric acid were separated; oleic, linoleic and linolenic acids were also identified (Bishop, Harwood and Purves,

1950). Douglas fir (*Pseudotsuga taxifolia*) bark is exceptionally waxy; lignoceryl alcohol, lignoceric acid and ferulic (4-hydroxy-3-methoxycinnamic) acid were identified in the products of hydrolysis of the wax (Kurth, 1950; Kurth and Kiefer, 1950). The bark wax of privet (*Ligustrum vulgare*) contains esters of palmitic, behenic and arachidic acids, that of sweet birch (*Betula lenta*) betulin, and that of the red mulberry (*Morus rubra*) α-amyrin.

The cork of the cork oak (*Quercus suber*) may yield up to 20% of wax depending upon the solvent used for extraction. The wax contains friedelin and triterpene alcohols and ketones (Zetzsche and Sonderegger, 1931; Drake and Jacobsen, 1935; Šorm and Bazant, 1950). The *n*-alkanes and *n*-primary alcohols of cork wax have been identified by Bescansa-López and Ribas-Marqués (1966) and Bescansa-López, Gil Curbera and Ribas-Marqués (1966). The *n*-alkanes are present to the extent of 0·6% in the cork and consist of all odd and even carbon-numbered compounds from C_{16}–C_{34}, with C_{22}–C_{32} most abundant. Alcohols in the neutral fraction of the wax contain odd carbon-numbered members, especially C_{21} and C_{23}; the alcohols liberated by hydrolysis consist of all the even carbon-numbered compounds from C_{20} to C_{26}, with C_{22} and C_{24} predominating. Cork wax contains 17% of ω-hydroxy acids chiefly phellonic and 24-hydroxytetracosanoic, 1·5% of dioic acids and 0·6% of betulin (Duhamel, 1963).

Terpenoid constituents of barks

Cyclic compounds are important constituents of barks and recent work has considerably extended our knowledge of terpenoid components. The outer bark of silver birch (*Betula verrucosa*) owes its white colour to an unusually high content (*c*. 30%) of betulin, which also occurs, in smaller amounts, in the barks of beech and hazel (*Corylus avelana*). Betulic acid occurs in the bark of *Syncarpia laurifolia* and other trees (Ralph and White, 1949), and oleanolic acid acetate in the bark of *Eucalyptus calophylla* (White and Zampatti, 1952). Other isolations made include oleanolic acid, betulin and β-amyrenone from elder (*Sambucus nigra*) bark (Lawrie, McLean and Paton, 1964), friedelin, epifriedelinol, ursolic acid and oleanolic acid from *Olearia paniculata* (Corbett, Young and Wilson, 1964) and morolic acid from *Eucalyptus papuana* (Hart and Lamberton, 1965). The bark of *Leptospermum ericoides* contains ursolic acid acetate, betulic acid and a triterpene acid $C_{30}H_{48}O_4$ (Corbett and McCraw, 1959). The outer bark of *L. scoparium* contains a complex mixture of terpenoid compounds, including ursolic acid acetate, oleanolic acid and betulic acid, together with aliphatic esters, *p*-coumaric esters, aliphatic alcohols and β-sitosterol. The aliphatic esters comprise a major component of the neutral fraction and consist of *n*-C_{24}, C_{26} and C_{28} fatty acids esterified with *n*-C_{18}, C_{20}, C_{22}, C_{24}, C_{26} and C_{28} primary alcohols. The *p*-coumaric esters, previously unreported, contain the same alcohols as the aliphatic esters (Corbett and McDowall, 1958; Corbett, McDowall and Wyllie, 1964). Other isolations include a new triterpene 3-epimoretenol from the bark of *Sapium sebiferum* (Khastgir, Pradhan, Duffield and Durham, 1967), friedelin, epifriedelinol and other compounds from *Syzygium cordatum* (Candy, McGarry and Pegel, 1968) and lupeol acetate, β-amyrin acetate, erythrodiol caprylate and other triterpenoids from *Madhuca latifolia* (Awasthi and Mitra, 1968) Diterpene acids (resin acids) also occur in barks.

Phenolic and other constituents of barks

Phenolic acids are localised in the cork cells. Hergert and Kurth (1952) found that phenolic acids were liberated on saponifying the cork of Douglas fir, and Hergert (1958) obtained ferulic acid and high-molecular phenolic acids by saponifying white fir (*Abies concolor*) cork. Another phenolic acid, sinapic (4-hydroxy-3,5-dimethoxycinnamic) acid, was obtained from the cork of *Quercus suber*. Hergert suggests that the phenolic acids are involved in the suberin complex by esterification with the hydroxy-fatty acids, and that suberin should therefore be defined as a hydroxy fatty-phenolic acid ester. This work was of further interest in suggesting that the suberin of gymnosperms is composed of hydroxy-fatty acids of shorter chain lengths than those of the suberin of angiosperms. Guillemonat and Traynard (1963) isolated from saponified oak cork a crystalline material named phellochryseine which was identified as the ferulic acid ester of methyl phellonate. They suggest that caffeic acid (3,4-dihydroxycinnamic) and not ferulic exists in the cork esterified with phellonic and that the phellochryseine was formed during the saponification with methanolic alkali. A phlobaphene-like material was obtained by Kurth and Hubbard (1951) from the alcohol-soluble wax of *Pinus ponderosa* bark, and phenolic acids were also obtained by Hergert and Kurth (1953) from white fir wax. Kurth and Hubbard suggest that the phlobaphene-like material exists as a complex with the fatty acids of the wax.

Wattle and mangrove barks (*Acacia* spp. and *Rhizophora* spp.) contain large amounts (up to 40%) of tannins and with many others have long been used as tanning materials. The tannins are mostly of the condensed type. Simpler gallotannins and ellagitannins which can be hydrolysed to give gallic and ellagic acids and sugar are frequent. Hathway (1958, 1959) has examined the sequence of events leading to the deposition of phlobatannin. Simple phenolic compounds are formed in the leaves and translocated through the sieve tubes of the inner bark to the cambium where they are enzymically oxidised to phlobatannin which is stored in the bark. The phenolic acids, tannins and phlobaphenes are probably polymers of catechins and related substances; other polyphenolic components may contain both catechyl and guaiacyl (methylcatechyl) nuclei (Fujii and Kurth, 1966). The phenolic acids and other phenolic derivatives may play a part in conferring upon barks resistance to microbial attack.

Chalconaringenin-2′-glucoside is the chief flavonoid of young *Salix purpurea* bark and is accompanied by naringenin-5-glucoside in old bark (Jarrett and Williams, 1967). Salicin, salicyl alcohol, 2,6-dimethoxy-*p*-benzoquinone, azaleic acid (*n*-heptane-1,7-dioic acid), gentisyl alcohol (2,5-dihydroxybenzyl alcohol) and other compounds were isolated by Pearl and Darling (1968) from *Populus balsamifera* bark; azaleic acid is reported for the first time in any bark and gentisyl alcohol in any higher plant material. Rahman and Bhatnagar (1968) found persicogenin (5,3′-dihydroxy-7,4′-dimethoxy-flavanone) and its glycoside persiconin among other flavonoids in peach (*Prunus persica*) bark. Swan (1968) obtained octadecanedioic, 18-hydroxy-octadecanoic, 3,4-dimethoxycinnamic and ferulic acids from western red cedar (*Thuja plicata*) bark; no phellochryseine or phloionic acid derivatives were found.

Derivatives of quinone and, more commonly, of anthraquinone are known to be present. Briggs and Locker (1951) identified four flavonols and xanthoxyletin (a substituted coumarin) in the bark of *Melicope ternata*. Chatterjee and Mitra (1949) obtained three crystalline constituents of the bark of *Aegle marmalos*: a compound identical with γ-fagarine (4,8-dimethoxyfuro(2,3,6) quinoline), a substituted dihydrofurocoumarin, and umbelliferone (7-hydroxycoumarin). The umbelliferone may have been derived from another coumarin compound. Shimada (1952) obtained fraxin (the glucoside of fraxetin, 7,8-dihydroxy-6-methoxycoumarin) from the bark of *Aesculus turbinata* and fraxetin, aesculetin (6,7-dihydroxycoumarin), fraxinol (6-hydroxy-5,7-dimethoxycoumarin) and syringin (the β-D-glucopyranoside of syringenin, 4-hydroxy-3,5-dimethoxycinnamyl alcohol) from the barks of *Fraxinus* spp. The coumarin and especially any furocoumarin derivatives may contribute to the resistance of the barks to fungal attack. Lignans have also been isolated from barks. Hughes and Ritchie (1954), for example, obtained the lignans galbulin, galcatin, galbacin and galgravin from the barks of *Himantandra* spp. The naturally occurring lignans have been comprehensively reviewed by Hearon and MacGregor (1955). Barks of commercial importance are reviewed in Chapter 11. The chemistry of barks has been described in detail by Jensen, Fremer, Sierilä and Wartiovaara (1963).

Suberin

Suberin constitutes up to 40% of the cork of *Quercus suber* or the outer bark of silver birch (*Betula verrucosa*). Other constituents of cork are wax, lignin, cellulose and other polysaccharides, and tannin. Glycerine is present, in a bound form, to the extent of 6–7% (Ribas and Blasco, 1940). Cork contains up to 4% of ash containing potassium, calcium, magnesium and other metals. The constituent long-chain hydroxy and dicarboxylic acids of suberin are inter-esterified to form a cross-linked polyester admirably suited to its function as the structural unit of the tough outer coating of trees (Downing, 1961). The polyester, of large molecular weight, has an affinity for basic dyes which suggests the presence of free carboxyl groups.

Kurth (1967) points out that the alkaline saponification of cork gives a complex mixture which consists chiefly of hydroxy-fatty acids, fatty acids and amorphous dark-coloured polymeric phenolic acids.

The early work on the isolation of constituent hydroxy-fatty and dibasic fatty acids of suberin has been discussed in Chapter 3. Phellonic acid obtained from the suberin of oak cork and birch bark is 22-hydroxydocosanoic acid ($CH_2OH(CH_2)_{20}$ COOH). This structure was reported by Jensen (1950a), Guillemonat and Strich (1950), Ribas and Gil Curbera (1951) and Seoane,

Gil Curbera and Ribas (1953) and has been confirmed by Dupont, Dulou and Chicoisne (1956) and Chicoisne, Dupont and Dulou (1957). Phellogenic acid is the corresponding docosanedioic acid ($COOH(CH_2)_{20}COOH$); its structure was confirmed also by Guillemonat and Strich (1950). Phloionic acid (9,10-dihydroxyoctadecanedioic acid) and phloionolic acid (9,10,18-trihydroxyoctadecanoic acid) occur in oak cork and birch bark suberin in differing proportions; oak cork suberin is rich in phloionic acid and birch suberin in phloionolic. Seoane and Ribas (1951) showed that oak cork of good quality contains 3·6–3·7% of phloionic acid and 0·6–0·8% of phloionolic; Jensen and Östman (1954) and Jensen and Rinne (1954) report more phloionolic acid than phloionic in birch. The structure of phloionic acid was confirmed by Guillemonat and Cesaire (1949), Ribas (1951) and Dupont, Dulou and Cohen (1955, 1956).

Unsaturated acids are known to occur in the suberin of oak cork and birch bark. Jensen (1950b, c) obtained phellonic, phellogenic and an unknown unsaturated acid from these sources. Ribas and Seoane (1954) isolated from the liquid acid fraction of oak cork suberin the *cis* form of 18-hydroxyoctadec-9-enoic acid and the *cis* and *trans* isomers of octadec-9-ene-1,18-dioic acid. These acids were also obtained by Jensen and Tinnis (1957) from the suberin of both oak cork and birch bark. The *cis* structure of 18-hydroxyoctadec-9-enoic acid was confirmed by Duhamel (1965). The acids of oak cork were further elucidated by Gonzalez-Gonzalez, Sola, Iglesias-Martin, Rivas-Paris and Ribas-Marqués (1968). 9-Hydroxyoctadecane-1,18-dioic acid was isolated for the first time and its identity confirmed by synthesis. Major acids obtained were *trans*-18-hydroxyoctadec-9-enoic (also synthesised), *trans*-octadec-9-ene-1,18-dioic, 9-hydroxyoctadecane-1,18-dioic and phloionic; acids present in smaller amounts were *cis*-octadec-9-ene-1,18-dioic, *cis*-18-hydroxyoctadec-9-enoic, phloionolic and phellogenic. Another major constituent, possibly a homologue of phloionic or phloionolic acid, and unidentified volatile acids were also found.

RELATIONSHIP BETWEEN SUBERIN AND CUTIN

Detailed information on the identities of the constituent acids of suberin and cutin has been obtained from the examination of comparatively few materials, mostly the suberin of *Quercus suber* and *Betula verrucosa* and the cutin of *Agave* and apple. Suberin and cutin from different sources have the same basic polyester structure, but differ in the relative proportions of constituent acids. Phellonic and phellogenic acids occur in oak cork and birch suberin, while phloionolic acid is prominent in birch but not in oak cork suberin. Phloionolic acid, on the other hand, is a major acid of *Agave* and apple cutin, but no hydroxy acids of C_{22} chain length have been detected in them. In some other cutins, phloionolic acid occurs in small amounts and the predominant acid is 10,16-dihydroxyhexadecanoic. No equivalent C_{16} acid has yet been identified in suberin. Brieskorn and Böss (1964) draw attention to the occurrence of pairs of related ω-hydroxy and dicarboxylic acids in cutin and suberin; the

C_{16} juniperic and thapsic acids are present in apple cutin and their C_{22} homologues phellonic and phellogenic acids in cork suberin.

With the isolation by Matic (1956) of phloionolic acid and 18-hydroxy-octadec-9-enoic acid from *Agave* cutin, the occurrence of acids common to cutin and suberin was established. Matic showed that his unsaturated acid has a *cis* configuration. The identical nature of the phloionolic acids of suberin and cutin has been confirmed; Matic established a *threo* configuration for the acid of *Agave* cutin; Seoane, Ribas and Fandino (1957, 1959) showed that the acid of oak cork suberin also is the *threo* form; and Eglinton and Hunneman (1968) found that the acid is present in apple cutin in both *threo* and *erythro* forms, with the *threo* predominating. Alvarez-Vazquez and Ribas-Marqués (1968) have separated the optical isomers of phloionic acid and have investigated the stereochemistry of phloionic and phloionolic acids. The 18-hydroxyoctadec-9-enoic acid and the *cis* and *trans* octadec-9-ene-1,18-dioic acids of oak cork and birch suberin also occur in apple cutin (Eglinton and Hunneman, 1968). The similarity between suberin and cutin was further indicated by the identification, by Eglinton and Hunneman, of phloionic acid in apple cutin.

From an examination of the suberised outer cell walls of roots Martin and Fisher (1966) obtained further indications that, like cutin, suberin from different sources may show differences in its hydroxy-fatty acid composition. Guillemonat and Triaca (1968) have suggested that the composition of the suberin of different botanical species may be of taxonomic interest. In a preliminary examination, the cork of *Kielmeyera coriacea* gave 46% of suberin which yielded phloionic, hexadec-8-ene-1,16-dioic and octacosanoic acids; of these, only phloionic acid has been detected in oak cork suberin. At present it seems that cutin is composed largely of acids of chain lengths C_{16}–C_{18} and that suberin differs from cutin by possessing a significant proportion of acids of C_{22} chain length. Detailed information on the composition of suberin, however, is scanty; a considerable proportion of the hydroxy acids of the polymer is not accounted for. With the further detailed examination of cutin and suberin from other plants, the ranges of acids present in the two substances may be found to overlap to a greater extent than is now indicated and a chemical differentiation between them would then disappear.

Association between the components of cuticles or barks

A close association exists between cutin or suberin and other components of the protective coverings. Whether this involves chemical linkage or physical occlusion is not clear. Huelin and Gallop (1951a) found during their fractionation of the apple fruit cuticle some evidence of an association between lipoid material and cellulose; the cellulose became available to the solubilising Schweizer's reagent only after the saponification of the skin. Huelin (1959) suggested that the carbohydrates and other components of the apple cuticle may be bound in the cutin complex. Crisp (1965) advanced the interesting view that linkage of the cutin to the uronic acid moiety of pectin may be a mechanism by which the cutin is cemented to the outer cell wall. Heinen and de Vries (1966b) in a study of the enzymic breakdown of the cutin membrane found evidence of the presence of relatively loosely bound fatty acids on its outer surfaces. They suggest that the fatty acids may serve as a link with the wax compounds at the outside of the membrane and with the pectic layer at the inside.

Brieskorn and Böss (1964) suggest a probable linkage between the hydroxyl groups of cutin and cellulose. The tomato fruit cuticle contains a polysaccharide which yields glucose, xylose and arabinose on hydrolysis; part of the cutin is firmly occluded by the polysaccharide, as evidenced by renewed solubility of the cutin in alcoholic alkali after treatment of the partly saponified membrane with Schweizer's reagent (Brieskorn and Reinartz, 1967b). The possibility exists of a linkage between cutin and the condensed tannin component of cuticles; this, however, is unlikely because of the ease with which the tannin is extractable. No significant quantities of phenolic acids have been found in association with cutin. There is evidence, on the other hand, that phenolic acids are chemically bound in the suberin complex.

SPOROPOLLENIN

A substance of unusual interest related to cutin and suberin occurs in the outer membranes of moss and fern spores and pollen. Zetzsche and his co-workers (see for example Zetzsche, Kalt, Liechti and Ziegler, 1937) showed that spore and pollen membranes, including those from fossil specimens from brown coal, have two layers, an inner one consisting chiefly of cellulose, known as the intine, and an outer one containing a material to which the name sporopollenin was given, the exine. Zetzsche concluded that sporopollenin, which was difficult to degrade, may be a type of polymeric triterpenoid. Kwiatkowski and Luliner-Mianowska (1957) reported sporopollenin in the spores of a moss, two pteridophytes, three gymnosperms and eleven angiosperms and in the pollen of conifers, and suggested that the stability of pollen in geological deposits depends upon the content of sporopollenin in its membrane. Heslop-Harrison (1963b) refers to the important role of sporopollenin as a highly resistant spore wall material in *Silene pendula* and *Cannabis sativa*; he suggests that the tapetal cells, the nutritive cells lining the sporangium, contain enzyme systems capable of mobilising sporopollenin, which is otherwise extremely resistant to decay. More recent reviews are given by Godwin (1968) and Heslop-Harrison (1968).

Comparatively little is known of the chemistry of sporopollenin. Shaw and Yeadon (1966) examined the walls of the spores of the common clubmoss (*Lycopodium clavatum*) and of the pollen of Scots pine (*Pinus sylvestris*). The membranes were obtained, in approximately 25% yield, by boiling the spores or pollen in 6% aqueous potassium hydroxide solution. The membranes from the two sources were similar and contained an almost pure cellulose intine (10–15%), an ill-defined xylan fraction (10%), a lignin-like fraction (10-15%) and an exine fraction containing the sporopollenin (55–65%). The exine on oxidative degradation yielded a mixture of non-branched mono- and dicarboxylic acids containing 18 carbon atoms or less. Hexadecanedioic acid, 7-hydroxyhexadecanedioic acid and 6,11-diketohexadecanedioic acid were obtained from the *L. clavatum* membrane; acids of chain lengths less than C_{16} were possibly breakdown products of compounds related to the hydroxy- and keto-hexadecanedioic acids isolated. Shaw and Yeadon suggest that the precursors of the dicarboxylic acids found could be acting as cross-linking units in a macromolecular complex.

Because of the extreme resistance of pollen exine to chemical degradation other than oxidation, Brooks and Shaw (1968) examined the parallel development of pollen exine and other substances in the anthers. The anthers from *Lilium henryii* plants were removed at intervals and solvent extracted, and pollen was taken from mature plants and their membranes were isolated. The formation of exine was found to be accompanied by a parallel formation of ***carotenoids***. The carotenoids from the anthers consisted of a mixture of free carotenoids and carotenoid esters which on saponification gave 90% of fatty acids mostly straight chain C_{16} compounds. Brooks and Shaw examined the polymerisation of the carotenoids obtained from *L. henryii* and also of the related ***β-carotene*** and ***Vitamin A palmitate*** under the catalytic action of boron trifluoride in the presence of oxygen. Insoluble oxygen-containing polymers were formed which had molecular formulae very similar to the formula of the pollen exine. Degradation of the polymers with ozone yielded straight- and branched-chain mono- and dicarboxylic acids which were similar, qualitatively and quantitatively, to those obtained in the same way from the pollen exine. Brooks and Shaw suggest that the pollen exine is formed by an oxidative polymerisation of the mixture of carotenoids and carotenoid esters contained in the anthers, and postulate an interesting new function for carotenoids in nature. They suggest that the lignin-like fraction of the wall found by Shaw and Yeadon may consist of phenolic acids derived from the polymerised materials. Carotenoids possess a large number of unsaturated linkages with the possibility of the formation of hydroperoxides in the presence of oxygen. The peroxides can form radicals, initiate polymerisation and give rise to hydroxyl, epoxy and carbonyl compounds (see p. 174). The insoluble polymers obtained by Brooks and Shaw contained appreciable proportions of oxygen; polymerisation may therefore occur by the hydroperoxide route. Heslop-Harrison (1968) and Heslop-Harrison and Dickinson (1969) show that the synthesis of coating materials of *Lilium* pollen takes place in the tapetum, that of sporopollenin beginning during the tetrad phase. At maturity the recesses in the surface of the exine are filled with an adhesive lipid material containing α-carotene-5,6-epoxide and other carotenoid pigments. The synthesis of the sporopollenin in the anthers is virtually complete before that of the pigmented substance begins. They therefore suggest that the carotenoids are unlikely to be the precursors of sporopollenin, but do not preclude the possibility that colourless precursors similar to carotenoids may give rise to the sporopollenin.

Cutin, suberin and sporopollenin are clearly similar polymeric structures. From the evidence so far available, the units of sporopollenin are of shorter chain lengths than those of cutin and suberin. Von Höhnel (1878) and Zetzsche (1932) observed that cutin is more resistant to degradation by alkali than suberin, and sporopollenin is regarded as more resistant than cutin; this may be a reflection of the closer cross-linking in cutin than suberin, and in sporopollenin than cutin, due to the progressive diminution of the chain lengths of the constituents from suberin to sporopollenin. The morphology and chemistry of suberin, cutin and sporopollenin have been discussed by Koljo and Sitte (1957), Sitte (1957) and Koljo (1957).

6

The Biosynthesis and Development of Cuticles

BIOSYNTHESIS

The outer wall of the epidermal cell consists of microfibrils of cellulose, 8–30 nm wide, which reinforce a soft amorphous matrix of hemicelluloses and 'pectinaceous' material. Hemicelluloses are defined chemically as carbohydrates that, unlike cellulose, are soluble in concentrated aqueous alkali solution (e.g. 17·5% sodium hydroxide). Whether the pectinaceous material in the wall is indeed pectin is unknown; hemicelluloses containing uronic acid residues give similar staining reactions, e.g. with ruthenium red. As the young epidermal cell grows, wall formation takes place over the entire surface of the protoplast, the deposition of part of the amorphous matrix preceding the formation of the microfibrils. The formation of the wall is thought to be controlled by the plasmalemma, a granular membranous structure which bounds the protoplast and is concerned with secretion from the cell. The carbohydrates constituting the matrix are formed in the Golgi vesicles of the cell and are secreted through the plasmalemma, and the cellulose microfibrils are synthesised *in situ* in contact with the plasmalemma.

To reach the surface the precursors of the wax components and of cutin must pass through the plasmalemma and the outer wall of the epidermal cell (see Figs. 1.1 and 1.2). The plasmalemma, about 7·5 nm thick, is built up of lipid molecules held in a framework of lipo-protein macromolecules. The passage of the cuticle precursors through the plasmalemma is a process of diffusion, solubility in the lipids of the plasmalemma facilitating the movement of large molecules. Bolliger (1959) observed the passive diffusion of droplets, 8–20 nm in diameter, of unsaturated fatty acids through the outer cellulosic wall of *Philodendron scandens*, and Frey-Wyssling and Mühlethaler (1959, 1965) reported a similar migration of droplets of what they thought to be a precursor of cutin, which they called pro-cutin, through the matrix of the epidermal wall of *Echeveria secunda*. The cuticle precursors are formed in the cells and both epidermal and parenchymatous cells are probably involved. Mazliak and Pommier-Miard (1963) suggest that the components of the cuticle of the apple fruit are synthesised by the epidermal cells and the outer cells of the cortex, from whence they migrate to the surface. The capacity of chloroplasts to synthesise fatty acids is known (Smirnov, 1960; Stumpf and James, 1962, 1963); the involvement of fatty acids in the formation of components of the cuticle suggests that the palisade layer of leaves plays a prominent part in the biosynthetic process. The cuticle is built up by accretion from within and its outer zone, being formed first, is in a state of tension as demonstrated by the curling of membranes isolated from the upper surfaces of leaves or from fruits.

Synthesis of wax components

Interest in the biosynthesis of wax components stems from the work of Channon and Chibnall (1929) who isolated nonacosane and nonacosan-15-one from cabbage leaves and assumed that the compounds were interrelated metabolically because of their structural similarity. These workers suggested that a condensation of two molecules of pentadecanoic acid ($CH_3(CH_2)_{13}COOH$) occurred, followed by a decarboxylation to give nonacosan-15-one which in turn was reduced to nonacosane.

$$\begin{matrix} CH_3(CH_2)_{13}COOH \\ CH_3(CH_2)_{13}COOH \end{matrix} \rightarrow \begin{matrix} CH_3(CH_2)_{13} \\ CH_3(CH_2)_{13} \end{matrix}\!\!>\!CO \rightarrow \begin{matrix} CH_3(CH_2)_{13} \\ CH_3(CH_2)_{13} \end{matrix}\!\!>\!CH_2$$

pentadecanoic acid   nonacosan-15-one   nonacosane

The position, however, was left in doubt because the naturally occurring fatty acids up to and including stearic acid (C_{18}) contained even numbers of carbon atoms and pentadecanoic acid was not known to occur in plants. Clenshaw and Smedley-Maclean (1929) obtained hentriacontane (C_{31}) from spinach leaves and thought that it might be derived from a similar condensation of two molecules of palmitic acid (C_{16}).

Other compounds, including secondary alcohols, were then obtained from plant waxes. Nonacosan-10-ol was isolated from apple wax (Chibnall, Piper,

Pollard, Smith and Williams, 1931) and nonacosan-15-ol from Brussels sprout wax (Sahai and Chibnall, 1932). The compounds were all regarded as end-products of metabolism and a modification of the biosynthetic route suggested by Channon and Chibnall was necessary to account for the presence of the secondary alcohols. This was made by Chibnall and Piper (1934) in a detailed examination of the metabolism of both plant and insect waxes. The condensation of two molecules of saturated fatty acids in association with decarboxylation was thought to give a ketone, which was then reduced to give successively a secondary alcohol, an olefin and a paraffin; this route was suggested in addition to the direct reduction of the ketone to the paraffin.

$$\begin{matrix} R_1CH_2COOH \\ R_2COOH \end{matrix} \quad \begin{matrix} R_1CH_2 \\ \quad \rangle CO \\ R_2 \\ \text{ketone} \end{matrix} \rightarrow \begin{matrix} R_1CH_2 \\ \quad \rangle CHOH \\ R_2 \\ \text{secondary alcohol} \end{matrix} \rightarrow \begin{matrix} R_1CH \\ \quad \rangle\!\!= CH \\ R_2 \\ \text{olefin} \end{matrix}$$

$$\text{secondary alcohol} \searrow \quad \text{olefin} \downarrow \quad \begin{matrix} R_1CH_2 \\ \quad \rangle CH_2 \\ R_2 \\ \text{paraffin} \end{matrix}$$

Nonacosan-10-ol in apple wax was assumed to be derived from eicosanoic and decanoic acids, both found in natural fats.

$$\begin{matrix} CH_3(CH_2)_{17}CH_2COOH \\ \text{eicosanoic} \\ CH_3(CH_2)_8COOH \\ \text{decanoic} \end{matrix} \rightarrow \begin{matrix} CH_3(CH_2)_{17}CH_2 \\ \quad \rangle CO \\ CH_3(CH_2)_8 \\ \text{nonacosan-10-one} \end{matrix} \rightarrow \begin{matrix} CH_3(CH_2)_{17}CH_2 \\ \quad \rangle CHOH \\ CH_3(CH_2)_8 \\ \text{nonacasan-10-ol} \end{matrix}$$

There was, however, still a difficulty in accounting for the presence in Brussels sprout wax of nonacosan-15-ol which required as starting compounds two molecules of pentadecanoic acid.

Other work by Chibnall and his colleagues on leaf waxes had shown that secondary alcohols were limited in distribution and that their importance had probably been over-emphasised. Many leaf waxes consisted in large part of primary alcohols with subordinate amounts of paraffins and fatty acids of similar chain lengths. The primary alcohols and fatty acids were believed to be essentially even carbon-numbered compounds, and the paraffins, secondary alcohols and ketones odd carbon-numbered compounds. Chibnall and Piper concluded that the paraffins were formed from the long-chain acids associated with them in waxes and not from the shorter acids (e.g. C_{15})

as earlier suggested. They then proposed a general scheme to account for the formation of primary alcohols and of other related compounds of known chain lengths. Unsaturated acids, of undetermined identities, but possibly of the type $CH_3(CH_2)_{13}CH{=}CH(CH_2)_2CH{=}CH(CH_2)_{2n+1}COOH$ were believed to be reduced to saturated acids and these to primary alcohols.

$$RCH{=}CH(CH_2)_2CH{=}CH(CH_2)_{2n+1}COOH \rightarrow R_1CH_2CH_2COOH$$

unsaturated fatty acid → saturated fatty acid

$$\downarrow$$

$$R_1CH_2CH_2CH_2OH$$

primary alcohol

An alternative route from the saturated acid was proposed for the formation of a paraffin with one carbon atom less than the original acid. Preliminary oxidation was believed to give a β-keto acid, which by hydrolysis and loss of CH_3COOH could give a fatty acid with two carbon atoms less or by decarboxylation a methyl ketone with one carbon atom less. The new fatty acid could be oxidised to give a primary alcohol or could undergo further shortening of its chain; the methyl ketone could yield a paraffin.

$$RCH_2CH_2COOH \rightarrow RCOCH_2COOH \begin{cases} \nearrow RCOOH \rightarrow RCH_2OH \\ \searrow RCOCH_3 \rightarrow RCH_2CH_3 \end{cases}$$

saturated fatty acid; β-keto acid; RCOOH: fatty acid; RCH_2OH: primary alcohol; $RCOCH_3$: methyl ketone; RCH_2CH_3: paraffin

The scheme proposed that unsaturated fatty acids were the starting compounds, that primary alcohols were formed by the reduction of the corresponding acids, and that the paraffins, secondary alcohols and ketones arose from the indirect decarboxylation of the corresponding C_{n+1} acids. The odd carbon-numbered paraffins resulted from the β-oxidation of even carbon-numbered acids, followed by decarboxylation and reduction of the odd carbon-numbered ketones. Starting from the series of unsaturated acids, the scheme accounted for all the long-chain wax constituents whose structures were definitely known at the time. Brenner and Fiora (1960) suggested that the dominance of C_{30} and C_{32} alcohols and acids and of C_{29} and C_{31} hydrocarbons in *Bulnesia retamo* wax could be explained by the biosynthetic mechanism put forward by Chibnall and Piper.

Kreger (1948) thought that a possible biosynthetic pathway consisted of the condensation of two molecules of palmitic acid with loss of carbon dioxide to give palmitone ($CH_3(CH_2)_{13}CH_2COCH_2(CH_2)_{13}CH_3$) from which two methyl-carbon atoms were subsequently removed by ω-oxidation and decarboxylation; this was a modification of the earlier pentadecanoic acid pathway. The complexity of the problem was further exposed by

Wanless, King and Ritter (1955) who found that pyrethrum (*Chrysanthemum coccineum*) cuticle wax contained all the paraffins from C_{24} to C_{36}, both odd and even carbon-numbered. According to the scheme of Chibnall and Piper, the odd carbon-numbered paraffins were synthesised from even carbon-numbered fatty acids, known to be the dominant acids in plants. The formation of the even carbon-numbered paraffins appeared to require the existence, at least temporarily, of some odd carbon-numbered acids. Wanless, King and Ritter suggested that a pathway occurs in plants analogous to the ***co-enzyme A (CoA)*** route of synthesis of fatty acids already postulated for animal tissues. They proposed modifications of the pathway which would account for the production of odd as well as even carbon-numbered fatty acids, and suggested a mechanism for the synthesis of both odd and even carbon-numbered paraffins beginning only with the known even carbon-numbered fatty acids. This involved alternate α- and β-oxidation to account for the formation of a minor proportion of odd carbon-numbered acids.

The synthesis of saturated fatty acids is controlled by various enzyme systems collectively known as fatty acid synthetase. The starting material is acetate which condenses with CoA to give the thiol ester acetyl-S-CoA. Under the influence of acetyl-CoA carboxylase and in the presence of biotin, acetyl-S-CoA is converted to malonyl-S-CoA. Energy is supplied by the hydrolytic splitting-off of phosphate from adenosine triphosphate. Malonate, if already present in the tissue, may esterify directly with CoA. An acyl carrier protein (ACP) then mediates the interaction of the acyl derivatives. The malonyl-S-CoA is converted to malonyl-S-ACP by transacylase and the acetyl-S-CoA to acetyl-S-ACP and these interact, with loss of carbon dioxide, to form acetoacetyl-S-ACP. This is converted by the action of reduced nicotinamide adenine dinucleotide phosphate (NADPH) to β-hydroxy-butyryl-S-ACP which loses water by the action of dehydrase to form crotonyl-S-ACP. Reduction of the crotonyl-S-ACP gives butyryl-S-ACP. The sequence is repeated from butyrate, with the addition of two carbon atoms to the chain at each stage. The route is well documented for the synthesis of fatty acids in bacteria, yeasts and animals and mounting evidence indicates a similar route in higher plants. Among the saturated acids, palmitic and stearic are the major immediate products of synthesis. Oleic acid is formed under aerobic conditions from acetate, but direct conversion of stearic to oleic has not been shown. Separate pathways exist, from some point intermediate in the chain, for the production of stearic and oleic acids. Linoleic and linolenic acids are successively formed by desaturation from oleic. Mudd (1967) has discussed in detail the present position concerning fatty acid synthesis and has compared the systems in higher plants and other organisms.

The synthesis of wax components from acetate is now generally accepted. Matsuda (1962) found that labelled carbon from acetate-1-^{14}C was readily incorporated into components of candelilla (*Euphorbia cerifera*) and jojoba (*Simmondsia californica*) waxes. Adenosine triphosphate was the source of energy in the wax synthesis. Examination of the waxes from young and old plant tissues indicated that the rates of metabolism of the different wax components varied, but that the mechanism of formation of paraffins was constant throughout the life of the plant. No precursor-product relationship was found between any two of the wax components. Mazliak (1963b, d) and Mazliak and Pommier-Miard (1963) supplied acetate – ^{14}C to the peel of ripe apple fruits, isolated the wax components 48 hr later and determined their radioactivities. The fatty acids (saturated and unsaturated) had the highest specific activities, the alcohols rather less, and the diols and hydrocarbons were only very weakly active, suggesting that the paraffins were formed by a different mechanism from that of the fatty acids. The initial formation of short chain (C_4–C_6) acids was envisaged, but acids of chain length shorter than C_{16} did not accumulate. In general, the short-chain acids were more radio-active than the long-chain, indicating a progressive alignment of the aliphatic molecules by condensation in units of two carbon atoms. As the period of exposure was extended, the long-chain acids incorporated more and more of the radioactivity. Mazliak (1963d) proposed different biosynthetic pathways for the major wax fractions. The fatty acids were formed from acetate, the unsaturated members (oleic, linoleic and linolenic) by a separate route from that of the saturated (myristic, palmitic and stearic), and the possibility that the saturated acids arose from the progressive saturation of polyunsaturated acids was suggested. The hydroxy acids were thought to be derived from the acids by oxidation and were removed from the waxy phase when they inter-esterified to form the cutin. The alcohols arose from the saturated fatty acids by reduction. Doubt was cast on the formation of the paraffins by decarboxylation of the corresponding C_{n+1} acids; if this occurred the paraffins would theoretically show radioactivities comparable with those of the acids, whereas the activities of the paraffins were negligible. The ursolic acid fraction of the wax showed considerable radioactivity and was probably formed, from original acetate, by the condensation and cyclisation of isoprene radicals derived from precursors of the mevalonic acid type (see p. 171). Separate pathways of formation of saturated and unsaturated fatty acids had earlier been suggested by Davenport (1960) to account for the presence of a wide range of saturated acids and the virtual absence of unsaturated acids of chain length other than C_{18} in the cuticle oil of the apple. Further evidence of the existence of separate routes of formation of saturated and unsaturated acids was provided by Mazliak (1965c) in work on their synthesis by the parenchyma of the fruit.

Purdy and Truter (1963b,c) quantitatively determined the classes of components of cabbage wax. The wax contained 36% of hydrocarbons (of which 93% was nonacosane), 20% of alcohols (primary and secondary), 14% of ketones (nonacosan-15-one) and about 1% of ketols (nonacosan-10-ol-15-one and nonacosan-10-one-15-ol). Occasional members of every series in the wax were missing or present only in small amounts; this, they considered, undermined the belief in the systematic lengthening of the molecular chain by units containing two carbon atoms. They assumed that nonacosane is the end-product of a series of biochemical reactions, and suggested the possibility of a biosynthetic pathway from a hypothetical 10,15-diketone *via* the ketol, ketone and alcohol to nonacosane.

$$CH_3(CH_2)_8CHOH(CH_2)_4CO(CH_2)_{13}CH_3 \rightarrow CH_3(CH_2)_{13}CO(CH_2)_{13}CH_3$$

nonacosan-10-ol-15-one nonacosan-15-one

$$\rightarrow CH_3(CH_2)_{13}CHOH(CH_2)_{13}CH_3 \rightarrow CH_3(CH_2)_{27}CH_3$$

nonacosan-15-ol nonacosane

Horn, Kranz and Lamberton (1964) confirmed the identity of the cabbage ketone as nonacosan-15-one. They showed that many *Eucalyptus* waxes contain long-chain β-diketones, the most commonly occurring member being tritriacontan-16,18-dione, $C_{15}H_{31}COCH_2COC_{15}H_{31}$. Unlike the concordance in chain length between nonacosan-15-one and nonacosane in cabbage wax, no simple relationship was found between the chain lengths of the *Eucalyptus* β-diketones and the alkanes occurring with them. This suggested that the biogenetic origins of the long-chain ketones and β-diketones are different. The position of the keto groups in the β-diketones suggested their derivation from a series of methylene keto groups built up from acetate units.

Channon and Chibnall (1929) thought that nonacosan-15-one might be derived from pentadecanoic acid, but failure to detect the acid in plants prompted Chibnall and Piper (1934) to discard this suggestion. The discovery by Martin and Stumpf (1959) of an α-oxidation mechanism especially in young leaves, however, raised the possibility that α-oxidation may produce pentadecanoic acid which could be utilised for the synthesis of C_{29} compounds. Stumpf (1965) has suggested that an α-oxidation enzyme system found in plant leaves may be involved in the synthesis of hydrocarbons from saturated fatty acids, possibly by the following route:

$$RCH_2CH_2COOH \rightarrow RCH_2CH(OH)COOH \rightarrow RCH_2CHO$$
$$\rightarrow RCH_2CH_2OH \rightarrow RCH{=}CH_2 \rightarrow RCH_2CH_3$$

By this mechanism, an even carbon-numbered acid gives rise to the corresponding odd carbon-numbered paraffin.

Nonacosane (C_{29}) comprises over 90% of the hydrocarbon fraction of cabbage wax, and C_{29} compounds as a whole account for over 60% of the wax. Interest has therefore again centred on the mechanism of formation of the C_{29} compounds, and especially nonacosane, in brassica wax. Palmitic acid (C_{16}) is clearly an intermediary, and its formation *via* the acetate route is accepted; the question therefore arises of the mechanism by which it gives rise to the C_{29} compounds. This has been investigated in considerable detail by Kolattukudy (1965; 1966a, b; 1967a, b; 1968 a, b, c, d).

Kolattukudy (1965) showed that radioactive acetate was readily assimilated into the wax of growing brassica leaves, both carbon atoms of the acetate being incorporated into the various wax components at equal rates. Other short-chain fatty acid units (pentanoate and hexanoate) contributed carbon to the wax almost as efficiently as acetate. Chemical degradation of the nonacosan-15-one isolated from the wax showed that its carbonyl carbon originated predominantly from the methyl carbon of the acetate and not from the carboxyl carbon; this supported the hypothesis that α-oxidation (of palmitic acid to pentadecanoic) was involved in the synthesis of the ketone. Trichloroacetate effectively reduced the amount of wax on the surface, the components most severely suppressed being the paraffins, ketones and secondary alcohols. This effect was believed to be exerted on the biosynthetic process rather than on secretion from the interior of the leaf (see p. 112). On the other hand, trichloroacetate did not inhibit the synthesis of the internal fatty acids, suggesting that either the site of synthesis of the surface wax, or its biosynthetic pathway, differed from that of the internal lipids. Measurements of the specific radioactivities of the wax components at different times of acetate incorporation indicated that the C_{29} compounds and the esters may be synthesised at different sites in the leaf. The pattern of radioactive labelling of the wax components was thought to provide support for the suggestion of Channon and Chibnall (1929) of a pentadecanoic acid pathway of synthesis.

This and other possible routes were further examined by Kolattukudy (1966a, b). Chopped broccoli leaves or discs cut from the leaves were incubated with various radioactive fatty acids, and the surface waxes and internal lipids were isolated, fractionated by column and thin-layer chromatography and the activities of their components determined. Capric (C_{10}), lauric (C_{12}), myristic (C_{14}), palmitic (C_{16}) and stearic (C_{18}) acids were utilised as precursors of the wax components, stearic acid being at least twice as efficient a precursor as palmitic. The possibility of the incorporation of a C_3 unit (propionyl) with thirteen C_2 (acetate) units in the synthesis of the C_{29} compounds was ruled out because of the observed derivation of the carbonyl carbon of nonacosan-15-one from the methyl carbon of acetate. Three other routes were considered: (a) the pathway involving the condensation of two molecules of pentadecanoic acid suggested by Channon and Chibnall (1929), (b) the pathway involving the condensation of two molecules of palmitic acid proposed by Kreger (1948) and (c) an elongation route by which the chain length of the intermediate fatty acid is extended, by the progressive addition of C_2 units, to a C_{30} unit which is then decarboxylated. The condensation mechanism (b) operates in the conversion of palmitic acid to palmitone in the synthesis of bacterial lipid (Gastambide-Odier and Lederer, 1959). The route from two acid molecules involves the formation, by ***'head to head' condensation,*** of an α-substituted β-ketoacid which can be reductively decarboxylated to give the hydrocarbon with one carbon atom less than the sum of the carbon atoms of the two precursor acids.

$$\mathrm{RCH_2COOH} + \underset{\displaystyle \mathrm{R_1}}{\underset{|}{\mathrm{CH_2}}}\mathrm{COOH} \rightarrow \mathrm{RCH_2CO}\underset{\displaystyle \mathrm{R_1}}{\underset{|}{\mathrm{CH}}}\mathrm{COOH}$$

$$\rightarrow \mathrm{RCH_2CH(OH)}\underset{\displaystyle \mathrm{R_1}}{\underset{|}{\mathrm{CH}}}\mathrm{COOH} \rightarrow \mathrm{RCH_2CH_2CH_2R_1}$$

The pentadecanoic acid route involved nonacosan-15-one and nonacosan-15-ol as precursors of nonacosane. Comparisons of the specific activities of the three compounds after different intervals of time following the administration of labelled acetate showed no precursor-product relationship between them and indicated that the ketone, like the paraffin, is an end-product of synthesis rather than an intermediate. No conversion to the alcohol or paraffin was detected when the labelled ketone was added to leaves or leaf homogenates. Tests with specifically labelled pentadecanoic and palmitic acids cast further doubt upon the validity of the pentadecanoic acid pathway. Pentadecanoic acid-1-^{14}C was incorporated into the C_{29} compounds less rapidly than palmitic acid-1-^{14}C or uniformly labelled palmitic acid-^{14}C, and the differently labelled palmitic acid preparations were equally efficient in contributing their activities to the compounds. No evidence was found that palmitic acid undergoes breakdown. The pentadecanoic acid pathway assumes that the acid is formed from palmitic acid by α-oxidation, but imidazole, known to be an inhibitor of α-oxidation in plants, did not prevent the incorporation of acetate into the C_{29} compounds. The results indicated that palmitic acid is not converted to pentadecanoic, and that the most likely synthetic route involves the incorporation of palmitic acid as a unit into the C_{29} compounds.

According to the hypothesis of Kreger (1948) two molecules of palmitic acid condense with loss of carbon dioxide to form palmitone which subsequently loses two methyl carbons by ω-oxidation and decarboxylation to give nonacosane. Kolattukudy's finding that palmitic acid-l-^{14}C and uniformly labelled palmitic acid-^{14}C are equally efficient in incorporating radioactivity into the C_{29} compounds also cast doubt upon the validity of this route. If Kreger's route were operative, one radioactive carbon atom would be lost in the condensation and decarboxylation, and uniformly labelled palmitic acid-^{14}C would be more efficient in contributing ^{14}C to the compounds than palmitic acid-1-^{14}C. The greater efficiency of stearic acid than palmitic as precursor also prejudiced Kreger's hypothesis.

Kolattukudy concluded that the C_{29} compounds are synthesised by the elongation of a preformed fatty acid with subsequent decarboxylation. Nonacosane is thought to be formed by the decarboxylation of the C_{30} acid built up by the progressive addition of seven C_2 units to palmitic. The synthetic process is visualised to comprise two phases; acetate is incorporated into the C_{16} acid which becomes the substrate for elongation to the C_{30} unit.

$$\begin{array}{cccccccc|c} C_2 \rightarrow & \nearrow & \rightarrow & \nearrow & \rightarrow & \nearrow & C_{16} & & \rightarrow [C_{30}] \rightarrow C_{29} \\ & C_{10} & & C_{12} & & C_{14} & \downarrow & & \\ & & & & & & C_{18} & & \end{array}$$

The formation of fatty acids from acetate was affected more severely by the herbicide CMU (3-(4-chlorophenyl)-1,1-dimethylurea) than the formation of paraffins, whereas the incorporation of palmitate into paraffins was unaffected by CMU although, it was suggested, the process was sensitive to trichloroacetate. According to the hypothesis, the elongation of the fatty acid chain takes place within the leaf cellular tissue; the incorporation of labelled acetate or palmitate into the paraffin is independent of light and so, unlike the synthesis of internal lipids, is not associated with photosynthetic processes. Kolattukudy suggests that the difference in photo-dependence may be due to the synthesis of the paraffin taking place at a different site from that of the internal lipids.

The scheme proposed by Kolattukudy deals primarily with the route of synthesis of nonacosane, but that of related compounds was also considered. Palmitic acid-1-^{14}C and uniformly labelled palmitic acid-^{14}C were equally efficient in contributing radioactivity to nonacosan-15-one and nonacosan-15-ol; the pentadecanoic acid pathway was therefore not involved in their synthesis. An elongation mechanism of formation was proposed for these compounds and possibly for non-C_{29} components of the wax. The secondary alcohol may be derived from the ketone but was not considered to be an intermediate in the synthesis of the paraffin. Acids derived from palmitic acid by the elongation process could be converted into primary alcohols which could condense with other acid units to give long-chain esters.

More evidence in support of the elongation pathway was obtained by Kolattukudy (1967a). Stearic acid was incorporated as a unit, without degradation, into nonacosane. Fatty acids of chain length up to C_{26}, derived from stearic by the same (or a similar) mechanism which synthesises the paraffins, were found in the leaves esterified in the phospholipids and triglycerides; no precursor-product relationship between the long-chain fatty acids and the paraffins was observed. It was suggested that the enzyme system responsible for elongation is somewhat like fatty acid synthetase and that the elongation-decarboxylation complex is situated in the cells other than in the chloroplasts, which are accepted as a site of fatty acid synthesis. Zill and Harmon (1962) have shown that in spinach the chloroplasts lack the ability to synthesise waxes. No free intermediates, including the C_{30} unit, were detected by Kolattukudy during paraffin synthesis; the C_{30} acid may possibly be an enzyme-bound β-keto acid which could undergo decarboxylation and reduction to give the C_{29} compound. Inability to detect the C_{30} acid may be due to its binding on the enzyme.

The synthesis of esters in broccoli leaves was also examined by Kolattukudy (1967b). The direct esterification of free fatty acids such as palmitic and stearic with alcohols, by a reversal of esterase reaction, was suggested by Kolattukudy (1966a). Other mechanisms of ester formation were identified. Chopped broccoli leaves and a powder suspension prepared from the leaves readily converted stearyl alcohol into esters whose components formed an homologous series with palmitate as the major member. Esterification was at a maximum at pH 5. Some of the palmitate moiety involved in the formation of ester was found to be derived from the cell phospholipids. When a radioactive powder suspension prepared from leaves that had metabolised palmitic acid-1-^{14}C was incubated in buffer solution at pH 5, radioactivity in the phospholipid fraction decreased and there was a corresponding increase in the free fatty acid and ester fractions. The phospholipid could participate in the formation of ester either by contributing free fatty acid or by the direct transference of the acyl moiety to the alcohol (trans-esterification due to the transacylase activity of the hydrolytic enzyme). Both mechanisms were considered to be operative. Triglyceride (tripalmitin) did not appear to be involved to any great extent in ester synthesis. Palmityl-CoA served as a substrate for esterification induced by a protein fraction obtained by precipitation with ammonium sulphate from a buffer extract of the leaf powder preparation. Purification of the enzyme by gel filtration enhanced esterification and the product was chiefly stearyl palmitate.

$$CH_3(CH_2)_{14}COCoA + CH_3(CH_2)_{17}OH \rightarrow CH_3(CH_2)_{14}COO(CH_2)_{17}CH_3$$

The optimum pH extended over a broad band above 6. Kolattukudy concluded that the broccoli leaf contains enzymes capable of synthesising esters from alcohols by direct esterification with fatty acids and by acyl transfer from phospholipids and acyl-CoA.

Work by Kaneda (1967, 1968) again drew attention to the possibility that a 'head to head' condensation mechanism operates in the synthesis of the wax hydrocarbons. He found (Kaneda, 1966) that L-valine, L-leucine or L-isoleucine are incorporated by *Bacillus subtilis* into specific branched-chain fatty acids. The amino acid is first deaminated to form the α-keto acid which is oxidatively decarboxylated to give the corresponding branched-chain acyl-CoA derivative. Elongation follows by the normal CoA route of fatty acid synthesis. The content of branched-chain (*iso* and *anteiso*) alkanes in tobacco wax almost equals that of the straight chain. Kaneda (1967) showed that when uniformly labelled (but not carboxyl-labelled) L-valine, L-leucine or L-isoleucine is fed into the growing plant, the radioactivity is found chiefly in the branched-chain alkanes. L-valine is incorporated chiefly into the C_{31} and C_{33} *iso*-alkanes, L-leucine into the C_{30} and C_{32} *iso*-alkanes and L-isoleucine into the C_{30} and C_{32} *anteiso*-alkanes. A substrate labelled on the carboxyl carbon conferred no radioactivity on the hydrocarbon fraction, the carbon being lost in the decarboxylation. Kaneda postulated in the tobacco leaf a pathway to the branched-chain fatty acid essentially identical with that in *B. subtilis*; conversion to the branched-chain alkane could be either by elongation by successive C_2 units followed by decarboxylation, or condensation with a straight-chain acid also with decarboxylation. The route from L-valine to the *iso*-C_{31} alkane, *via* the C_{14} *iso*-acid, is depicted as follows:

$$\underset{\displaystyle CH_3}{CH_3\overset{|}{C}HCH(NH_2)COOH} \rightarrow \underset{\displaystyle CH_3}{CH_3\overset{|}{C}HCOCOOH} \rightarrow \underset{\displaystyle CH_3}{CH_3\overset{|}{C}HCO\text{-}S\text{-}CoA}$$

$$\rightarrow \underset{\displaystyle CH_3}{CH_3\overset{|}{C}HCH_2(CH_2)_9COOH} \rightarrow \underset{\displaystyle CH_3}{CH_3\overset{|}{C}HCH_2(CH_2)_{26}CH_3}$$

iso-C_{31} alkane

The final stage involves elongation by nine C_2 units, or condensation with the n-C_{18} acid. The route from L-isoleucine to the *anteiso*-C_{30} alkane *via* the corresponding C_{14} acid is as follows:

$$\underset{\displaystyle CH_3}{CH_3CH_2\overset{|}{C}HCH(NH_2)COOH} \rightarrow \underset{\displaystyle CH_3}{CH_3CH_2\overset{|}{C}HCOCOOH} \rightarrow \underset{\displaystyle CH_3}{CH_3CH_2\overset{|}{C}HCO\text{-}S\text{-}CoA}$$

$$\rightarrow \underset{\displaystyle CH_3}{CH_3CH_2\overset{|}{C}HCH_2(CH_2)_9COOH} \rightarrow \underset{\displaystyle CH_3}{CH_3CH_2\overset{|}{C}HCH_2(CH_2)_{24}CH_3}$$

anteiso-C_{30} alkane

Here, elongation by eight C_2 units or condensation with the n-C_{16} acid is involved. Among the n-fatty acids in the leaf, palmitic acid (C_{16}, 70%) and stearic (C_{18}, 14%) predominated, and the *iso* and *anteiso* fatty acids necessary for the formation of the branched chain alkanes were also present. The young leaf was richer in fatty acids and had a higher proportion of branched members than the old; the young leaf also had a higher content of branched alkanes than the old.

Further work with the radioactive n-fatty acid substrates fed into excised tobacco leaves gave evidence that the condensation of two long-chain fatty acids, at least one being a normal fatty acid, is more likely to be involved than elongation by C_2 units in the later stage of synthesis. Short-chain acids were incorporated into both straight- and branched-chain alkanes; had the elongation mechanism operated, incorporation only into the straight-chain members would have been expected. Kaneda postulates two steps in the process of synthesis of the known alkanes in tobacco. The first is assumed to involve the condensation of a biologically active form of palmitic acid, the predominant fatty acid in tobacco, with another saturated fatty acid, either straight chain or branched, in the range C_{14}–C_{20}. The second consists of the conversion of the product of condensation to the straight- or branched-chain alkane by decarboxylation and reduction possibly *via* an intermediate ketone.

$$\begin{matrix} CH_3(CH_2)_{13}CH_2COOH \\ RCH_2COOH \end{matrix} \rightarrow \begin{matrix} CH_3(CH_2)_{13}CHCOOH \\ | \\ RCH_2CO \end{matrix}$$

$$\rightarrow \left[\begin{matrix} CH_3(CH_2)_{13}CH_2 \\ | \\ RCH_2CO \end{matrix} \right] \rightarrow \begin{matrix} CH_3(CH_2)_{13}CH_2 \\ | \\ RCH_2CH_2 \end{matrix}$$

represents a straight or branched chain. The first step simulates the condensation mechanism proposed by Kreger (1948) and Gastambide-Odier and Lederer (1959). Additional evidence obtained from the observed incorporation of differently labelled caprylate into both straight and branched-chain alkanes supported the condensation hypothesis. The high proportion of branched-chain compounds in the alkane fraction of the wax, relative to the low proportion of branched-chain acids in the acid fraction may, it was thought, be due to selectivity by the enzyme system in favour of the branched-chain precursors. Kaneda suggests that the utilisation of the amino acids to form unreactive *iso*- and *anteiso*-paraffins is a device to control the amounts of the acids in the plant.

Further evidence in support of the elongation mechanism of synthesis was obtained by Kolattukudy (1968a, b) who confirmed that valine and isobutyrate fed to excised tobacco leaves produce branched-chain C_{29}, C_{31} and C_{33} paraffins, and isoleucine produces branched-chain C_{30} and C_{32} paraffins. Furthermore, isobutyrate gives rise to branched-chain C_{16} to C_{26} fatty acids tentatively identified as belonging to the *iso* series, and isoleucine to C_{17} to C_{25} fatty acids of the *anteiso* series. He suggests that valine yields a C_4 unit which by the addition of C_2 units by the CoA route gives the *iso*-C_{16} acid; this by elongation and decarboxylation produces the *iso*-C_{29}, C_{31} and C_{33} paraffins. Similarly, isoleucine yields an *anteiso*-C_5 starter which gives in turn the *anteiso*-C_{17} acid and the *anteiso*-C_{30} and C_{32} paraffins.

Kolattukudy (1968a) reconsidered the condensation routes proposed by Channon and Chibnall (1929) and Kreger (1948) in the synthesis of nonacosane and concluded that they cannot be substantiated. Each route postulates the condensation of two molecules of an individual acid; pentadecanoic acid was suggested by Channon and Chibnall and palmitic by Kreger. The elongation route from the C_{16} acid to nonacosane involves the successive addition of C_2 units to the chain, but no direct proof of this exists. Units longer than C_2 may condense with the C_{16} acid; the C_{14} acid could condense with the C_{16}, with the formation of the paraffin by the process of reductive decarboxylation. Similarly, the C_{12} acid could condense with the C_{18}. Kolattukudy examined this possibility in broccoli leaves by using as substrate the C_{12} acid labelled with ^{14}C in the carboxyl carbon and tritium (^{3}H) in the methylene carbon. Paraffins isolated from the leaves contained the same ratio of ^{14}C to tritium as in the substrate; had condensation occurred, ^{14}C would have been lost in the decarboxylation. This finding and other observations led Kolattukudy to believe that condensation is unlikely.

Kolattukudy (1968c) has examined the possibility that a 'head to head' condensation of two molecules of palmitic acid accounts for the production of *n*-hentriacontane in pea and spinach waxes. He produces evidence that condensation involving two molecules of C_{16} acid, or stearic acid and a C_{14} acid, followed by decarboxylation is untenable. A process which involves malonate dialkylation is, however, put forward by Kolattukudy as a possible alternative to the elongation-decarboxylation mechanism for the synthesis of paraffins. This hypothesis suggests a condensation between a fatty acyl-CoA and malonyl enzyme, the reduction of the carbonyl group of the acyl-CoA to give an alkyl malonyl enzyme, the condensation of this with a second acyl-CoA with reduction as before, and the loss of carboxyl carbon of the malonate moiety to give the paraffin. Such a route could account for the production of diketones but not, in general, of monoketones. The condensation of two molecules of C_{16} acid with a malonyl moiety would give, for example, *n*-tritriacontane-16,18-dione found in *Eucalyptus urnigera* wax. Kolattukudy (1968d) has also investigated species specificity in the biosynthesis of branched paraffins. In the broccoli leaf, unlike the tobacco leaf, the paraffin synthesising system is specific for straight chain precursors although the fatty acid synthetase is not.

Possible routes of synthesis of nonacosan-15-one in broccoli leaves have been re-examined by Kolattukudy, Jaeger and Robinson (1968). In tests with carboxyl labelled and uniformly labelled (^{14}C) stearic acid as substrates, the carboxyl carbon was incorporated as efficiently as the other carbon atoms into nonacosan-15-one and other components of the wax, and nonacosane and nonacosan-15-one isolated from leaves that had metabolised doubly labelled (^{14}C, ^{3}H) stearic acid showed the same isotopic ratio as the starting acid. The stearic acid molecule was thus incorporated as an intact unit into nonacosan-15-one as well as into nonacosane. While accepting that several alternative routes are possible they conclude that the simplest hypothesis to explain the formation of nonacosan-15-one from both intact palmitic acid and intact stearic acid is that the nonacosane chain is first formed and is then oxidised selectively in the 15 position.

Eglinton and Hamilton (1967) suggest that if there is a simple decarboxylation of acids to the hydrocarbons, in the absence of an intermediary pool of acids or their derivatives the maximum of the distribution curve for the hydrocarbons might be expected to occur at a chain length one carbon atom less than the maximum for the acids. This is not found; many waxes contain major paraffins of chain length C_{29}–C_{31} and major acids up to and including C_{26}. No quantitative relationship, however, need exist between a precursor and its product in a biochemical system; the abundance of components in the surface wax merely represents an accumulation in metabolically inert pools (Kolattukudy, 1968a).

Kolattukudy rules out the pentadecanoic acid condensation pathway and suggests that the long-chain alkanes and fatty acids of *Brassica* wax are formed by elongation from acids of medium chain length especially palmitic, and that the ketones and secondary alcohols result from specific oxidation of preformed saturated chains. The observations of Macey and Barber (1969), however, again raise the possibility that the pentadecanoic acid condensation mechanism operates in the formation of C_{29} compounds in *Brassica* wax. These workers found that pentadecanoic acid is a significant component of the free fatty acid fraction of the wax of normal glaucous plants whereas it is difficult to detect in mutant lines and the contents of C_{29} compounds, including alkane, ketone and secondary alcohol, are drastically reduced. Alternatively α-oxidation (p. 163) might be involved. The presence of 2,(ω-1)-dimethylalkanes in *Marrubium vulgare* wax (Brieskorn and Feilner, 1968) further suggests that a 'head to head' condensation mechanism operates. Wollrab (1969) found that the alkanes and secondary alcohols in the wax of any individual organ of a member of the Rosaceae occur in similar homologous series, indicating a biogenetic relationship between them; he suggests that classes of compounds with odd carbon-numbered members predominant are synthesised by a different pathway from that of classes with even, and that secondary alcohols are formed by a different route from that of primary. Barber and Netting (1968) in a study of the chemical genetics of wheat showed that the production of primary alcohols is correlated with a suppression of β-diketones and suggest that the alcohols are formed from a precursor of the β-diketones. Streibl and Stránský (1968) presume that *n*-alkenes arise from unsaturated acids by a process analogous to the formation of *n*-alkanes from saturated acids.

The site of synthesis of the wax components has also received attention (Kolattukudy, 1968a, b). The wax is synthesised mostly, if not entirely, in the epidermis. The epidermal layer of cells peeled from the leaves of *Senecio odoris* incorporated acetate into both paraffins (chiefly n-C_{31}) and long-chain ($> C_{18}$) fatty acids, whereas the mesophyll tissue incorporated acetate chiefly into the C_{16} acid. The epidermal layer also readily converted added stearyl alcohol into ester. Nearly all of the paraffin fraction present in the leaf is located on the surface; the ease and rapidity with which it reaches the surface indicates a site of synthesis near the cuticle. Marekov, Stoïanova-Ivanova, Mondeshky and Zolotovitch (1968) found that acetate and also formate were incorporated into the alkanes of the flowers of the Damask rose (*Rosa damascena*), and concluded that the most probable route of synthesis is the one suggested by Kolattukudy.

Synthesis of terpenoid compounds

The sesquiterpene farnesene, diterpene compounds such as phyllocladene and lindleyol, and triterpene derivatives such as β-amyrin, lupeol, betulin and oleanolic and ursolic acids have been identified as constituents of plant waxes. The terpenes belong to a larger class of compounds known as the isoprenoids, so-called because the carbon skeleton of the hydrocarbon isoprene forms the basic structural unit of their molecules.

$$\begin{array}{ccc} CH_3 & & \\ & \diagdown & \\ & C{-}CH & \\ // & & \diagdown\diagdown \\ CH_2 & & CH_2 \end{array} \qquad \text{isoprene}$$

The route of synthesis, known as the isoprenoid pathway, is well documented. The starting material is again acetate. By successive enzymically controlled acetyl-CoA condensations acetoacetyl-CoA, β-hydroxy-β-methyl-glutaryl-CoA and then mevalonic acid are formed.

$$CH_3COCoA + CH_3CO\ CoA \rightarrow CH_3COCH_2COCoA$$

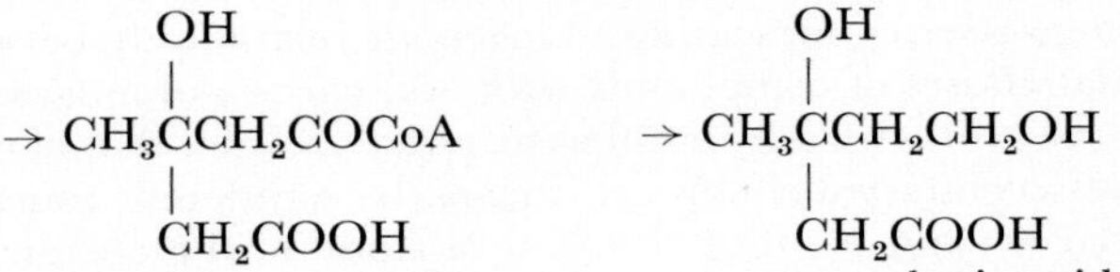

$$\rightarrow \begin{array}{c} OH \\ | \\ CH_3CCH_2COCoA \\ | \\ CH_2COOH \end{array} \rightarrow \begin{array}{c} OH \\ | \\ CH_3CCH_2CH_2OH \\ | \\ CH_2COOH \\ \text{mevalonic acid} \end{array}$$

The mevalonic acid is then transformed, through the agency of adenosine triphosphate and the intermediate mevalonic acid pyrophosphate, to *iso*pentenyl pyrophosphate which is the monomer from which all isoprenoids are derived.

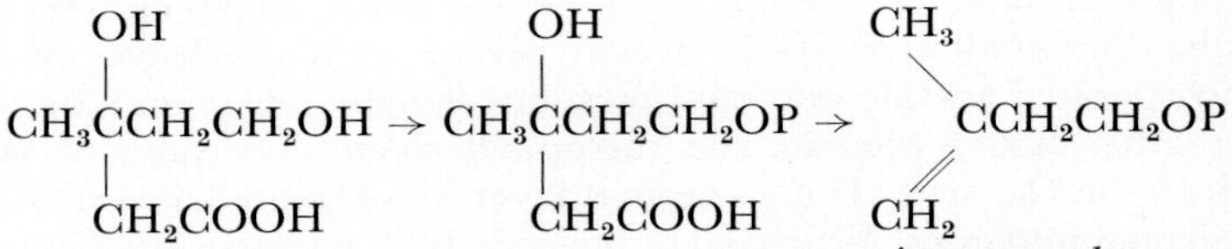

$$\begin{array}{c} OH \\ | \\ CH_3CCH_2CH_2OH \\ | \\ CH_2COOH \end{array} \rightarrow \begin{array}{c} OH \\ | \\ CH_3CCH_2CH_2OP \\ | \\ CH_2COOH \end{array} \rightarrow \begin{array}{l} CH_3 \\ \quad \diagdown \\ \qquad CCH_2CH_2OP \\ \quad /\!/ \\ CH_2 \\ \textit{iso}\text{pentenyl} \\ \text{pyrophosphate} \end{array}$$

The *iso*pentenyl pyrophosphate exists along with its isomer dimethylallyl pyrophosphate formed by enzymic action.

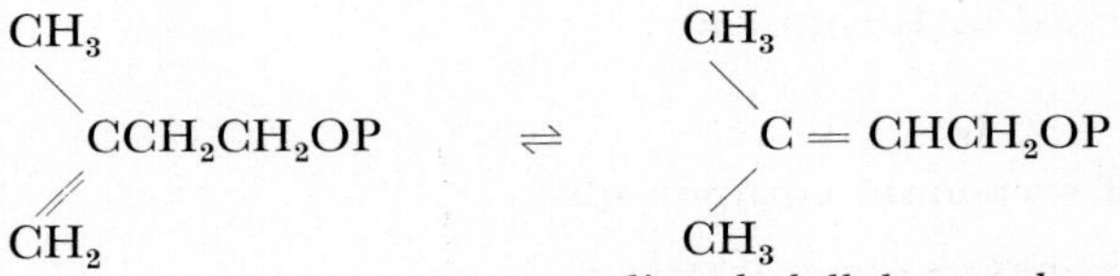

$$\begin{array}{l} CH_3 \\ \quad \diagdown \\ \qquad CCH_2CH_2OP \\ \quad /\!/ \\ CH_2 \end{array} \rightleftharpoons \begin{array}{l} CH_3 \\ \quad \diagdown \\ \qquad C = CHCH_2OP \\ \quad / \\ CH_3 \\ \text{dimethylallyl pyrophosphate} \end{array}$$

By the linking of one molecule of *iso*pentenyl pyrophosphate with one molecule of its isomer, a di-isoprenoid containing ten carbon atoms (geraniol pyrophosphate) is produced. The isoprenoid chain is extended by further successive additions of *iso*pentenyl pyrophosphate units; at each step, the chain is lengthened by five carbon atoms. The cyclic terpenes are assumed to be formed by cyclisation of the isoprenoid chains.

The addition of *iso*pentenyl pyrophosphate to geraniol pyrophosphate gives the pyrophosphate of the sesquiterpene alcohol farnesol. The corresponding hydrocarbon is farnesene, a tri-isoprenoid (C_{15}).

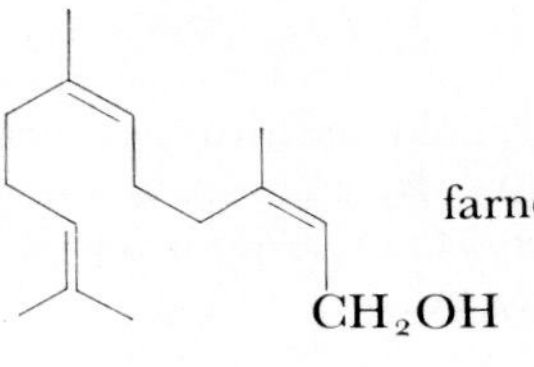

farnesol

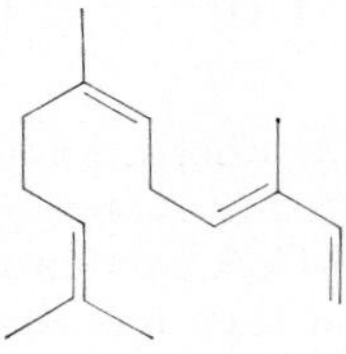

CH_2OH

farnesene

The diterpenes or tetra-isoprenoids are derived from the C_{20} homologue geranyl-geraniol pyrophosphate, formed by the linking of two units of geraniol pyrophosphate.

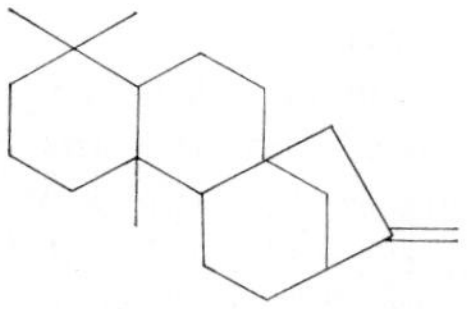

phyllocladene

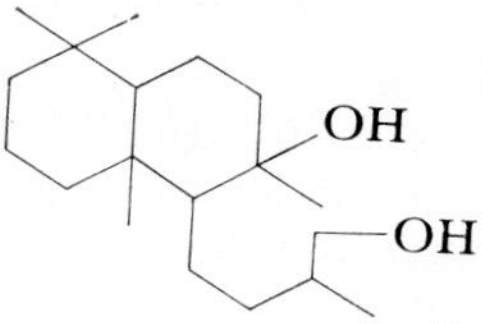

OH
OH
lindeyol

The triterpenoids or hexa-isoprenoids (C_{30}) are derived from the basic unit squalene which is formed from two molecules of farnesol pyrophosphate. The six isoprene units which make up squalene are shown

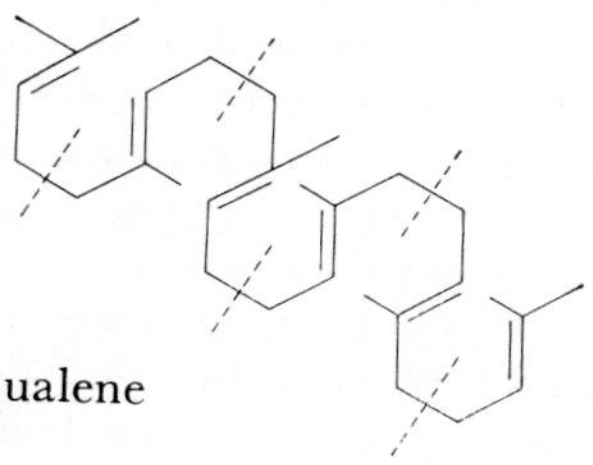

squalene

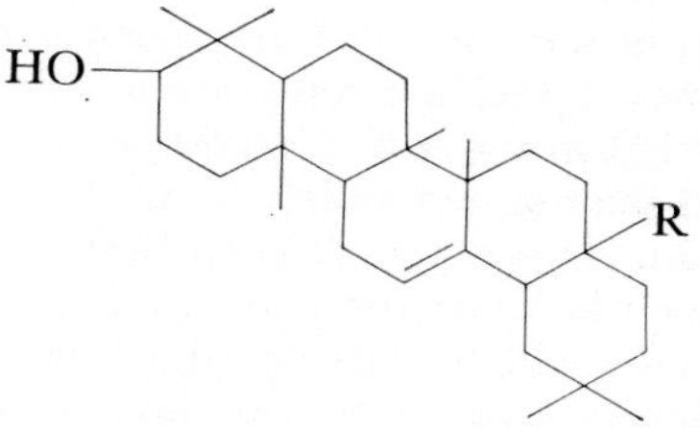

HO
R
β-amyrin R=CH_3

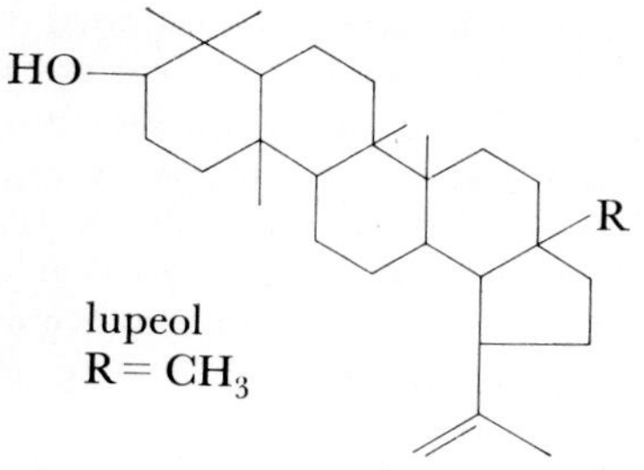

HO
R
lupeol
R= CH_3

HO
R

betulin R= CH_2OH

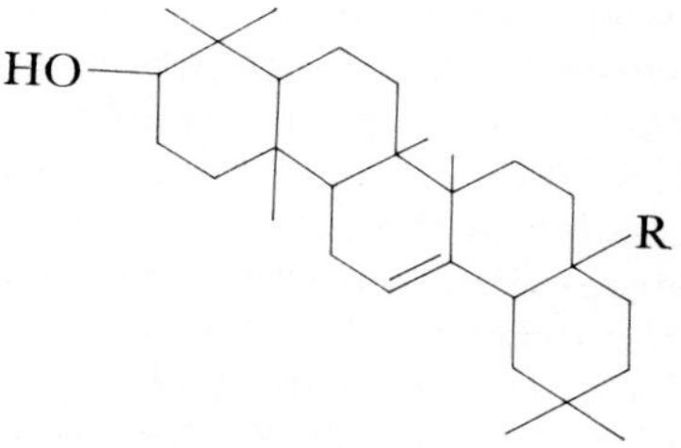

oleanolic acid R=COOH

Ursolic acid is an isomer of oleanolic acid; it is a member of the group of triterpenoids which include α-amyrin, an isomer of β-amyrin.

The leaves and inflorescence of *Nicotiana tabacum* are covered with a gummy layer, part of which appears to be an exudate from the leaf hairs. The secretion contains non-volatile terpenes. Isolated cuticles bearing trichomes, or trichomes alone, when incubated with radioactive acetate or mevalonate produced labelled terpenoids. The radioactivity of labelled squalene applied to the surface of the leaf was chiefly incorporated into sterols, but part was transferred to the terpenoid compounds (Michie and Reid, 1968). The routes of synthesis outlined have recently been discussed in detail by Porter and Anderson (1967) in connection with the formation of carotenes.

Synthesis of cutin

Huelin and Gallop (1951a) showed that hydroxy-fatty acids are major components of cutin, and Huelin (1959) suggested that the process of synthesis occurs under the control of a specific enzyme localised near the surface. The nature of this enzyme was first postulated by Siddiqi and Tappel (1956) who suggested that lipoxidase may be involved in the formation of the surface layer of plants. They thought that fatty substances secreted by the cytoplasm are oxidised to monohydroperoxides which migrate along the cell wall to the outside where they polymerise to form cutin. This was an amplification of the view expressed by Priestley (1943) that fatty materials are oxidised and polymerised at the surface. The participation of lipoxidase was confirmed by Heinen and van den Brand (1963) who studied the regeneration of the cutin of *Gasteria* leaf after injury. Three fatty acid oxidising enzymes, lipoxidase and stearic and oleic acid oxidases, were shown to be present in the leaf. Lipoxidase was the first enzyme to respond to wounding and maintained a high order of activity during the period of regeneration of cutin. Heinen and van den Brand concluded that lipoxidase occurs in or near the cutin layer and that it initiates the synthesis by inducing a series of reactions which activates the other enzymes involved. The function of the oxidases, which were probably located in the underlying tissue, was believed to be the provision of a suitable substrate. Bredemeijer and Heinen (1968) examined the production, transport and utilisation of fatty acids in cutin re-synthesis after injury to the leaves. Free fatty acids are first consumed due to wound respiration. The formation of fatty acids within the leaf is then intensified with an accompanying increase in protein, part of which consists of enzyme involved in fatty acid synthesis, and the acids are transported to the site of injury. The acids in the region of the wound then decline due to their utilisation in the cutin polymerisation process. It is also suggested that the mature cutin around the wound participates in the regeneration by becoming partly depolymerised and so providing an additional supply of fatty acid units for the new cutin layer.

Lipoxidase catalyses the oxidation of unsaturated fatty acids such as linoleic and linolenic to form reactive hydroperoxides. In the plant the hydroperoxides may undergo molecular rearrangement to form keto derivatives; the hydroperoxy groups are labile, effective oxidising agents and react readily with themselves and other units. Hydroperoxide formation is involved in the oxidative polymerisation of various types of fatty material; linolenic acid, for example, undergoes slow polymerisation to dimers and trimers linked together by peroxide bonds or, if irradiated with ultra-violet light, by both carbon-carbon and peroxide bonds. Siddiqi and Tappel found that a mixture of linseed oil and a lipoxidase extracted from peas, when exposed to air, hardened to a film which resembled cutin.

Heinen and van den Brand postulated that cutin synthesis depends upon the formation, under the influence of lipoxidase, of the hydroperoxide of linoleic and possibly linolenic acid. The oxidation involves a shift in position of a double bond in linoleic acid.

$$CH_3(CH_2)_4CH{=}CH{-}CH_2{-}CH{=}CH(CH_2)_7COOH$$

$$\downarrow$$

$$CH_3(CH_2)_4CH{=}CH{-}CH{=}CH{-}\underset{\substack{|\\ O-OH}}{CH}(CH_2)_7COOH$$

linoleic-9-hydroperoxide
(9-hydroperoxyoctadec-10,12-dienoic acid)

They suggest that stearic acid, formed from acetate by the acetyl-CoA route, is converted to oleic by the stearic acid oxidase, and that the oleic is converted to linoleic by the oleic acid oxidase. They put forward a series of possible interactions involving the linoleic-9-hydroperoxide and the fatty acids which could lead to the formation of various saturated and unsaturated hydroxy-fatty acids. Oleic acid, for example, may interact with the hydroperoxide to give 9-hydroxylinoleic acid and, through an epoxy intermediate, 9,10-dihydroxystearic acid.

$$-\underset{\substack{|\\ O-OH}}{CH}- \; + \; -CH{=}CH- \longrightarrow -\underset{\substack{|\\ OH}}{CH}- \; + \; -\underbrace{CH{-}CH}_{O}-$$

$$\rightarrow -\underset{\substack{|\\ OH}}{CH}-\underset{\substack{|\\ OH}}{CH}-$$

Interaction of the hydroperoxide with linoleic acid could give 9-hydroxylinoleic acid and 12,13-dihydroxyoleic acid.

$$-\underset{\substack{|\\ O-OH}}{CH} \; + \; -CH{=}CH{-}CH_2{-}CH{=}CH- \rightarrow -\underset{\substack{|\\ OH}}{CH}-$$

$$+ \; -\underbrace{CH{-}CH}_{O}{-}CH_2{-}CH{=}CH- \rightarrow -\underset{\substack{|\\ OH}}{CH}-\underset{\substack{|\\ OH}}{CH}-CH_2{-}CH{=}CH-$$

The hydroperoxide may react with a saturated fatty acid to give two monohydroxy derivatives.

$$-\underset{\substack{|\\ O-OH}}{CH}- \; + \; -CH_2{-}CH_2- \longrightarrow -\underset{\substack{|\\ OH}}{CH}- \; + \; -\underset{\substack{|\\ OH}}{CH}-CH_2-$$

A dihydroxy unsaturated acid such as 12,13-dihydroxyoleic acid may be converted by the lipoxidase to the hydroperoxide, which could lose water to form the ketone.

$$-\underset{OH}{\underset{|}{CH}}-\underset{OH}{\underset{|}{CH}}-CH_2-CH{=}CH- \longrightarrow -\underset{OH}{\underset{|}{CH}}-\underset{OH}{\underset{|}{CH}}-CH{=}CH-\underset{O-OH}{\underset{|}{CH}}-$$

$$\longrightarrow -\underset{OH}{\underset{|}{CH}}-\underset{OH}{\underset{|}{CH}}-CH{=}CH-\underset{O}{\underset{\|}{C}}-$$

The ketone may then react with a saturated fatty acid to give mono- and trihydroxy acids.

$$-\underset{OH}{\underset{|}{CH}}-\underset{OH}{\underset{|}{CH}}-CH{=}CH-\underset{O}{\underset{\|}{C}}- + -CH_2-CH_2- \rightarrow -\underset{OH}{\underset{|}{CH}}-CH_2-$$

$$+ -\underset{OH}{\underset{|}{CH}}-\underset{OH}{\underset{|}{CH}}-CH{=}CH-\underset{OH}{\underset{|}{CH}}-$$

As a result of the reactions of the primary peroxides a large number of C_{18} acids can thus arise, differing in the numbers of hydroxyl groups and double bonds.

The scheme put forward by Heinen and van den Brand suggests that the synthesis of cutin from the acids probably involves a preliminary coupling of fatty acid chains through peroxide bonds followed by esterification between hydroxyl and carboxyl groups.

$$\begin{array}{ccc}
\begin{array}{l}
R-\underset{OH}{\underset{|}{CH}}-CH_2-\underset{\underset{\underset{O}{|}}{O}}{\underset{|}{CH}}---COOH \\
\quad \quad \quad \quad \quad \quad | \\
HOOC-CH_2-CH-CH_2-\overset{OH}{\overset{|}{CH}}-R_1
\end{array}
& \longrightarrow &
\begin{array}{l}
R-CH-CH_2-\underset{\underset{\underset{O}{|}}{O}}{\underset{|}{CH}}---C{=}O \\
\quad\ \ | \quad \quad \quad \quad \quad | \quad \quad \quad \ \ | \\
\quad\ \ O \quad \quad \quad \quad \quad O \quad \quad \quad \ \ O \\
\quad\ \ | \quad \quad \quad \quad \quad | \quad \quad \quad \ \ | \\
O{=}C-CH_2-CH-CH_2-CH-R_1
\end{array}
\end{array}$$

The later stages of synthesis are probably under the control of a number of enzymes; catalase for example could play a part by reacting with peroxidised linoleic acid to form alkoxy and hydroxy radicals by splitting the peroxide linkage. Polymerisation may set in by the combination of radicals or their attachment to double bonds; the process probably proceeds by several intermediate stages. It is suggested that the final esterification step is controlled by the synthetic activity of cutinesterase, an enzyme reported by Heinen (1962) to be responsible for cutin degradation.

Further information on the biosynthesis of cutin was provided in a detailed study of *Agave* cutin by Crisp (1965). His scheme supports that proposed by Heinen and van den Brand by showing the crucial part played by lipoxidase in producing alkyl hydroperoxides. Crisp suggests that oleic, linoleic and linolenic are the most likely acids to be involved in cutin synthesis and that the key intermediates are the ω-hydroxyalkyl hydroperoxides. His scheme, however, extends and differs from the earlier one in several ways. In accordance with the generally accepted view, the later stage of synthesis of both C_{16} and C_{18} mono-, di- and tri-enoic acids takes place by a separate pathway from that of the saturated acids. Stepwise synthesis from acetate gives the C_8, C_{10} and C_{12} saturated acids, which are built up to the enoic acids by the addition of C_2 units and enzymic dehydrogenation. Stearic acid is thus eliminated as a direct precursor of oleic. The enoic acids then undergo ω-hydroxylation and this is followed by the formation of the hydroperoxides. Crisp makes the interesting suggestion that the ω-hydroxylated unsaturated fatty acids comprise the 'pro-cutin' that other workers have observed migrating, as droplets, from the cellular tissues to the surface. The unsaturated acids are formed in the chloroplasts; ω-hydroxylation is believed to occur in the cytoplasm or in association with cellular structures such as ribosomes, and hydroperoxidation in or near the cutin layer.

Crisp points out that the hydroperoxy groups may then react in a number of ways. Dimers (R_1—O—O—R_2) may be formed by the interaction of the groups with unsaturated bonds of other molecules, and the process can be repeated to form polymeric structures. Enzymically catalysed cleavage of the hydroperoxy group, of the type suggested by Heinen and van den Brand, yields free alkyloxy and hydroxyl radicals which by recombination give epoxy, hydroxyl and carbonyl compounds as well as polymerisation products. Molecular rearrangement may take place, with the formation of similar compounds. The formation of ester linkages between carboxyl and hydroxyl groups in terminal positions or within the chains dominates the polymerisation process; peroxide linkages and a small proportion of ether linkages also occur. Crisp depicts the later stages of cutin synthesis as follows:

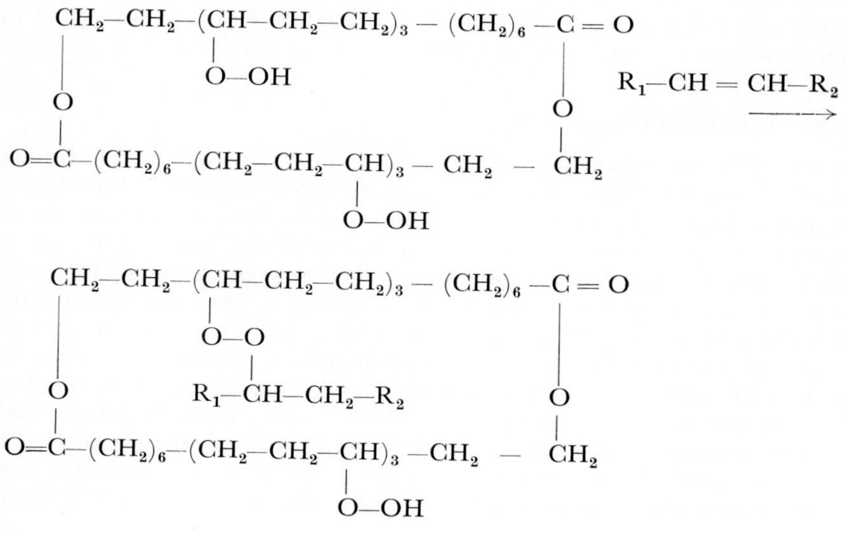

$$\begin{array}{l}
CH_2{-}CH_2{-}(\underset{\displaystyle |\atop \displaystyle O{-}OH}{CH}{-}CH_2{-}CH_2)_3 - (CH_2)_6 - C{=}O \\
\quad | \hspace{17em} | \\
\quad O \hspace{16.5em} O \qquad \xrightarrow{R_1{-}CH\,=\,CH{-}R_2} \\
\quad | \hspace{17em} | \\
O{=}C{-}(CH_2)_6{-}(CH_2{-}CH_2{-}\underset{\displaystyle |\atop \displaystyle O{-}OH}{CH})_3 - CH_2 - CH_2
\end{array}$$

$$\begin{array}{l}
CH_2{-}CH_2{-}(\underset{\displaystyle |\atop \displaystyle O{-}O \atop \displaystyle | \atop \displaystyle R_1{-}CH{-}CH_2{-}R_2}{CH}{-}CH_2{-}CH_2)_3 - (CH_2)_6 - C{=}O \\
\quad | \hspace{17em} | \\
\quad O \hspace{16.5em} O \\
\quad | \hspace{17em} | \\
O{=}C{-}(CH_2)_6{-}(CH_2{-}CH_2{-}\underset{\displaystyle |\atop \displaystyle O{-}OH}{CH})_3 - CH_2 - CH_2
\end{array}$$

Polymerisation continues by the repeated addition of other chains. The ester linkages in association with the peroxy cross-linkages induce stability of the high molecular weight polymer. Heinen and van den Brand suggest that the later stages of formation of cutin are controlled enzymically. Crisp found that the production of free radicals from hydroperoxides is enhanced by ultra-violet irradiation, and in tests on *Agave* plants grown under controlled conditions, ultra-violet light promoted the formation of peroxide linkages and the yield of polymer was increased in the presence of oxygen. Crisp therefore believes that an auto-oxidative polymerisation process, dependent upon the inherent instability of hydroperoxy groups of the pro-cutin and stimulated by ultra-violet irradiation, may play an important role in cutin synthesis. This, he suggests, may explain the development of the thicker cuticle found on the upper surface of leaves than on the lower, and the thickening of the cuticle of alpine plants.

DEVELOPMENT

The cuticular substances are secreted through the outer walls of the epidermal cells. Sometimes the cells are cutinised on all sides, when a so-called cuticular epithelium is formed. Cells lying immediately below the epidermis may also become cutinised (see p. 93). If cutin is deposited in excess, swellings or cutin cystoliths may be formed in the middle lamellae (Fritz, 1937) (see Fig. 4.1). Cuticles develop during the early stages of growth of leaves and fruits, and it has long been thought that the cuticular material gradually hardens due to continuing processes of oxidation and polymerisation. Schieferstein and Loomis (1959) suggested that surface wax is extruded only through the fragile cuticle of young leaves, later extrusion being prevented by the thickening of the cuticular layer; plasmodesmata-like channels through which the wax may reach the surface of young leaves may be blocked in the thickened cuticle (see, however, p. 84).

The cuticles of leaves

Kurtz (1950) determined cuticle thickness and the chemical characteristics of the waxes of the leaves or stems of thirteen species of plants indigenous to southern Arizona in relation to plant age. Cuticle thickness varied with the species, ranging from virtually nil in *Beloperone californica* to 5·3 μm in *Baccharis sarathroides*. In some cases, cuticle thickness increased with increasing age; in others, it hardly changed or even diminished. The waxes were isolated by solvent extraction of the dried plant material, and fractionated into 'hard' and 'soft' wax components. The amount of acids in the hard wax decreased rapidly in young plants and then slowly increased as the plants matured, an indication of the early synthesis of wax esters. Kurtz concluded that the yield of the hard wax fraction varies directly with the thickness of the cuticle. The concept that thick cuticles are well waxed, however, is not invariably true. Some leaves are characterised by exceptionally well-developed cuticular membranes with comparatively little wax. It has generally been thought that plants indigenous to arid and hot regions have waxy cuticles designed to aid the conservation of water in the plant. Kurtz (1958) produced evidence that this view is untenable. He examined the waxiness of 42 species growing in Southern Arizona under conditions of high temperatures, intense light and low rainfall. A few species, e.g. of the genera *Juniperus* and *Asclepias*, produce considerable amounts of wax (0·38–2·0%), but the majority of succulents and xerophytes only small amounts.

Sifton (1963) made a detailed investigation of the development of the cuticle of leaves and hairs of Labrador tea (*Ledum groenlandicum*). The small leaves in buds have an incipient cuticle which, although very thin, can be isolated from the cellular tissue by steeping the leaves in a hot 2% ammonium hydroxide solution. The isolated layer gave a positive stain with Sudan IV, indicative of the presence of cutin. The membrane, however, quickly disintegrated in a hot dilute solution of sodium hydroxide, showing that the forging of linkages in the formation of cutin in the young leaves was not complete. With increasing age of the leaf, the resistance to disintegration increased, showing that progressive polymerisation occurred in the cutin complex. Polymerisation of the hydroxy-fatty acids to form cutin clearly begins at an early stage in the development of the leaf. The observations supported the suggestion by Watson (1942) that the gradual hardening of the cuticle is responsible for the wavy outline of the epidermal cells in leaves. Frederiksen (1957) examined by analytical methods the progressive development of wax on the leaves and stems of young fibre-flax plants. Wax was virtually absent on the seedlings, but increased to 0·02 mg/cm^2 in 12 days after germination, falling to 0·01 mg in 19 days and then rising to 0·02 mg in 31 days. The development of the radially striped cuticular wall of the leaves and stems of *Ilex integra* was studied by Hülsbruch (1966) using cytochemistry and polarising and fluorescence microscopy. At first, a cuticularised primary wall 0·5 μm thick encloses a cellulosic layer 0·2 μm thick. As the cuticle thickens, diminutive tubes grow in a radial direction from the cellulosic wall outwards. Cuticular wax is deposited on the sides of the fine tubes.

Cunze (1926) observed that plants grown in humid atmospheres under bell jars tended to produce waxier and thinner cuticles than when grown under normal conditions in the open. The cuticles of varieties of cranberry fruit (*Vaccinium macrocarpon*) varied in thickness in different years, but the differences could not be correlated with any variations in the weather during the growing seasons (Stevens, 1932). Ivy plants (*Hedera helix*) growing in the sun have small, thick, leathery leaves whereas those in the shade have larger, thin, succulent leaves. More wax and cutin are produced on the sun leaves than on the shade. In a cool environment *Nicotiana glauca* plants produce less surface lipoidal material than in a warm; at the lower temperature less wax is deposited compared with cutin. Plants grown in direct sunlight or under water stress are less easily wetted (Skoss, 1955). A reduction in sunlight (by artificial shade) influenced the morphological development of lucerne plants, and the wetting capacity of applied chemical sprays was then increased

(Dorschner and Buchholtz, 1956). Daly (1964) isolated the wax from the leaves of thirteen field populations of the New Zealand tussock grass (*Poa colensoi*). The wax on the leaves was highly and negatively correlated with the aridity of their environments and slightly positively correlated with temperature. Other evidence suggests that the waxiness of leaves is an adaptation to an arid or unfavourable climate and has an important physiological significance. Populations of *Eucalyptus gigantea* are usually uniformly green at the lower altitudinal range and uniformly glaucous at the upper. The green phenotypes of *Eucalyptus* are found in the more sheltered habitats and the glaucous in the more exposed (Barber and Jackson, 1957). Mazliak (1963c) observed that the waxes of plants of warm climates contain constituents that are more saturated than those of plants of temperate climates.

Hull (1958) observed differential responses to herbicides between glasshouse and outdoor plants which he thought might be caused, at least in part, by differences in their cuticles due to environmental factors. Seedlings of mesquite (*Prosopis juliflora*) were grown for 97 days under closely controlled glasshouse conditions with a constant light regime (8 hr of daylight, 8 hr of artificial light at 1300 ft candles [121 lux] and 8 hr of darkness) and 6 different conditions of day and night temperatures ranging from 17 °C to 30 °C. The total wax contents of the leaflets and the degree of unsaturation of the wax were highest in the plants grown at the higher temperatures both day and night. Field-grown mesquite had a cuticle up to 8–10 μm thick on both surfaces of the leaflets, but the glasshouse-grown material usually possessed no visible cuticle. The stems of the glasshouse-grown plants had a cuticle about 2–5 μm thick. Troughton and Hall (1967) found differences in ease of wetting of the upper and lower leaf surfaces of glasshouse and outside wheat plants. The development of the wax on *Eucalyptus* leaves is also dependent on growth conditions. Cold temperatures appear to stimulate the development of the wax when the genotype allows it. For example *E. risdoni* remains green and does not develop any wax if grown in a heated glasshouse during the winter. If grown outside and allowed to experience mild frost it becomes heavily glaucous (Barber, 1955). Work cited by Barber (1955) states that the degree of development of wax in *Nicotiana* is also influenced by the degree of available soil moisture, drought tending to increase the amount of cuticular wax. Tribe (1967) grew *Hordeum vulgare* var. *distichon* under controlled conditions. The effect of light was investigated using illuminations of 750, 1200 and 3000 ft-candles (70, 112 and 280 lux) intensities. With increased illumination more cuticle was secreted, but no differences were detected in the lipid pattern on the lamina under different light regimes. The effects of 15, 40 and 85% relative humidities also were tested; under more humid conditions more cuticle was formed, also of identical pattern. Increasing the temperature gave the same effect. Only slight differences were found in the relative proportion of hydrocarbons, alcohols, esters, fatty acids and glycerides in the waxes of the *Hordeum* plants of different age. Hull (1964) suggests that a glass house may filter out light rays, presumably ultra-violet, necessary for the proper development of the cuticle. It would seem to be desirable that this possibility and also the effect on the cuticle of growing plants under artificial light in controlled environment rooms should be more fully examined.

Pits and ectodesmata have been suggested as pathways for the transport of wax to the surface, but neither extend into the cuticle (see p. 84 and Figs. 4.6 and 4.7). Submicroscopic canals, the openings of which have been observed as pores beneath deposits of wax, are a more likely route (Hall, 1967a, b) but no evidence has yet been obtained that the canals run right through the cuticular layer (Fig. 2.12). Mueller, Carr and Loomis (1954) observed shallow pit-like depressions on the leaves of *Musa* sp. before wax deposition began, but found that wax developed near the pits but not from them. Removal of the wax from older leaves with organic solvent exposed the unchanged depressions and revealed no breaks in the cuticle from which the wax may have emerged. No canals were detected by O'Brien (1967), Hallam (1967) or Ledbetter (1967) in a wide range of plant cuticles. Hallam found that wax removed from the *Eucalyptus* leaf and allowed to recrystallise slowly on glass had a similar physical form to that on the leaf surface. This suggests that the pattern of the wax deposit on the leaf is more dependent on the chemical composition of the wax than on the way in which it is exuded.

Wax formation takes place at an early stage in the development of the leaf and continues during the period of leaf expansion. The very young leaves of *Pisum* and of some *Eucalyptus* spp. show prominent crystalline or tubular wax deposits almost identical in appearance with those on the mature leaves (Juniper, 1960a, b; Hallam, 1967). On the other hand, the very young leaves of *E. ovata*, a species which produces plate-like wax deposits, show plates which are small and undifferentiated, and only on the fourth and older leaves is the fully developed plate-like structure apparent (Hallam, 1967). Juniper (1960a, b) examined the sequence of development of the wax on *Pisum sativum* var. Alaska leaves. Plants were grown in sand in darkness at 18 °C for 8 days and then subjected to light intensity just below 5000 ft candles (466 lux) (just below full English summer daylight). Carbon replicas were immediately taken from the surfaces of the minute leaves which had already developed and successive replicas were taken from corresponding leaves every 24 hr. The results are shown in Fig. 6.1 A–D. The appearance and wettability of the surfaces of the pea leaf grown in the dark are very similar to those of normally grown *Rumex obtusifolius*, *Beta vulgaris* or other broad-leaved species which have no apparent surface wax projections. After 24 hr in the light, however, the adaxial and abaxial surfaces of the pea leaf differed in appearance from one another and from the *Rumex* surface. The development of wax on the abaxial surface, away from the light, was retarded at this stage. Hallam (1967) studied the formation of wax on *Eucalyptus* in a similar experiment and found no such difference between the adaxial and abaxial surfaces.

Study of the progressive development of the cuticles of plants has been greatly facilitated in recent years by the availability of analytical methods for the determination of wax, cutin and other cuticle components. Methods for the fractionation of the wax permit an examination of the change in composition with increasing age. The formation of the cuticles of apple leaves was studied quantitatively by Richmond and Martin (1959) and Baker, Batt,

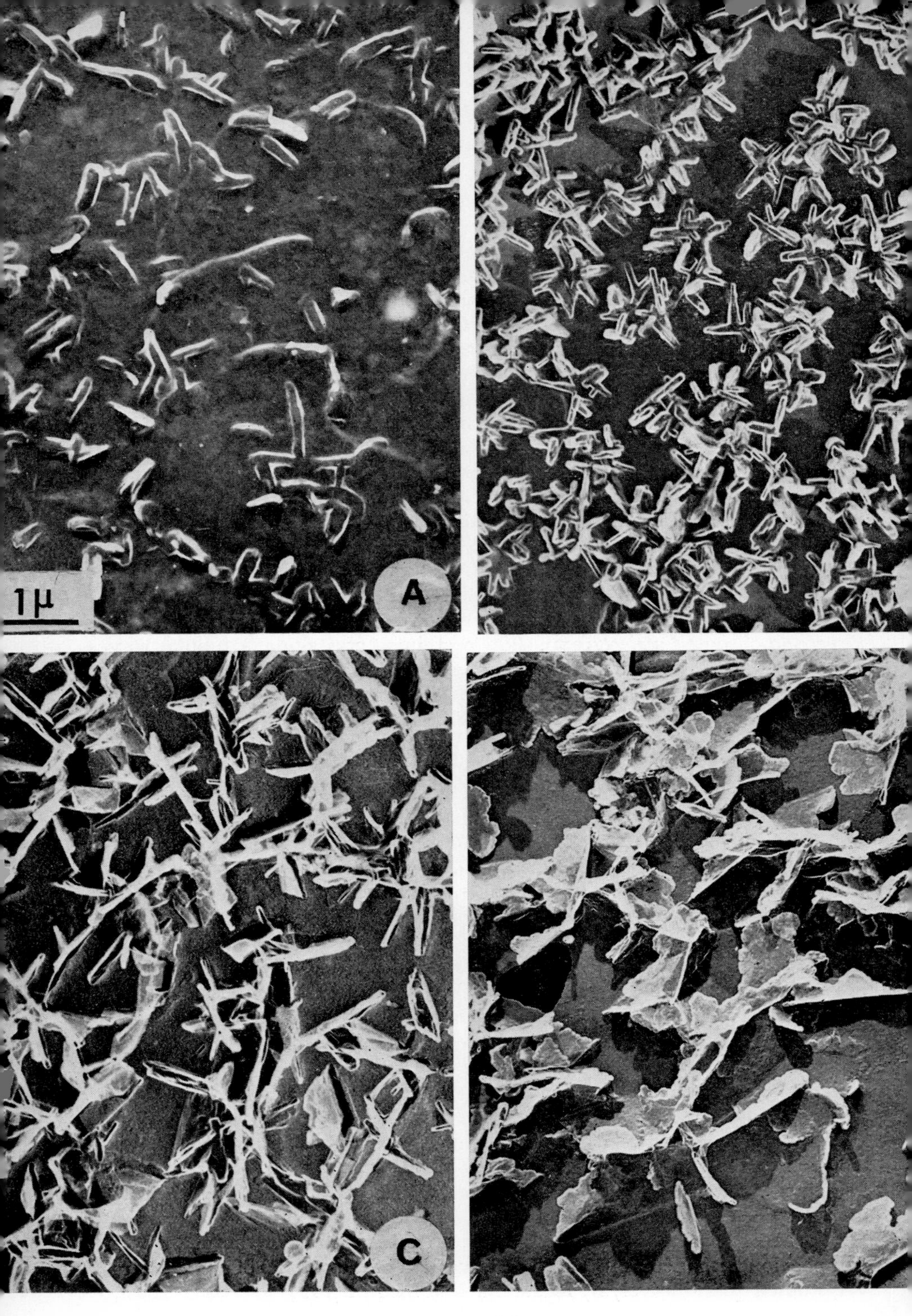

FIG 6.1

Roberts and Martin (1962); often the young leaves have as much wax and cutin on a unit area of surface as the fully grown, but during the period of leaf expansion the deposits decline slightly which suggests that the secretion of wax and the deposition of cutin just fail to keep pace with the expanding surface. The amounts of wax and cutin produced by each leaf increase steadily, attaining up to 2·5 mg of wax and 3·5 mg of cutin in the fully expanded leaf. The most active production of wax and cutin occurs during the later stages of leaf expansion. When the leaf is fully grown, its cuticle has 0·4–0·6 mg/cm² of wax and about 0·7 mg/cm² of cutin. The cuticle is fully formed soon after the leaf attains its maximum size; thereafter, the cutin deposit remains constant, but the wax deposit tends to decline presumably due to the effect of weathering.

The development of the cuticle of the blackcurrant leaf differs from that of the apple. The young blackcurrant leaf is appreciably waxy (0·05 mg/cm²) but as the leaf expands, the deposit declines. On the other hand, the cuticular membrane and the cutin build up rapidly, outstripping the expansion of the surface. When the leaf is half-grown, the cuticular membrane weighs 0·25 mg/cm² with a cutin content of 30%; when fully expanded, the membrane weighs 0·4 mg/cm² and contains 78% of cutin (Baker, Dawkins and Smith, 1968).

The leaf of the ornamental shrub *Euonymus japonicus* carries little wax (0·02 mg/cm²) throughout its life, but quickly develops an exceptionally heavy cuticular membrane (Roberts, Martin and Peries, 1961). The very young leaf (2 cm² in size) already has 0·22 mg/cm² cutin in its adaxial membrane and 0·19 mg/cm² in its abaxial. Further rapid deposition of cutin leads to 0·64 mg/cm² (adaxial) and 0·54 mg/cm² (abaxial) in the cuticle of the fully expanded leaf. By the time the leaf is mature it has produced less than 1 mg of superficial wax but up to 20 mg of cutin. The leaf of citrus lime (*Citrus aurantifolia*) also shows a rapid formation of cutin (Roberts and Martin, 1963). The mature leaf contains 0·25 mg/cm² cutin in its adaxial surface and 0·16 mg/cm² in its abaxial. The fully grown apple, *Euonymus* and lime leaves are of similar size (*c.* 25 cm²). The apple leaf has more surface wax than the others, but the *Euonymus* leaf produces six times more cutin than the apple and at least twice more cutin than the lime.

The leaf of the para rubber tree (*Hevea*) also has a heavy cuticular membrane with comparatively little surface wax. In the young leaf (45 cm² in size) the adaxial membrane weighs 0·45 mg/cm² and the abaxial 0·30 mg/cm². As the leaf expands, the membrane becomes slightly heavier and its cutin content increases from about 50% to nearly 80%. Other plants which have heavy leaf cuticular membranes are *Prunus laurocerasus* (laurel) (0·5 mg/cm²) and *Rhododendron* spp. (0·4 mg/cm²). The adaxial surface of the leaf

---

FIG. 6.1 Electron micrographs of carbon replicas of the adaxial surfaces of the leaves of peas (*Pisum sativum* var. *Alaska*). Pea plants were grown in total darkness to the second-leaf stage; they were then transferred to full daylight in a greehouse. **A**, a leaf 24 hr after transfer to the light. **B**, a similar leaf 48 hr after transfer. **C**, a similar leaf 72 hr after transfer. **D**, a similar leaf 7 days after transfer. Cr/Au/Pd

of *Rhododendron falconeri* is heavily waxed (0·2 mg/cm^2) and the wax-free membrane amounts to 0·9 mg/cm^2 (Batt and Martin, unpublished). The leaf of *Ginkgo biloba* also has a cuticular membrane of 0·9 mg/cm^2 (Johnston and Sproston, 1965). The cuticles of *R. falconeri* and *G. biloba* are the heaviest of the leaf cuticles so far assayed; it seems unlikely that the leaf of any species has a cuticle greatly in excess of 1 mg/cm^2.

The cuticles of the leaves of weed plants are in general quite delicate. Chickweed (*Stellaria media*), wild radish (*Raphanus raphanistrum*), henbit (*Lamium amplexicaule*), pennycress (*Thlaspi arvense*), dandelion (*Taraxacum officinale*), daisy (*Bellis perennis*), plantain (*Plantago major*), nipplewort (*Lapsana communis*), fat hen (*Chenopodium album*), ground elder (*Aegopodium podagraria*), and bindweed (*Convolvulus arvensis*) have little surface wax (*c.* 0·02 mg/cm^2) and cuticular membranes not exceeding in weight 0·25 mg/cm^2. No direct relationship exists between the heaviness of the cuticle and resistance to the hormone herbicides. Plantain, for example, has a comparatively well-developed membrane (0·25 mg/cm^2) and is susceptible to MCPA (4-chloro-2-methylphenoxyacetic acid), 2,4-D (2,4-dichlorophenoxyacetic acid) and other herbicides, whereas henbit with a lighter membrane (0·08 mg/cm^2) is moderately resistant.

The leaves of many vegetable and fruit plants, including potato (*Solanum tuberosum*) red beet (*Beta vulgaris*), turnip (*Brassica campestris*), broad bean (*Vicia faba*), runner bean (*Phaseolus coccineus*), gooseberry (*Ribes uva-crispa*), strawberry (*Fragaria* sp.) and tomato, have delicate cuticles with small deposits of surface wax. (<0·05 mg/cm^2) and cuticular membranes not in excess of 0·2 mg/cm^2. The fully expanded cabbage leaf is more heavily waxed (0·08 mg/cm^2) and has a membrane which weighs 0·17 mg/cm^2. The skin of the potato tuber has a small deposit of wax (0·02–0·03 mg/cm^2) and when freed from attached cellular tissue weighs 0·9–1·1 mg/cm^2; it contains a relatively small proportion of material which from its behaviour on saponification resembles cutin. Baker and Martin (1967) assessed the development and composition of the cuticles of members of the Saxifragaceae, Rosaceae and Leguminosae in relation to botanical classification. Wide differences occur in the relative proportions of surface wax and cuticular membrane in leaves even of species of one genus. The leaf of *Ribes glaciale*, for example, has a much heavier membrane than the equally slightly waxed leaf of *R. sativum*. The leaf of *R. viscosissimum* is exceptional in that it has a heavier deposit of wax than of membrane, and the leaf of *R. cereum* is heavily waxed and has a well-developed membrane. The cutin content in the wax-free cuticular membrane of the leaves of six species of *Ribes* varied between 30 and 50%.

The adaxial surface of leaves usually has a heavier cuticular membrane than the abaxial. By contrast, more wax often is present on the abaxial surface than on the adaxial. Purdy and Truter (1963a) showed that the lipid deposit on cabbage differs in amount from leaf to leaf and that 25% more lipid may be recovered from the abaxial surface. At all stages of growth of apple leaves, more wax is obtained from the abaxial surface than from the adaxial (Silva Fernandes, Batt and Martin, 1964). The leaf of *Eucalyptus globulus* yields more wax from its abaxial surface and a denser mat of crystals on this surface is revealed by the electron microscope. Furthermore, the waxes isolated from the adaxial and abaxial surfaces of leaves may differ in chemical composition. Wulff and Stahl (1960) found differences in the proportions of materials derived from the adaxial and abaxial surfaces of leaves of *Acorus calamus*. The waxes from the two surfaces of banana or rose show wide variations in their content of paraffins (Silva Fernandes, 1964). The surface wax of the canes of the *Latham* and *Malling Exploit* cultivars and of a reputedly wax-free cultivar of raspberry in successive years had low contents ($<10\%$) of paraffins whereas that of the *Norfolk Giant* cultivar was rich in paraffins (30–30%); the wax from the base of the *Norfolk Giant* canes contained more paraffins than that from the top (Baker, Batt, Silva Fernandes and Martin, 1964). The wax of *Norfolk Giant* has a crystalline form which may be due to its high paraffin content. The wide variation in the content of paraffins in plant waxes is of special interest because of the strongly hydrophobic properties of these components (Holloway, 1969b).

The physical characteristics of a leaf cannot be depended upon to give an indication of the extent of development of its cuticle. A wax bloom, due to the scattering of light at the surface, is an expression of the configuration of the wax rather than the amount. A soft, yielding leaf may have a well-developed cuticular membrane and a hard, tough leaf a poorly developed one. The pressure required to puncture a leaf is not necessarily an indication of cuticle thickness; the size and compactness of the epidermal cells and the thickness of their walls contribute, with the cuticle, to resistance to puncture.

The cuticles of fruits

The cuticles of fruits are invariably much waxier and heavier than those of the corresponding leaves (Baker and Martin, 1967). The young fruits have well-developed waxy membranes, and wax and cutin are rapidly deposited as the fruits increase in size. Unlike the cuticles of some leaves, those of fruits often become heavier per unit area of surface as the fruits grow, the deposition of cuticle outstripping the expansion of the surface. Little further development of the cuticular membrane may occur after the fruits are fully grown. The formation of cuticular material in both leaves and fruits is thus associated with the active growth of the cellular tissues.

The progressive development of the cuticles of cider, dessert and culinary apple fruits from the fruitlet stage in late June to harvest in September was studied quantitatively by Baker, Batt, Roberts and Martin (1962) and Baker, Batt, Silva Fernandes and Martin (1963). Chloroform-soluble surface and membrane-occluded wax and the weight of the wax-free cuticular membrane and its content of cutin were determined. The weights of the cuticles were assessed by summation of the weights of wax and wax-free membranes on the assumption that no significant amount of cuticular material was lost during the enzymic isolation of the membrane. As the fruits grow, all the components steadily increase in amount per unit area of surface. The fruitlets (*c.* 10 g weight) have about 0·4 mg/cm^2 surface wax, 0·2 mg/cm^2 sub-surface wax and wax-free cuticular membranes which weigh 0·6 mg/cm² or more. The fully grown fruits carry up to 1 mg/cm² total wax and wax-free membranes which weigh up to 1·5 mg/cm². Irrespective of the type or size of the fruits, surface wax varies between 20 and 30% of the complete cuticle, total wax between 30 and 45% and cutin between 30 and 40%. Cellulose does not exceed 5% of the cuticle and the remaining cuticular material consists chiefly of tannin. The large culinary *Bramley* apple produces one of the heaviest of cuticles (2·5 mg/cm²) found in many varieties examined; the fully grown fruit of average size (*c.* 150 g in weight) secretes about 150 mg each of wax and cutin.

During the growth of apple fruits the surface wax changes in composition; the content of paraffins remains constant at 15–20% but the ester fraction increases in proportion and the ursolic acid fraction decreases. The wax embedded in the membrane differs considerably from the wax on the surface; the sub-surface wax is almost devoid of paraffins and is richer in triterpenoid compounds. It is of some interest that the paraffins, to which the fruit owes much of its water-repellent and water conservation properties, are expelled to the surface. Mazliak (1963d) found that the wax in the membrane of the apple contains a higher proportion of unsaturated compounds than that on the surface.

Fully grown pear fruits carry lighter cuticles than those of apples. One of the heaviest of pear cuticles, that of the variety *Conference*, amounts to 1·8 mg/cm². Unlike apple, the pear cuticle shows no marked increase in weight as the fruit grows; the deposition of wax, cutin and other cuticle components keeps slightly ahead of surface expansion (Silva Fernandes, Batt and Martin, 1964). The plum, variety *Giant Prune*, has 0·12 mg/cm² of surface wax and a cuticular membrane which weighs 0·17 mg/cm². The fruit of *Cotoneaster cornubea* carries 0·24 mg/cm² of wax and a membrane of 0·53 mg/cm². The fruits of species of *Ribes* show widely varying ratios of surface wax to cuticular membrane; the *R. glaciale* fruit, for example, has a lighter deposit of wax and a much heavier membrane than *R. nigrum*, and the *R. sativum* fruit is waxier than *R. aureum* but has an equally well-developed membrane. The pods of members of the Leguminosae differ appreciably in waxiness; the *Russell* lupin pod is well waxed externally but the pods of broad (*Vicia faba*) and runner (*Phaseolus coccineus*) bean have comparatively little wax. The lupin pod also has, by comparison with the bean pods, a well-formed external cuticular membrane. The internal cavities of the broad and runner bean pods appear to be devoid of a cuticular membrane (Baker and Martin, 1967).

The tomato fruit has comparatively little surface wax, but a well-developed cuticular membrane. As the fruit grows, a substantial skin containing about 1 mg/cm² cutin is formed. The peel of the orange fruit consists of an outer pigmented zone, the flavedo, and an inner white zone, the albedo. The epidermis, or outer wall of the flavedo, is cutinised and contains a few stomata; oil glands are uniformly distributed within the flavedo and extend into the albedo. Well-formed cuticular membranes may be isolated by enzymic or chemical treatment of the rind; the membranes of fruits obtained from different citrus-producing countries varied in weight between 0·35 and 0·75 mg/cm² and contained 60–70% of cutin. The banana fruit is well waxed (0·16 mg/cm²) and has a cuticle which may weigh 3 mg/cm² (Batt and Martin, unpublished). The values reported for wax, cuticular membrane and cutin are intended to give indications of the relative extent of development of cuticles in fruits; the values may differ appreciably according to species, analytical methods and other factors.

Apart from apples, little information is as yet available on the progressive development of the cuticles as fruits increase in age. It has long been known that changes occur in the composition of the wax of apple fruits during their growth and storage. The early method of analysis fractionated the wax into a hard or true wax and a soft wax or oil. Markley and Sando (1931) showed that the oil fraction of apple wax increased at a greater rate than the hard wax fraction during the growth and storage of the fruits. An increase in the oil fraction of the cuticle of apples was associated by Hackney (1943) with an increase in the resistance of the fruits to gaseous diffusion. Huelin and Gallop (1951b) found that the proportion of esters in apple wax, especially the volatile ones derived from the shorter-chain alcohols and acids, increased during storage of the fruits. The changes in the composition of the wax of apple fruits during the growth and storage periods were examined by Mazliak (1958) and by Mazliak and Pommier-Miard (1963). When the fruits are on the tree, the hard wax fraction increases in amount more rapidly than the oil, whereas during later storage in the cold the hard wax remains constant, but the oil increases. The maximum oil to hard wax ratio occurs at the peak of the respiration rate. The later stages of storage are marked by a degeneration of the wax and especially by a reduction in the amount of the oil. Short-chain acids, aldehydes and alcohols are formed in the wax and may be lost by volatilisation.

Morozova and Sal'kova (1966) found that during the storage of apples for 3 months, the amount of hard wax in the peel showed little change, whereas the soft wax fraction increased by 50–60%. The amounts of ursolic acid and cutin in the peel of one variety remained unchanged, but increased in the peel of another. As storage proceeded, the soft wax fraction of the cuticle of both varieties contained increasing amounts of unsaturated compounds. Morozova, Platonova and Sal'kova (1968) examined the development of the cuticles of apple fruits during the periods of growth and subsequent storage. During growth, soft wax was secreted on to the cuticle surface, and cutin and ursolic acid increased in amount within the cuticular membrane. The cuticle showed a slow increase in thickness during the period of active growth, but continued to develop during the ripening period and during storage of the ripened fruits. The fruits of some varieties of apple become noticeably greasy during storage. The surface wax of one variety (*Cox's Orange Pippin*) remains constant in amount and composition; the increasing greasiness of another (*Bramley*) is associated with an increase in the amount of surface wax and with an increase in the content of esters in the waxy covering (Baker, Batt, Silva Fernandes and Martin, 1963). Electron microscopy has revealed changes in the appearance of apple wax with increasing age of the fruit. As it approaches ripeness its surface becomes stickier, and the wax plates appear to soften and merge into one another (Skene, 1963 see Fig. 4.25). Fatty acid breakdown occurs during the later stages of *Cox's Orange Pippin* fruits on the tree (Meigh and Hulme, 1965); the comparatively short storage life of this apple is partly ascribed to the earlier formation and more rapid breakdown of the fatty acids of the peel (Schmitz, 1968).

Little is known of the development of the cuticle in seeds. Sanders (1955) found that a layer of cutin is present in sorghum (*Sorghum*) at the time of pollination and is transformed into the seed coat, by the thickening of the cutin layer, as the kernel matures. The cuticle protects seeds against loss of water and is broken in the process of germination, when a water-accumulating mechanism operates. A mucilaginous material deposited within the epidermis swells when the seeds are moistened and the epidermal cells are forced open.

Genetical control of the development of the superficial wax

Attention has been given to the genetical control of waxiness in plants, but so far the genetical control of other cuticular features has not been examined because ***phenotypic*** differences are difficult to detect. Abnormalities in waxiness are readily detected in plants in the field and can be measured by water repellency and other tests and studied in detail with the electron microscope.

Some of the first work on the genetical control of glaucousness was done by Hayes and Brewbaker in 1938 as part of an investigation into the genetics of maize (*Zea*). They found that at least three independently segregating genes Gl_1, Gl_2 and Gl_3 control the development of glaucousness. Non-glaucous (green or glossy) mutants, which they found amongst a large number of varieties of maize, have recessive genes gl_1 gl_1, gl_2 gl_2 and gl_3 gl_3. The presence of the homozygous recessive genes may be detected by sprinkling the young plants with water; *Gl Gl* types are water-repellent whereas the *gl gl* types are wetted.

Bianchi and Marchesi (1960) and Bianchi and Salamini (1964) worked on maize and *Triticum durum*. They confirmed the dominance of glaucousness, but observed it only on the first five leaves; older leaves were green regardless of the ***genotype***. They were unable to detect any difference, either macroscopically or with the electron microscope, between the surfaces of the *Gl Gl* and *Gl gl* types. They considered that glaucousness was controlled by five genes: Gl_1 (located in chromosome 7), Gl_2 (chromosome 2), Gl_3 (chromosome 4), Gl^H (chromosome 9) and *Cg* (chromosome 3). The effects of these on the development and ultrastructure of the wax of plants grown under controlled conditions were as shown in Table on p. 189.

The surface of the gl_3gl_3 plant appeared to be slightly glaucous, and the ultrastructure of the wax showed that the reduction in glaucousness was due not to fewer but to smaller wax projections. The early leaves of the gl_2gl_2 plants had wax deposits similar in appearance to those of the gl_3 series. The surfaces of the first and second leaves of the gl^H plants were identical with those of the *Gl* plants, whereas the third and fourth leaves resembled those of the Gl_3 series; the fifth leaf of the gl^H series was extremely glossy. The analysis by

Bianchi and Marchesi (1960) showed that wax formation occurred extremely slowly if at all on the leaves of the gl_1 plants.

Bianchi and Salamini (1964) produced, by irradiation, mutants of *Triticum durum* which differed visually from one another in waxiness. The macroscopic

Gene	*Gl-*	gl_1gl_1	gl_2gl_2	gl_3gl_3	gl^Hgl^H	*cgcg*
Leaf 1	fully glaucous	green	slightly glaucous with small projections	slightly glaucous with small projections	fully glaucous	mixed glaucous/ green
2	”	”	”	”	”	”
3	”	”	”	”	slightly glaucous with small projections	”
4	”	”	green	”	”	”
5	slightly glaucous	”	”	green	”	green

differences, however, were not confirmed at the ultrastructure level. They suggest that the mutations alter only the height of the wax projections (hence the differences in visual appearance) and not to any significant extent the amount of wax secreted.

Hall, Matus, Lamberton and Barber (1965) studied the genetics and the surface ultrastructure of several mutants of the pea (*Pisum*). They examined two sub-glaucous mutants *wa* and *wb* and one green mutant *wlo*, all with genes recessive to those of the normal glaucous form. The adaxial surface of the leaf of the normal glaucous *Kelvedon Monarch* pea has a wax deposit which consists of rods and irregularly shaped platelets. Visually, the surfaces of the *wa* plants resemble those of *wb*, but when examined under the electron microscope, differences appear. The wax of *wa* plants consists of large and small platelets, commonly arranged in concentric circles, and that of *wb* only of large platelets. The double mutant *wb wlo* shows only small spherical granules of wax with a few rods.

Certain strains of the castor-oil plant (*Ricinus communis*) can be distinguished by the presence or absence of a waxy bloom on the stem, petioles and capsules. Harland (1947) showed that a single factor *B* is involved, which is incompletely dominant in the heterozygous *Bb* form. If *b* occurs, there is an absence of bloom. Peat (1928) found that in some strains wax also occurred on the abaxial surface of the leaf, involving another factor which he called double bloom *C*. This, like *B*, is incompletely dominant in the heterozygous *Cc* form, and operates only in the presence of *B*. The factor *C* intensifies the effect of *B* on the stem, petioles and capsules; *c* precludes the formation of a bloom on the leaf. An additional factor *D* intensifies the bloom on the stem, petioles and capsules, but its effect is shown only in the presence of *B* and may be masked by *C*; *d* induces a light bloom on the stem. Harland (1947) found a correlation between the waxiness of *Ricinus* and climatic conditions. Around Lima, Peru, which has a hot, dry summer and a cold, wet and foggy winter, the *Ricinus* plants were almost completely the non-glaucous type. In experimental trials, *BB* genotype plants with a prominent bloom failed to produce fruit in the cold season, whereas *bb* plants fruited, although slowly. This was explained by the correlation between cold season sterility and bloom found in populations of *Ricinus*. Harland concludes that had the *BB* genotype been introduced to Lima, *Ricinus* would be an uncommon plant in that area. Lima lies at 340 ft (110 m) above sea level; with increasing elevation, with more sunlight and less fog, the proportion of *BB* plants increases until at 7760 ft (2600 m) all the plants are bloomed. Strong selective forces are obviously at work. As Harland points out, this is a good example of how climatic conditions can act differentially upon the two members of a pair of ***alleles*** (*bb*).

Barber and Jackson (1957) observed that the glaucousness of the leaves of Tasmanian eucalyptus (Fig. 6.2) was correlated with habitat. The populations of *Eucalyptus gigantea*, *E. pauciflora*, *E. salicifolia*, *E. gunnii* and *E. urnigera* at lower altitudes are usually uniformly non-glaucous while at higher levels they are glaucous. The change to glaucousness takes place at 400–500 ft (130–165 m). Seedlings were grown under identical conditions from seed collected from naturally pollinated mother trees growing at different altitudes. Twelve of 28 non-glaucous mother trees segregated glaucous seedlings and 17 of 39 glaucous mother trees non-glaucous seedlings. One possibility suggested by Barber and Jackson is that all the mother trees cross-pollinate and that glaucousness is dominant to greenness as shown by Hayes and Brewbaker, and Biancho and Marchesi in maize and Harland in castor-oil plants. The glaucous trees in the simplest possible system are either of genotype *GlGl* or *Gl gl* and the green trees *gl gl*. The glaucous mother trees *Gl gl* crossed with pollen from the green trees ($Gl\ gl \times gl\ gl$) or self-pollinating ($Gl\ gl \times Gl\ gl$) segregate to give a proportion of green seedlings. The green mother trees pollinated from the *Gl Gl* glaucous trees ($gl\ gl \times Gl\ Gl$) give 100% of glaucous seedlings and when pollinated from the *Gl gl* glaucous trees ($gl\ gl \times Gl\ gl$) 50% of glaucous seedlings; when self-pollinated they give only green progeny.

The other possibility is a system of multiple dominant inhibitors (I_1^G, I_2^G, etc.; recessives i_1^G, i_2^G, etc.) of glaucousness at *low* altitude, e.g.

> *Glgl*, $I_1^G\ i_1^G$, $I_2^G\ i_2^G$, $I_3^G\ i_3^G$. (green) segregating after selfing to give a proportion of *Glgl*, $i_1^G\ i_1^G$, $i_2^G\ i_2^G$, $i_3^G\ i_3^G$ (glaucous)

and multiple recessive genes at *high* altitudes, e.g.

> Gl_1gl_1, Gl_2gl_2, Gl_3gl_3, (glaucous) segregating after selfing to give a proportion of gl_1gl_1, gl_2gl_2, gl_3gl_3 (green).

Barber and Jackson think it probable that more than one pair of alleles controlling glaucousness are segregating along the cline. As an indication of this, some waxy seedling lose their wax as the leaves mature; wax persists on the mature leaves of others probably because they continue to secrete it. This suggests a control of glaucousness similar to that in maize. Barber and

Jackson assume also that all the trees within the cline are able freely to cross with one another, despite the fact that this means that a proportion of the progeny in each habitat is doomed to fail. That the glaucousness of eucalypts benefits the plants is suggested by the dimorphism shown by *E. urnigera* and

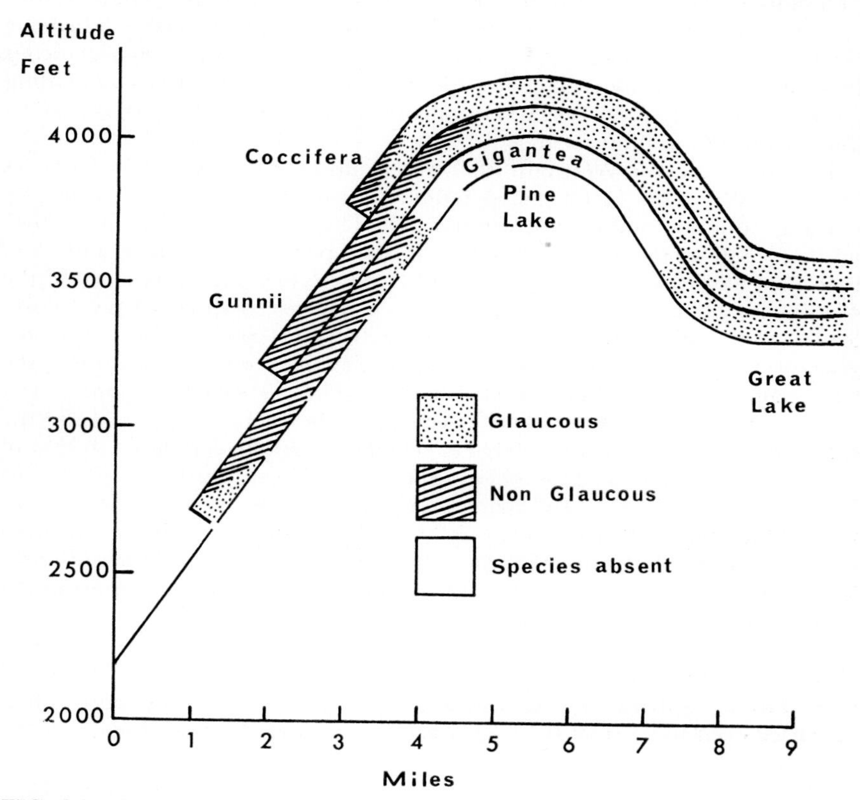

FIG. 6.2 Diagrammatic section in North Central Tasmania from above Golden Valley to Great Lake showing altitudinal zonation of *Eucalyptus* species and changes in the proportion of their glaucous to non-glaucous saplings. (Redrawn from Barber, 1955)

at least eight other species (Barber, 1955). Possible functions of the wax include the promotion of hardiness to frost and of resistance to pathogens.

Anstey and Moore (1954) studied the inheritance of non-glaucous foliage in broccoli (*Brassica oleracea* var. *italica*). The green mutant is distinguishable from the normal glaucous form when about six weeks old. The juvenile leaves of both glaucous and non-glaucous forms are green and the mutation is first noticeable in the first adult leaf. As in maize, the leaves of the non-glaucous form are easily wetted and those of the glaucous are not. The mutant appears to be a normal recessive *gl gl*. The genotype *Gl gl* is indistinguishable from the homozygous *Gl Gl* and crosses and back-crosses between pure lines of the normal and mutant forms produce normal ratios of individuals. Other examples of non-glaucous forms of *Brassica* are found in cabbage (*B. oleracea* var. *capitata*) (Bailey, 1949), marrow stem kale (*B. oleracea* var. *acephala*) (Thompson, 1963) and cauliflower (*B. oleracea* var. *botrytis*) (Hall, Matus, Lamberton and Barber, 1965). Macey and Barber (1969) have studied the biosynthesis of nonacosan-15-one and related compounds in a non-glaucous mutant of *Brassica oleracea*. They showed that C_{29} compounds, including paraffins, ketones and secondary alcohols, were drastically reduced in amounts in the mutant form.

Lundqvist and von Wettstein 1962) and Lundqvist, von Wettstein-Knowles and von Wettstein (1968) have induced wax-free mutants in barley by ionising radiation and chemical mutagens. They called these *eceriferum* (*cer*) mutants. 380 *cer* mutants were induced, corresponding to 44 loci, and 13 of these loci have been mapped on five of barley's seven chromosomes. Of the 44 *cer* loci, 11 in the phenotype produce bright green spikes, 8 give glossy leaves and 7 give bright green leaf sheaths and spikes. The other loci give seven other different types of phenotypic effects on some feature of the wax coating. Some of these induced mutants do not alter the amount of wax on the leaf blades, but, as can be seen from the carbon replicas, only the shape of the submicroscopic platelets. The effect can be compared to that of small doses of dalapon (2,2-dichloropropionic acid) on pea leaf surfaces (see Fig. 8.6 and page 245). Another mutant had 20% less wax on the surface; the mutant wax contained very little of the long-chain primary alcohols which are normally a principal constituent of the surface wax. Other mutants failed to form any of the long-chain ketones, the predominant constituent of the wild type's long submicroscopic wax needles.

Since there are at least eight lipid classes and 50 different long-chain aliphatic compounds contributing to the surface wax, as well as structural and transport apparatus to bring about the secretion, we may expect to find in barley even more than the 44 mutant *cer* loci already discovered.

7

The Cuticle in Action, I Physiological Functions

The cuticle has many roles to play; natural selection on any one group of plants in any one situation will have brought about a compromise in cuticle thickness which is a reflection of the various functions the cuticle performs. In the broadest terms the cuticle controls the movement of substances into or out of the plant. In particular, it exercises decisive control over the water relations of the plant. The cuticle on some plants has become adapted to trap and absorb water and on others to shed it, but its over-riding function is to limit the loss of water by transpiration.

Under systems of intensive agriculture, a wide variety of chemicals are applied to plants to protect or kill them. The cuticle influences the behaviour and efficiency of the chemicals; some are shed, some remain on the surface and many penetrate. The cuticle itself may be affected by some of them. The cuticle is the initial site of establishment of infection by pathogens and has long been thought to contribute to the plant's defence. Modifications of the cuticle ensure the secretion of nectar or of substances to entrap insects or pollen grains. Some structures on plant surfaces, e.g. the stomata, and some nectaries (Shuel, 1961), glands and stigmas are responsible for both egress and ingress.

The functions and activities of the cuticle may conveniently be subdivided into those that are concerned with physiological processes of the plant and those that are involved in the plant's interaction with chemicals or organisms. These are dealt with in this and the following two chapters.

The trapping and absorption of water

Gerarde in 1597 wrote:

> 'The leaves growe foorth of the jointes by couples, not onely opposite or set one right against another, but also compassing the stalke about, and fastened together; and so fastened that they hold deaw and raine water in manner of a little bason.'

He was speaking of the teazle (*Dipsacus fullonum*). A single teazle can trap up to half a pint (284 ml.) of water in its leaf axils (Lloyd, 1942). It is not known that the teazle derives any benefit from this mechanism, but it is more certain that other plants with similar mechanisms do.

The genus *Dischidia* is an epiphytic group restricted mainly to the tropical forests of Indonesia, Malaysia and Polynesia. Some of the species, of which the best known is *D. rafflesiana*, develop pitcher-like leaves which catch rain water and, on older trees, masses of decaying leaves and detritus. Ants sometimes take up residence in these pitchers and introduce more chopped leaf material and frass. None of the pitchers has any mechanism, like a true pitcher plant (see p. 272), for preventing the escape of an insect nor directly for the absorption of any nutrient. However, into these pitchers, both wet and dry, grow adventitious roots from the same plant developed from nearby stems and petioles. Since the plant is an obligate epiphyte these roots must draw from the pitcher most of their water and almost all of their nutrient supply, which consists of a mixture of rain washings, insect faeces and the decayed remains of insects. The pitchers themselves do not appear to absorb anything. They have a very thick waxy layer covering the whole of the leaf. The stomata, which occur inside the pitchers, are sunk at the base of a deep wax turret and cannot be penetrated by liquid water (Groom, 1893; Scott and Sargant, 1893).

Philips (1926, 1928) showed how vegetation could intercept and make available to itself rain and mist, not directly like *Dischidia*, but indirectly by drainage. He exposed two rain gauges under similar conditions. One was left unmodified, but over the other was erected a vertical screen of interwoven conifer branches complete with leaves. The screened gauge collected almost twice as much rain in a season as the control gauge. Much of this, of course, was rain which would have fallen on the lee side of the gauge, but some may have been mist converted to heavier droplets by the screen.

Maize (*Zea*) grown in certain areas of East Africa suffers long dry periods, sometimes up to about 180 days, which are broken only by light showers. These showers suffice only to wet the surface of the soil above the root zone and the gain in soil moisture is soon lost by evaporation. Glover and Gwynne (1962) have shown that *Zea* can concentrate the rainfall of a light shower by stem flow to such an extent that a significant increase in the amount of water is experienced by the roots close to the stem. The root system of normal plants shows a heavy exploitation of the area just around the stem probably in response to the enhanced rainfall experienced at this spot. It seems likely that both the posture of the leaves, the shape of the plant and the fact that young *Zea* leaves up to the fifth leaf are waxy and relatively unwettable contribute to this stem flow (see Fig. 4.23). Stem flow of a similar type is known to contribute to the water economy of many forest trees; in *Acacia anenia* up to 40% of the total rainfall (Slatyer, 1965). Stem flow in trees is reviewed by Penman (1963).

A number of xerophytes neither catch rain like *Dischidia* nor drain it like maize, but have complex water absorbing hairs. Examples of these are *Diplotaxis harra*, *Heliotropium luteum*, *Centaurea argentea* and *Convolvulus cneorum* (Haberlandt, 1914). The hairs of *C. cneorum* have already been described in Chapter 4. They cover both surfaces of the leaf and their structure is shown in Fig. 4.11. Each hair has a very thick-walled, elongated terminal cell, while the outer wall of the short disc-shaped absorbing cell (A) has a very thin wall. This cell is separated by only a thin wall from the embedded basal cell. It is thought that water is drawn by capillarity down the thick-walled distal tip of the hair and absorbed by the thin-walled area. There is no doubt about the leaf's efficiency in absorbing surface moisture. Haberlandt observed that a leaf of *Convolvulus cneorum* immersed in water for 24 hr was able to increase its weight by 10%. *C. cneorum* has a conventional root system and is thus not entirely dependent upon aerial irrigation. Most of the Bromeliaceae are entirely dependent for their water balance on leaf absorption and they have the most complex water-absorbing organs found on any leaves. Their structure, as studied by Haberlandt (1914) and Dolzman (1964, 1965), has already been described in Chapter 4 (see Fig. 4.18). When dry, the covering cells of the water-absorbing scale dry down and form an impermeable lid to the valve (Fig. 4.18B). When wetted, the cells rapidly absorb water, swell and thus raise the lid of the scale. Water can then flow by capillarity over the cuticular surface of the normal epidermal cells down towards the absorbent surfaces of the still living cells at the base of the scale (arrows Fig. 4.18A). Dolzman observed that the still living cells at the base of the scale are connected to one another by large numbers of plasmodesmata. He thinks that these plasmodesmata may assist the passage of absorbed water through the dome cells, through the gap in the cuticle into the cells of the leaf below. Although it is well known that the Bromeliaceae take up little, if any, water through their root systems (even when they are present) and use almost entirely the rain that falls on their leaves, it is not known whether they can use mist or dew in the same way.

Many pines are resistant to desiccation during drought and this is generally attributed to the thick cuticles and waxy surfaces of the needles. Uptake of water through the cuticle is likely to be limited by the waxy structures. Nevertheless a number of observations suggest that some pines can take up water like the Bromeliaceae. Stone (1958) reported that mist spraying at night prolonged the survival of *Pinus ponderosa* seedlings which were growing in dry soil. Schopmeyer (1961) showed that when radioactive phosphate was applied to the shoots of *Pinus banksiana* a substantial translocation took place to the roots. In both cases direct absorption through the needle cuticle was assumed. The highly repellent surface and deeply sunken stomata of pine needles almost completely preclude any significant uptake through the exposed needles. Observations by Leyton and Juniper (1963) on Scots pine (*Pinus sylvestris*) suggest that a mechanism for indirect uptake does exist, at least in Scots pine, and may be applicable to other species. The young dwarf shoot of Scots pine comprises a pair of needles wrapped at the base by a tightly enveloping sheath. This sheath degenerates as the needle ages, but usually persists as long as the needles are on the tree. Electron micrographs show very different surfaces on the exposed needle, the adaxial surface within the sheath and the abaxial within the sheath. The crystalline pattern of tubules and plates on the exposed needles (see Fig. 4.27) weathers slowly and gathers soot and dust, but much of it persists for the life of the needle. The same pattern of surface structure of the needle was found in all current and one-year-old leaves investigated. Contact angles for the four different surfaces were measured in the usual way (see p. 223) and are given below.

Contact angles on pine needle surfaces
Means of 40 measurements ± S.E.

Surface	August Adaxial	August Abaxial	October Adaxial	October Abaxial
Exposed	85·7 ± 0·9	77·3 ± 1·4	67·7 ± 0·5	70·5 ± 1·1
Inside sheath	49·3 ± 1·0	93·9 ± 0·9	55·3 ± 1·5	91·1 ± 1·4

Analysis of variance of the data showed a significantly lower contact angle (P >0·01) of the adaxial surface inside the sheath, independent of the sampling time. Measurements were also made of the relative permeability to water of different parts of the needles. At different times during the season dwarf shoots were cut from neighbouring positions on the same branch, weighed and immersed with their cut ends upwards in different depths of distilled water. The cut ends were sealed with a low melting point wax. After varying periods in the dark (usually about 24 hr) in sealed specimen tubes, the dwarf shoots were removed, surface dried with filter paper and re-weighed. The change in weight as a percentage of the initial fresh weight was taken as a measure of the relative permeability of that part of the cuticle immersed. It was assumed that, from any one sample, neighbouring shoots could be compared in this way. This assumption was supported by the fact that initial fresh weights, moisture contents and dimensions of the sampled needles did not differ by more than 3%.

The uptake of water by first-year needles (October) as a percentage of the initial fresh weight after 24 hr was as follows:

Depth of immersion (cm)	Uptake	S.E.
3	−0·16	±0·57
6	+0·61	±0·48
9 (1·5 cm below sheath)	+1·38	±0·42
12 (1·5 cm above sheath)	+6·53	±1·12
Through cut base	+7·30	±0·35
Through cut tips	+9·52	±0·47

Even if allowance is made for the effect of different hydrostatic pressures and areas of needles immersed, a greater amount of water was invariably taken up when the needle surface below the sheath was in contact with the water. The values given in the table suggest a more or less linear increase in uptake of water with the depth of immersion of the exposed needle surfaces. On this basis, uptake through the surface below the sheath is more than three times greater than would be expected, and of the same order as uptake through the cut bases of the dwarf shoots.

These results not only provide explanation for the passage of water (and presumably dissolved substances) through an apparently impermeable cuticle, but also suggest a path by which some pines may have become adapted to drought conditions and a way in which light rainfall and mist, and possibly dew fall, may be significant in the water economy of these species. During drought, the wax crystals on all exposed needle surfaces (including the abaxial below the sheath) effectively reduce cuticular transpiration; the only surface permeable to water remains protected between the needles and below the sheath. When the foliage is wetted by rain, mist or dew droplets, water will tend to run off the exposed inner surfaces into the gap between the needles and thence by capillary attraction to the surfaces within the sheath. These areas are easily wetted and relatively permeable. The Scots pine has a marked channel on the inner surface of the needle; this undoubtedly helps the conduction of water.

Yoshida (1953) noticed that the fog intercepted by broad-leaved trees such as *Betula ermani* and *Fraxinus sieboldiana* dripped directly from the foliage to the ground whereas much of the fog intercepted by *Picea glehni* reached the ground by stem flow. Whether this was due to the posture of the leaves or the properties of their surfaces was not determined. Kerfoot (1968) concluded that, at high altitudes at least, the coniferous type of needle is the most efficient condenser of fog droplets. Hoar frost, which comprises frozen fog droplets, may cover conifer needles, but only appears on the edges of the leaves of broad-leaved species.

Pinus radiata, a species of pine very different from *Pinus sylvestris* and growing in a very different habitat, has been studied by Leyton and Armitage (1968). Unlike *P. sylvestris* the whole of the needle surface, with the exception of the area within the sheath, is sparsely covered by tubules of wax. There are no platelets such as are visible on *P. sylvestris*, although there is probably a more or less continuous film of wax. *P. radiata* takes up water much more readily through its exposed needle surfaces than *P. sylvestris*. The fact that it is a three-needled pine, with needles very different in shape from those of *P. sylvestris*, suggests that it would not be able to lead intercepted water or condensed mist or dew to the protected surfaces below the sheath. *P. radiata* is possibly less efficient in conserving water during drought, but it has probably evolved in a region with an exceptionally humid climate: the Californian 'fog belt'.

There is no doubt that plant surfaces can concentrate light rainfall and convert it into significant amounts either through specialised pores or regions in their cuticle or *via* stem flow to the soil around the roots. There also seems no doubt, although the evidence is less well documented, that fog and mist can, through interception by vegetation, make a definite, if local contribution to the water economy of a plant.

Gilbert White (1776) was aware of the contribution made by fog to precipitation. He observed:

> 'In heavy fogs, *on elevated situations* especially, trees are perfect alembics and no one that has not attended to such matters can imagine how much water one tree will distill in a night's time, by condensing the vapour which trickles down the twigs and boughs, so as to make the ground below quite in a float. In Newton Lane in October 1775 a particular oak in leaf dropped so much water that the cartway stood in puddles and ruts ran with water, though the ground in general was dusty.'

White also observed that not only was the position of the tree important (see our italics above), but also that certain types of leaf were more efficient in influencing the amount of water collected. Philips (1928) claimed that mist interception by tall *Eucalyptus globulus* trees contributed 10–15 in. (25–37 cm) of water each year to the soil beneath trees. Johnston (1964) believes that *Pinus radiata* is able to maintain a satisfactory leaf water balance even under conditions of soil moisture deficiency through direct absorption of fog.

Penman (1963) has reviewed the whole problem of interception of water by vegetation. Observations cited by him confirm the work by Philips above on the ability of *Eucalyptus* trees to increase the apparent precipitation up to three times by condensation. However, Penman is very careful to emphasise that, except under exceptional conditions such as those on Table Mountain, South Africa, mist and fog precipitation is an edge effect. The front edge of a forest receiving a mist flow can demonstrate impressive gains in precipitation whereas the trees in the centre gain very little. The net gains under normal conditions are much less than is claimed.

The work of Ôura (1953) suggests that needle-leaved trees may be more effective than broad-leaved trees in capturing fog and converting it to precipitation. Whether this reflects the difference between unwettable and wettable leaf surfaces was not observed although as a general rule needle leaves are less wettable than those of broad-leaved trees. Ôura also claims that over a comparable area fog capture by a forest was six times as great as by a grass field.

Some plants in favoured conditions can obviously use mist or fog to augment their water supplies. Considerable confusion exists, however, concerning the importance of dew in the water balance of plants. An objective account of the phenomenon is given by Monteith (1963). Gindel (1965) has shown that artificial condensation surfaces which consisted of polythene sheets with a large surface area mounted at about 30° to the horizontal and with channels to lead the condensate to the plant roots were highly successful. Seedlings of *Pinus halepensis*, *P. brutia*, *Eucalyptus gomphocephala* and *Tamarix aphylla*, amongst others, were able to survive through the rainless August of Israel desert areas until the winter rains began, whereas control plants, unassisted, died. Stone (1958) concluded that under experimental conditions using an artificial dew system, seedlings of a number of conifers, e.g. *Pinus jeffreyi*, *Libocedrus decurrens* and *Abies concolor*, can benefit. He showed that they were capable of surviving for a considerable period of time after the soil moisture was reduced to wilting point. Artificial dew supplied at night extended their survival by 30 days, 20 days and 72 days respectively. In both of these experiments, however, the amount of water supplied by dew was much greater than that which could be condensed and made available to a plant under natural conditions. According to Monteith (1963), on the available evidence it is unlikely that dew is of any physiological importance either in humid climates or in the desert. A list of relevant publications is given by the Commonwealth Bureau of Soils Publication No. 691.

Fine rain, mist and dew, if they are to make any contribution to the plant's water supplies, are obviously affected by the properties of the surface on which they condense. Wettable and unwettable surfaces behave differently when acting as condensing surfaces under mist or dew conditions. Which surface is the more advantageous to the plant has not yet been demonstrated. A wettable leaf surface whose temperature is below ambient will, in a saturated atmosphere, soon be covered with a continuous film of water. An unwettable surface under the same conditions will be covered by small drops of water, many of which will run off to vegetation and the soil below or be trapped in the folds of the leaf or in axils. An engineer designing a condensing surface for maximum efficiency ensures that, so far as possible, condensation takes place in the form of drops and not as a film. To bring this about he coats the polished metal surfaces of the condenser with 'promoters' such as stearic, oleic, and linoleic acids and more recently polytetrafluorethylene and silicone resins. All these substances inhibit condensation as a film and thus do not seriously increase the resistance to heat transfer (Osment, 1963). It is obvious that many leaf surfaces approach very closely to the ideal condensing surface by providing a wax-coated surface upon which perfect droplet condensation can take place.

The control of loss of water by transpiration

Cuticular resistance to transpiration

The most obvious role of the cuticle in the shoot is the control of loss of water. From early days, attempts were made to evaluate the comparative significance of cuticular and stomatal transpiration. The prevalence of thick cuticles amongst xerophytes (Stålfelt, 1956a) might suggest that thickness makes a significant contribution to preventing transpiration, but many plants growing in desiccating conditions have comparatively thin cuticles (e.g. Bromeliaceae). On the other hand, thick cuticles occur in plants in some situations, e.g. peat bogs, where no water stress would be expected (Priestley, 1943).

Kamp (1930) made a comprehensive study of cuticle thickness in relation to loss of water. Many plant species, leaves of different ages and the effects of environmental conditions were examined. Water loss was not correlated with the thickness of the cuticle; the structure and chemical composition of the cuticle played a more important role. Old leaves transpired more freely than young despite the fact that the old leaves had thicker cuticles. Gäumann and Jaag (1935) showed that the hart's tongue fern (*Phyllitis scolopendrium*) has a cuticle of equal thickness to that of the conifer *Abies sibirica* yet it is 57 times more permeable to water. No correlation was found between the thickness of the cuticular walls and the cuticular transpiration rates in *Prunus*, *Agave* and *Sansevieria* (Sitte and Rennier, 1963) or in grapes (Radler, 1965c).

There are three main sites of evaporation within a leaf; one is the epithem surfaces below the hydathode pores, another is the mesophyll cells leading eventually to the substomatal cavity and the third is the outer epidermal cells themselves. The vapour, or sometimes liquid water, flows to the leaf surface through the hydathodes, through the stomata or through the cuticle. Since the number of hydathodes in comparison to the number of stomata is small the contribution of the hydathodes to vapour flow can probably be ignored. When stomata are open there is little resistance to vapour flow and a high proportion of the vapour lost from a plant flows through them. When they are closed no flow takes place through them and the only pathway is through the cuticle.

According to Slatyer (1967), since the cuticular and stomatal paths are in parallel, the resistance to flow they represent may be linked as follows.

$\frac{1}{r_1} = \frac{1}{r_c} + \frac{1}{r_s}$ where r_1 is the resistance of the surface layer as a whole to the diffusion of water and r_c and r_s are the resistances in the cuticular and stomatal paths respectively (in sec cm^{-1}). The resistance r_s itself comprises the following components: $r_s = (r_w + r_i + r_p)$ where r_w, r_j and r_p are the resistances in the mesophyll cell walls, intercellular spaces and stomatal pores respectively. The contribution of the stomata to transpiration lies outside the scope of this book, but a discussion will be found in Stålfelt (1956b), Slatyer (1967) and Gates (1968). Stålfelt (1956a) has also reviewed the earlier literature on the amount of water lost by cuticular transpiration. Gates (1968) gives tables showing the resistance to the transfer of water by diffusion through the cuticle alone, through the stomata and the total internal resistance to diffusion in a wide range of species.

The resistance of the cuticular path can be broken down into its constituents in similar fashion, e.g. the resistance of the epidermal cell walls, the cutinised layer, the cuticularised layer and the epicuticular wax layer $(r_w + r_{cn} + r_{cl} + r_{wa})$. The removal of any one of these alters the permeability of the cuticular path as a whole. The great difficulty in obtaining individual measurements for any one of these individual units of resistance results from the impossibility as yet of deciding where the liquid/air interface is. Some of these resistances will affect the liquid-phase transport for which the driving forces are the gradients of osmotic and hydrostatic pressure and some will affect the vapour phase transport for which the driving force is the gradient of water vapour concentration. All three components determine the water potential of the system.

This relatively simple mathematical treatment is complicated too by the fact that resistance to cuticular transpiration $\left(\frac{1}{r_c}\right)$ is not uniform. Cuticular transpiration comprises the loss of water through the cuticle of epidermal cells and peristomatal transpiration which means the loss of water through guard cells and their accessory cells. The cuticles and the epicuticular wax of guard cells are frequently different from those of the normal epidermal cells (see Fig. 2.7). Enhanced peristomatal transpiration has been demonstrated by Franke (1967b) and is further discussed by Stålfelt (1967). Franke has shown that the sites of highest concentration of ectodesmata, e.g. the guard cells in the leaves of *Zantedeschia aethiopica* (see Fig. 4.10), are also the sites of the greatest loss of water vapour. Franke suggests that the ectodesmata act as channels for the exit of water from the guard cells and their subsidiary cells and points out that this hypothesis is consistent with earlier reports on the connection between cuticular excretion of substances and the distribution of ectodesmata. Arens (1968) has also suggested that ectodesmata may be pathways of peristomatal transpiration. It should not be forgotten that ectodesmata do not pass through the cuticle, but only through the epidermal walls (Fig. 4.6). The final passage of water, if it is transmitted through the ectodesmata, must be through an intact cuticle.

The role of the cuticular wax in the conservation of water

Many workers have shown the importance of the wax layer in conserving water. Pieniazek (1944) studied the physical characters of the skin of apples in relation to transpirational loss (see Fig. 4.25). No correlation was found between the thickness of the cuticle and transpiration rate, but surface russeting greatly increased water loss. Wiping the fruit increased the rate of transpiration which was not reduced by the new layer of wax which formed on the surface. Hall (1966) found that the ease of removal of wax from apple fruits differs according to the variety, but removal by any method causes an increase in transpiration rate; abrasion by the wrapping paper in contact with the surface of the *Granny Smith* fruit is sufficient to impair the protective action of the wax layer. Exposure to petroleum ether vapour disorganised the wax structures on the fruits and leaves of the grape (*Vitis vinifera*) and markedly increased transpiration (Possingham, Chambers, Radler and Grncarevic, 1967).

Schieferstein and Loomis (1956, 1959) question whether there is a correlation between xeromorphic adaptation and the amount of surface wax and suggest that the subcuticular wax is the more important factor enabling plants to withstand dry habitats. The relationship between the ultrastructure of the epidermis and wax coverings and the conservation of water in species of S. American palms was studied by Machado (1958). When the wax deposits were scattered they were of little significance; when continuous, probably effective. Bolliger (1959) suggests that cuticular transpiration is greater than is usually thought, and raises doubt as to the efficiency of the cuticles of many plants as a protection against transpiration.

Pfeiffer, Dewey and Brunskill (1957) found that treatment of the soil with trichloroacetic acid suppressed the formation of wax on pea plants and that, as a consequence, transpiration from the leaves was greatly increased (see p. 245). Hall and Jones (1961) obtained a similar effect, not by preventing the wax layer from forming, but by damaging the leaf surface. They noticed that plants of white clover *Trifolium repens* (see Fig. 2.6) in the field, by contact with each other and with the ground, lost much of their epicuticular wax over the crowns of the epidermal cells. They simulated these conditions by brushing the surface of the leaf with a camel-hair brush. They then matched pairs of leaves, brushed and unbrushed, and plotted their loss of weight under the same conditions. Figure 7.1 shows a semi-logarithmic plot of the weight of

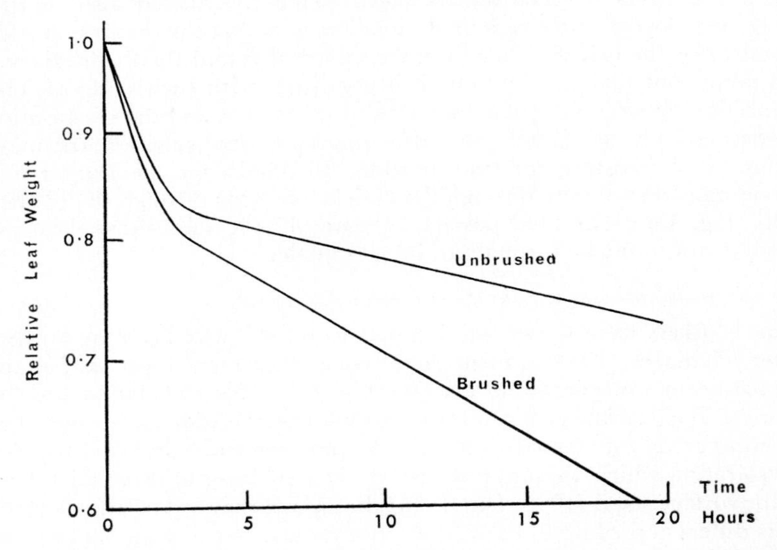

FIG. 7.1 Graph of the effect of damage on Moroccan red clover (*Trifolium pratense*). Semi-log plot. (Redrawn from Hall and Jones, 1961)

the leaves against time. By extending the weighings over a number of hours they found that the increased transpiration of the brushed leaves continued when the stomata were closed, indicating that enhanced cuticular transpiration was mainly responsible for the difference. Horrocks (1964) has shown the importance of the wax layer in inhibiting transpiration. He isolated areas of the cuticle from two varieties of apple differing greatly in their ability to withstand drying in storage and tested their cuticular permeabilities both intact and with the wax removed. *Granny Smith* showed a difference of 1 : 70 in the permeability of its cuticle in the complete and extracted form and *Golden Delicious* a difference of 1 : 30. The reason for the difference between the two varieties is not understood.

It is difficult to analyse these experiments in terms of the resistance of the constituent parts of the cuticle. The experiments of Pfeiffer and his colleagues eliminate the resistance due to the epicuticular wax (r_{wa}), but they have almost certainly modified the resistance due to the cuticularised layer (r_{cl}) and the cutinised layer (r_{cn}) as well. Hall and Jones' experiments have probably not modified r_{cl} or r_{cn}, but they will have partly removed and partly modified the epicuticular wax, thus reducing or eliminating r_{wa}, but they may also have converted the epicuticular wax layer from a zone through which water passes slowly as a vapour to a zone through which it can pass more quickly as a liquid. Horrocks' experiments will also have modified both r_{wa} and r_{cl}.

The problem of distinguishing between cuticular and stomatal transpiration was overcome by Martin and Stott (1957) who studied water loss in harvested grapes (*Vitis vinifera* var. *sultana*). Grape cuticles have no stomata; they must transpire either through the cuticle or the pedicel. The pedicel is thin and dry and can be ignored as a significant area of water loss. The loss of water in grape-drying to produce sultanas takes place by transfer through the parenchyma to the inside of the cuticle, diffusion through the cuticle, and evaporation from the outside surface or diffusion through the stationary air layer at the surface. Martin and Stott were able to show that the only limiting factor in the drying of a grape is the diffusion of water through its cuticle and wax layer. The transfer of heat determines the temperature of the grape and so the rate of drying varies with the size of the berry. The drying rate is inversely proportional to the amount of cuticular wax present as determined by extraction with chloroform. Any surface treatments to enhance drying are only successful if they modify the properties of the cuticle in some way. Another interesting phenomenon noticed by Martin and Stott is that the rate of drying depends very much on the conformation of the cuticle. With a water loss of up to 20% the grape retains its shape. As drying proceeds the cuticle thickens and wrinkles, thus reducing the water loss due to elastic deformation of the cuticle. This elastic deformation of the cuticle under water stress and its concomitant thickening is obviously an important feature of cuticular behaviour.

Grncarevic and Radler (1967) studied the action of the components of the wax of the grape cuticle by models to determine which of them is the active principle in inhibiting water loss. The grape cuticle wax comprises 30% of petroleum ether-soluble, low melting 'soft wax' consisting of long-chain alcohols, aldehydes, esters, free acids and hydrocarbons and about 70% of a 'hard wax' comprising mainly the triterpene oleanolic acid. The models were made by covering triacetate sheet, chosen because of its similarity in permeability to the isolated cuticle, with isolated wax fractions to a thickness similar to that found on the plant surface. They found that the wax components causing the greatest reduction of transpiration through these artificial membranes were the hydrocarbon, alcohol and aldehyde fractions. Their effect was comparable to that of the complete wax or to paraffin wax. The principal constituent of the natural wax, the oleanolic acid, did not affect evaporation in the artificial system, nor did free docosanoic acid (C_{22}). However, a mixture of free fatty acids, of which the main constituents were C_{24} and C_{26} from grape wax, did reduce evaporation slightly.

The dried grape industry is of considerable economic importance and artificial methods are used to speed up drying. The practice of grape-drying dates back to the ancient Greeks, who used olive oil and potash as drying accelerators. Preparations consisting of emulsions of free fatty acids, their ethyl esters and sulphonated compounds in alkaline solution are now used as dips to assist drying in the production of light-coloured sultanas (Radler, 1965c). In the production of prunes, the fruits are sprayed with a hot, dilute alkaline solution before they are dried. The treatment removes part of the surface wax, the structure of the residue is disorganised and a saving of drying time is achieved (Bain and McBean, 1967, 1969).

Hallam (1967) has shown that waxes isolated from leaf surfaces and allowed slowly to recrystallise again begin to take up the complex crystalline forms they held on the original surface. The restoration is, however, only a partial one. The fine structure of the wax on a grape cuticle is very complex (Chambers and Possingham, 1963) and it seems unlikely that the reconstituted model membranes created by Grncarevic and Radler began to approach in complexity the original surface. The waxy bloom of a grape comprises a series of wax platelets overlapping one another; they have very lobed and undulate margins and the plates are about 0·1 μm thick. Some of the platelet margins curve up so that they lie perpendicular to the plane of the grape cuticle. Chambers and Possingham were able to show that, although commercial dipping preparations accelerated drying while they were present, they could be washed off and apparently leave the surface undamaged. Moreover, the rate of drying also returned to normal, i.e. r_{wa} could be reversibly altered. Nor does dipping remove significant amounts of surface wax. In normal grape-drying, water diffuses in the liquid phase through the parenchyma, pectin and cuticle layers until it reaches the wax-platelet region. From here, they suggest, it moves as vapour around and between the platelets. Water droplets or films cannot form in these ultramicroscopic spaces because of the hydrophobic nature of the wax surfaces. The pathway for the water vapour must be many times longer than the thickness of the wax layer and the rate of water movement by diffusion therefore low. In a dipped grape the emulsion forms a continuous aqueous film over the surfaces of the wax plates. During drying, water leaves the cuticle in a liquid phase and enters the emulsion-filled capillaries of the wax bloom. It remains in the liquid phase as it moves through the capillaries pulled by the forces of capillarity. This continuous liquid capillarity continues as long as the wax platelets remain hydrophilic. This hypothesis suggests that the chemistry of the different wax layers is not the most significant factor in the rate of water loss, provided that they are basically hydrophobic, but rather their shape. The more complex their shape, and the longer and more tortuous the vapour path that they force the internal water to take, the more effective will be their waterproofing capacity.

If Chambers and Possingham are correct, many of the experiments modifying the wax bloom by damage or variation in the environment achieve their greatest effect by destroying the crystalline structure and not by reducing the total amount of wax on the surface.

Confirmation of the importance of the wax bloom in water relations is shown by the experiments already described on p. 196 of Leyton and Juniper (1963). The needles of *Pinus sylvestris* lose water very slowly. Oddly enough they also appear to take up water very readily either from rain, dew or mist or by immersion. However, very little is taken up through the bloomed area of the needle and virtually all of it is taken in through the unbloomed portion of the needle which lies below the sheath. The complex crystal structure on the surface of a pine needle (see Fig. 4.27) is obviously just as effective at keeping water out as keeping water in. Great care must be taken in interpreting results from permeability experiments on isolated leaves since, as Skoss (1955) showed, permeability depends very much from which part of the plant the leaf is harvested. Skoss demonstrated that wide variations in the amount of cutin, the amount of wax and the ratio of wax to cutin can be brought about in leaves from the same plant grown under different conditions. He has also

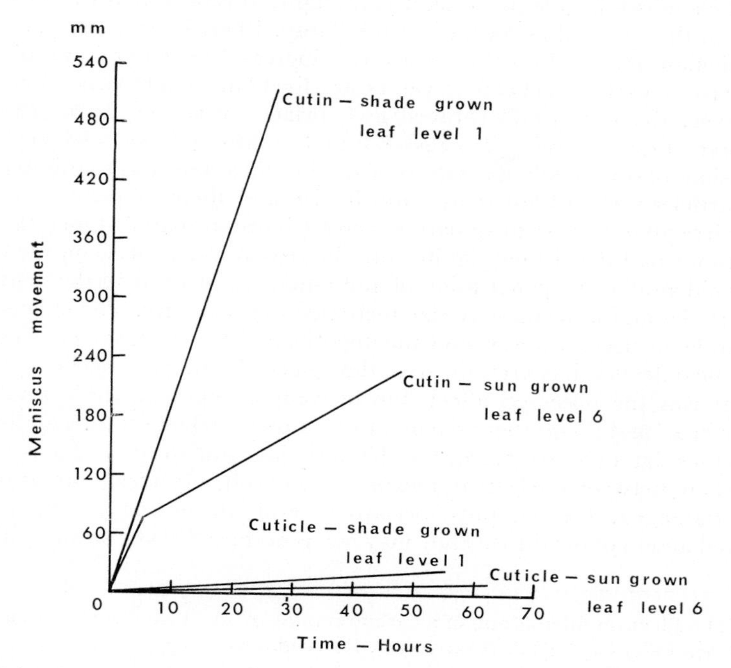

FIG. 7.2 Graph showing the permeability of isolated cuticles and cutin from different parts of a plant of *Hedera helix*. Leaves at level 1 were the youngest on the plant. Leaves at level 6 were the oldest on the plant. (Redrawn from Skoss, 1955)

shown that leaves from different positions on a single plant have very different permeabilities to water and solutions. Discs were punched from sun and shade ivy leaves, *Hedera helix*, the cuticles isolated and their permeabilities to water tested both before and after extraction of wax by refluxing in absolute alcohol for 1 hr. The results are given in Fig. 7.2. It appears, as before, that the principal resistance to the passage of water is provided by the wax within or on the cuticle. *Hedera* does not possess a visible bloom, there is no crystalline wax on the surface and that which is present is laid down in the form of laminae on or within the cuticle (Fig. 7.3). It might be expected that a leaf surface with a bloom would show an even greater permeability to water after extraction of the wax than that shown by *Hedera*. Skoss has also shown that certain plants, e.g. *Nicotiana glauca*, under continuous water stress develop up to twice the amount of cuticle as plants under an optimal water regime. The differential permeability of such cuticles does not appear to have been tested.

Sivadjian (1956) showed that when the leaves of *Phaseolus multiflorus* are treated with 1% glycerol solution, the cuticular permeability and consequently transpiration are greatly increased. The effect of the glycerol is probably comparable to that of the fatty material used to promote the drying of grapes; a hydrophilic pathway is established across the cuticle, thus facilitating the flow of water. Glycerol and other 'humectants' assist the penetration of chemicals into leaves due, it is thought, to retardation of the rate of drying of the spray films on the surface (Gray, 1956; Evans, 1968).

Particulate material on the surface can increase the rate of transpiration from leaves and the effect can, as yet, be ascribed only to a modification of the wax layer. Eveling (1967) sprayed leaf surfaces of runner bean (*Phaseolus multiflorus*), *Coleus blumei*, *Zebrina pendula* and potato (*Solanum tuberosum*) with suspensions of supposedly inert dusts of kaolin, silica, talc and titanium oxide. The particles were all less than 5 μm in size and their shapes varied from highly irregular (silica) to smooth spheres (titanium oxide). Once the dust suspensions had dried they significantly increased water loss from the leaves and, incidentally, the penetration of ammonia. An increase in the density of the particles and a decrease in size increased permeability. The effects were noticeable for up to 3 weeks after the deposits had dried; they were so severe on the bean leaves that necrotic areas developed. In only one species (*Zebrina pendula*) was the observed effect due to wedging open of the stomata; in runner bean and *Coleus* there seemed to be a direct effect on the cuticle with no obvious sign of damage. Most of the particles came to rest in the cuticular depressions between the cells and no one type of cell, e.g. a leaf hair, appeared to be damaged. Eveling puts forward several suggestions to explain the observed action of the dusts. They may act as minute 'wicks'; actually abrade

---

FIG. 7.3 Electron micrograph of a carbon replica of the adaxial surface of a leaf of ivy (*Hedera helix*) Cr/Au/Pd. (Courtesy P. R. Williams)

FIG. 7.4 Electron micrograph of a carbon replica of the adaxial leaf surface of pea (*Pisum sativum* var. *Alaska*) grown at 1500 ft candles (140 lux) light intensity in a wind tunnel at 25 m.p.h. (40 k.p.h.) constant speed. Cr/Au/Pd

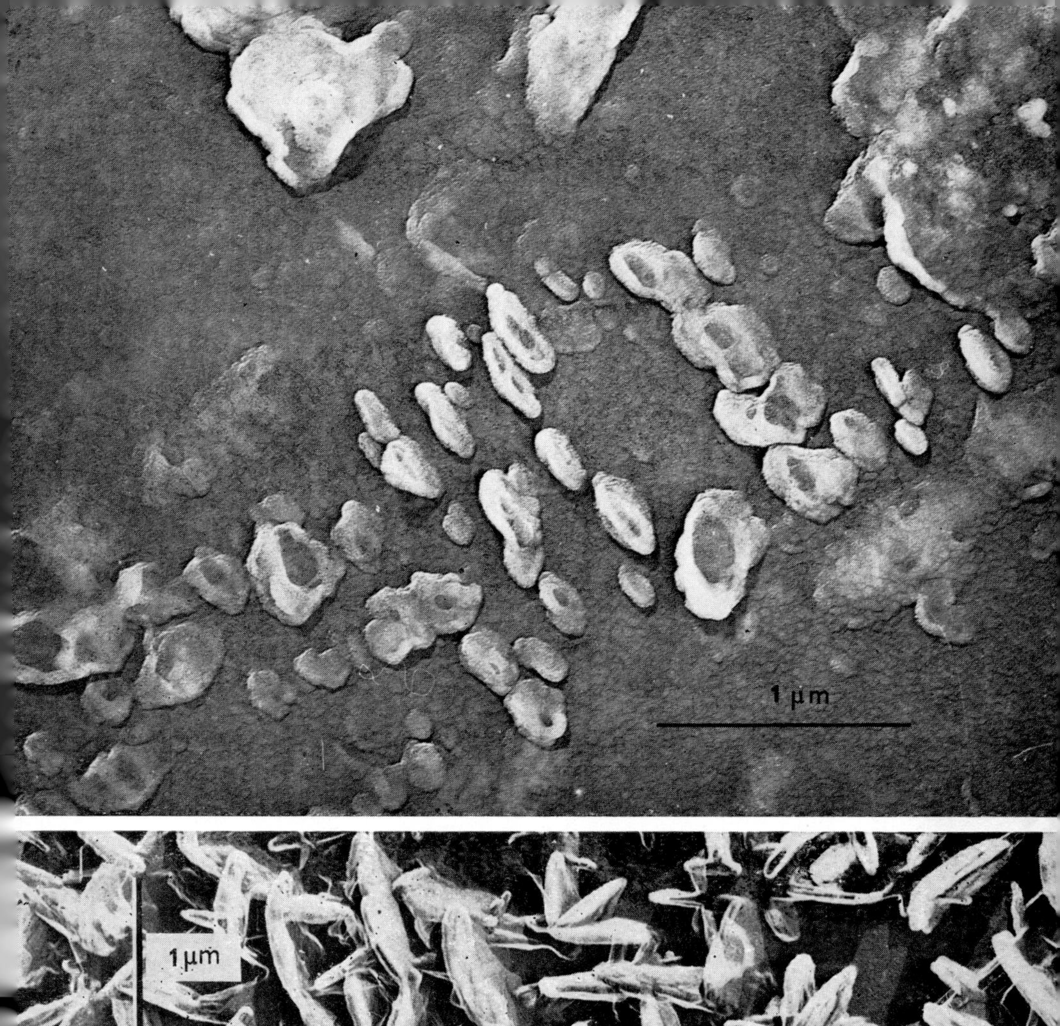
1 µm

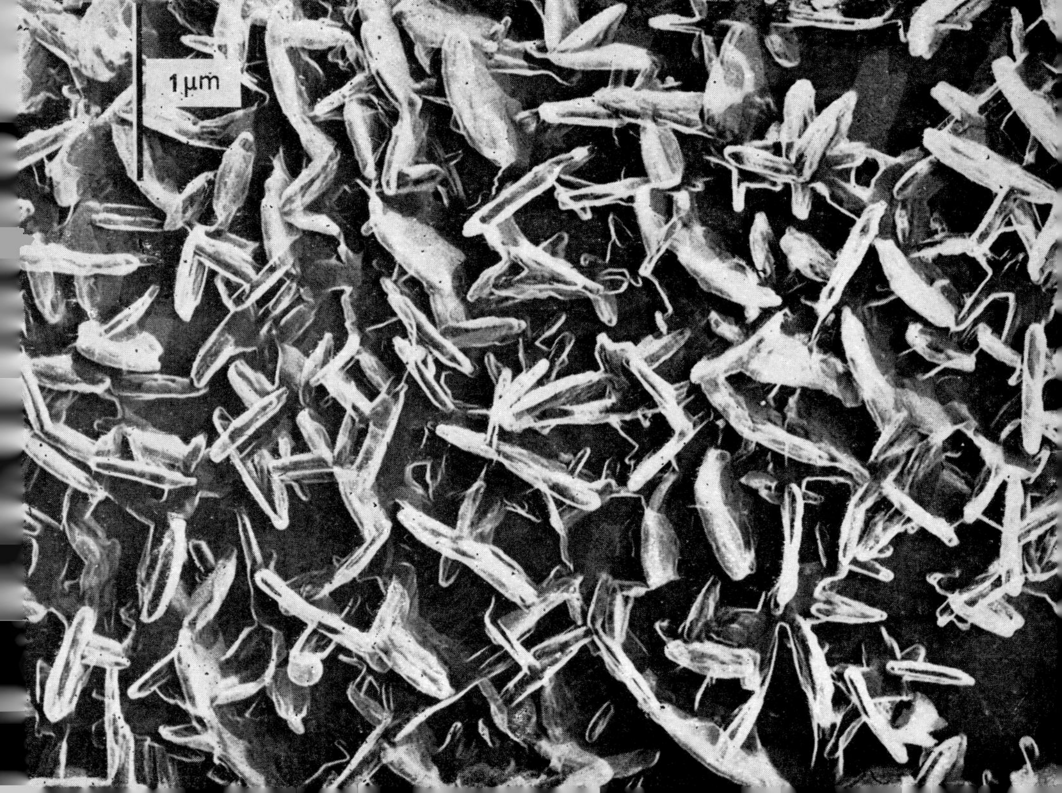
1 µm

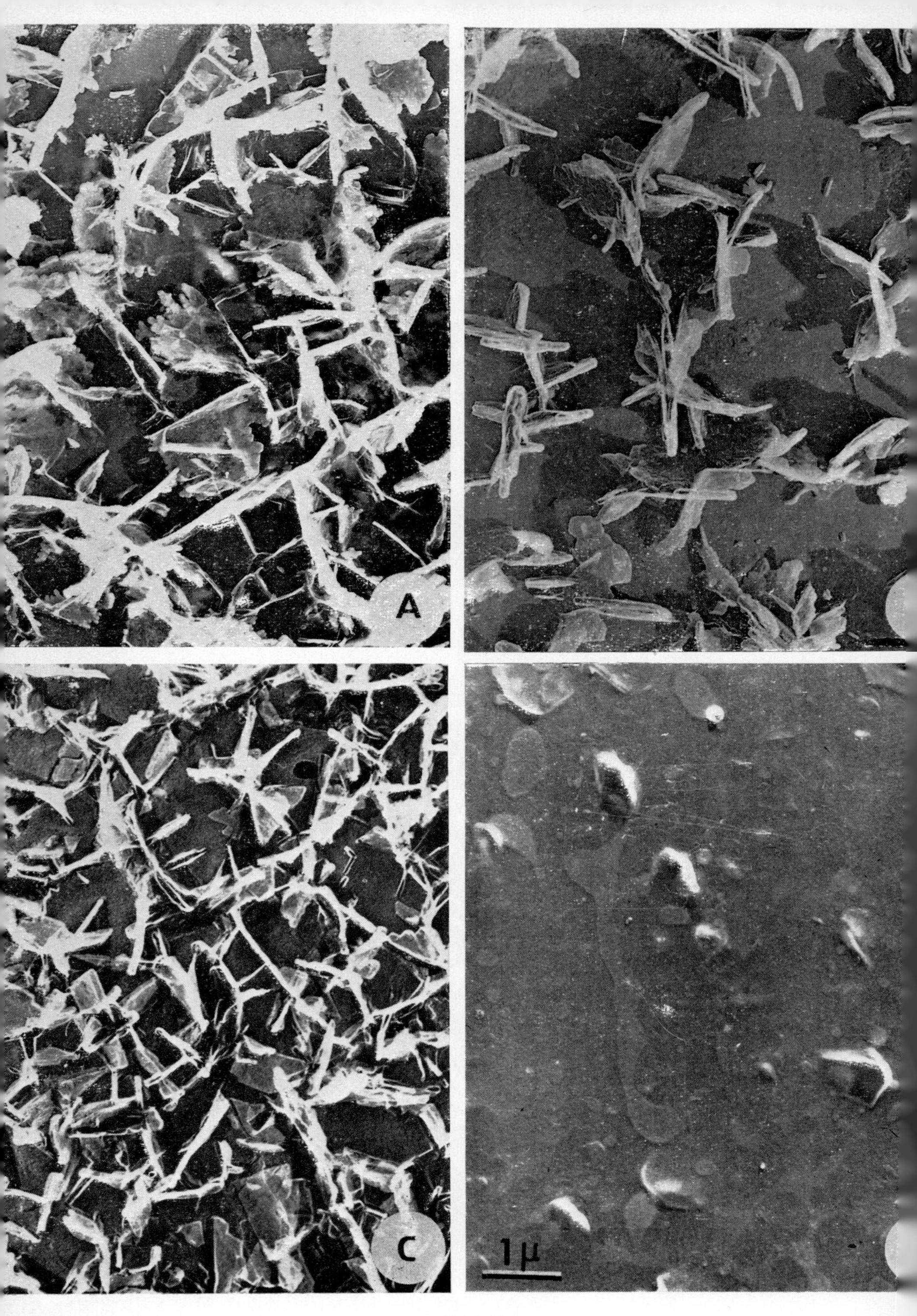

FIG. 7.5

the surface; alter its properties by surface tension effects; absorb certain components of the cuticle into themselves; or alter its properties by the reactive nature of the surfaces of freshly ground dusts. Eveling believes that no one explanation fits all the evidence. Similar results on the effect of dusts increasing the permeability of insect cuticle reported by Ebeling (1961) and Ebeling and Wagner (1961) are generally thought to be due to the adsorptive capacity of the dusts for certain waxy components of the insect cuticle.

Very little work has been done on the effect of windspeed on the development of the cuticle, but it is known that a high wind speed results in the development of a denser epicuticular wax layer than that of plants grown in still conditions (compare Figs. 4.21 and 7.4). This increased density of crystalline projections probably retards water loss by lengthening the vapour path to the surface.

The artificial suppression of transpiration

Some attempts have been made to augment existing cuticles and cut down transpiration by applying a wide range of transpiration suppressants such as latex and plastic compounds (Pallas, Bertrand, Harris, Elkins and Parks, 1962). Another approach has been to use salts such as phenyl mercuric acetate to bring about a premature closing of the stomata. However both of these techniques may cause leaf scorch under high light conditions and both of them reduce growth rate markedly. The work has been reviewed recently by Gale and Hagan (1966). The artificial closing of stomata is also discussed by Meidner and Mansfield (1968).

Transpiration through trichomes and emergences

Any trichome or emergence or any modification of the epidermis theoretically increases the surface area of the epidermis available for transpiration, in the same way that root hairs increase enormously the absorptive area of the root. The trichomes frequently modify the surface characteristics and alter the aerodynamic properties of the leaf and thus have an important indirect role in the water economy of the plant. Little transpiration appears to occur through them although, as Slatyer (1967) points out, this supposition needs quantitative confirmation.

Guttation

Guttation is the transfer of liquid water and dissolved solutes to the surface of the leaf and involves, according to Schnepf (1965), the temporary rupture of the cuticle. The tissues in which this phenomenon occurs have already been described in Chapter 4. Guttation occurs when conditions for absorption of water by roots are favourable and those for transpiration are unfavourable. It

---

FIG. 7.5 Electron micrographs of carbon replicas of the adaxial surfaces of the leaves of peas (*Pisum sativum* var. *Alaska*) grown under different light intensities. ***A***, grown at 5000 ft candles (406 lux). ***B***, grown at 1500 ft candles (140 lux). ***C***, grown at 900 ft candles (84 lux). ***D***, grown in darkness. Cr/Au/Pd

is thought that positive xylem pressure develops forcing water through the hydathodes. The most favourable conditions are cool mornings following a warm day. Most plants guttate, but only in some climates are conditions frequently favourable. Not only water may be exuded onto the leaf surface. The guttation fluid analysed from the surfaces of the leaves of tamarisk (*Tamarix*) contained calcium carbonate, magnesium sulphate, sodium chloride, sodium nitrate and sodium carbonate and a frequent component along with the salts are sugars. Much of the loss of substances from leaves is due to guttation and not a result of salt secretion through the intact cuticle. Sometimes this loss of material by guttation can be so severe that leaves can become coated with a white crust and the properties of the surface altered. A single large shrub of the crape myrtle (*Lagerstroemia* sp.) studied by Ivanoff (1963) lost more than 152 g of total solids, apart from that which dropped to the ground before being collected. Apparently as a result of the loss of salt the whole shrub defoliated.

Häusermann and Frey-Wyssling (1963) have shown that the hydathodes, like the nectaries, of a number of species of plant are characterised by a high phosphatase activity. This suggests that the secretion of water by hydathodes, like the secretion of sugars from nectaries, is an active process and not a passive process dependent entirely upon environmental conditions.

Schnepf (1965) has studied the structures associated with guttation in the chick-pea (*Cicer arietinum*) under the electron microscope. His findings support the idea of an active process. The young leaves of *Cicer* are covered with trichome hydathodes. These specialised hairs secrete water with some organic acids. In the gland cells the plasmalemma is separated from the wall and within the space so formed is a finely granular material; a similar material appears to fill vesicles which line the walls. These vesicles are also surrounded by a unit membrane like the plasmalemma and are thought to discharge their contents into the space between the plasmalemma and the wall. How they are formed is not known. Eventually the cuticle and the cutinised part of the wall separate from the cellulose layer by the intrusion of the secreted material. This sub-cuticular space is enlarged by the secretion. When the pressure reaches a certain level one or more pores open like a valve in the cutinised layer and a droplet is exuded. It then passes through a further pore in the cuticle proper to emerge finally on the surface. In thin section under the electron microscope these pores can be seen as apertures through both the cutinised and cuticularised layers (Fig. 7.6). The amount of liquid they can lose has been measured by Schnepf and one gland can exude 1000 μm^3 /min. Hafiz (1952) has suggested that some of the organic acids secreted along with the water from the trichomes of *Cicer* may confer disease resistance on certain varieties.

FIG. 7.6 Electron micrograph of a section through the cuticle of a hydathode pore of *Cicer arietinum*. Osmium fixation. (Courtesy E. Schnepf)

FIG. 7.7 Drawing of a nectary hair from a petal of *Abutilon striatum*. (Redrawn from Mercer and Rathgeber, 1962)

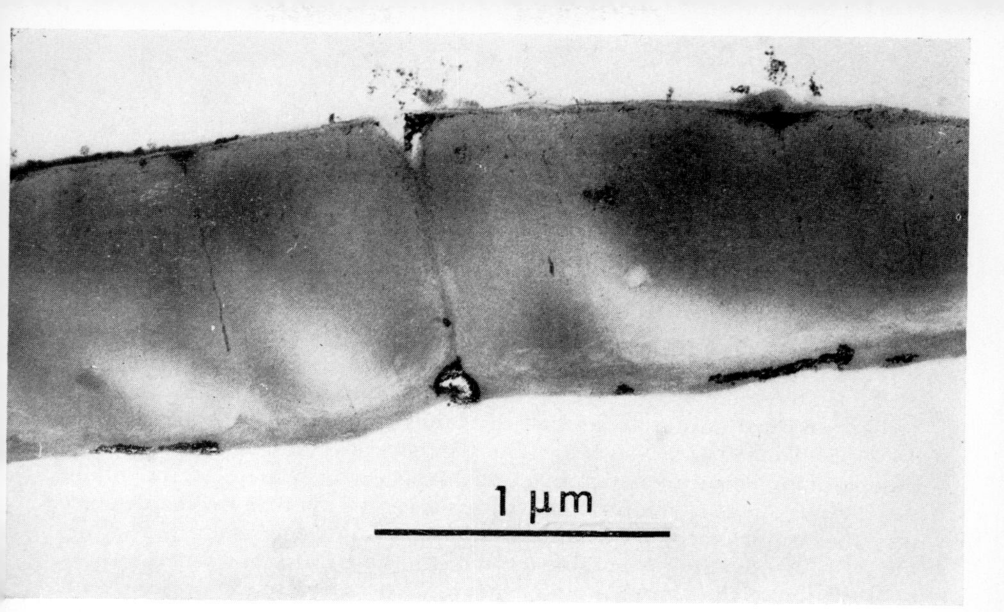

FIG. 7.6

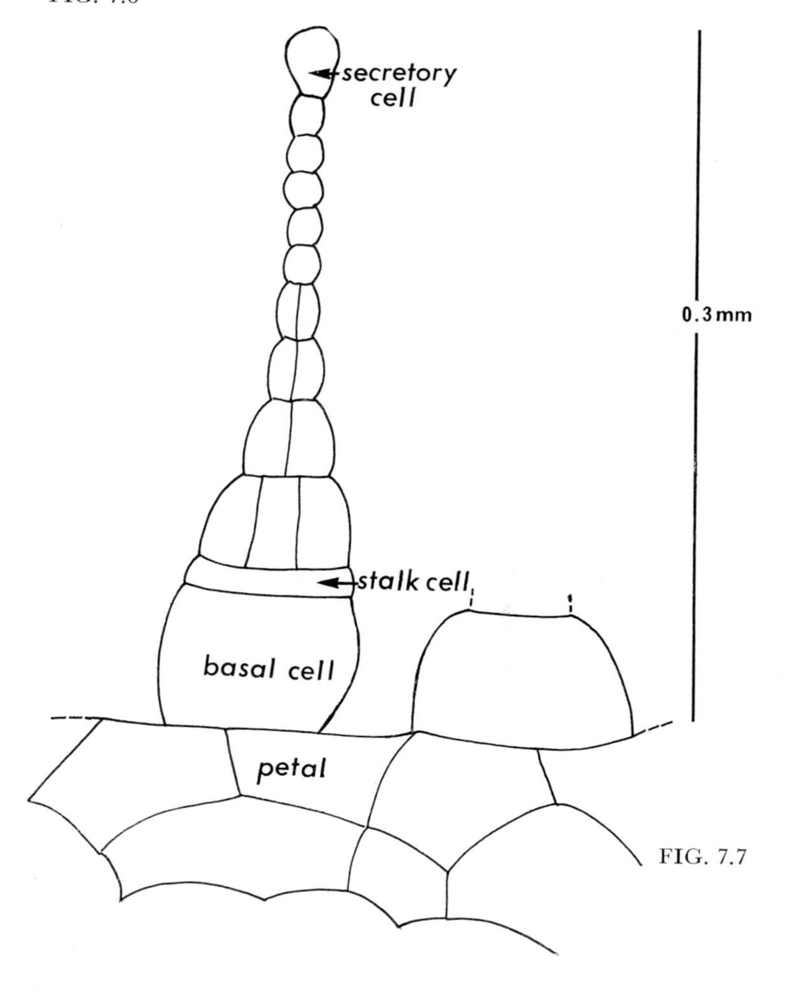

FIG. 7.7

Ivanoff (1963) discusses the organs and channels of guttation, the chemical composition of guttation drops and deposits, the losses of nutrients by guttation, the injuries caused by the action of concentrated guttation solutions and the guttation fluid as a medium for the growth of micro-organisms.

Other substances, such as gases and volatile organic compounds, e.g. ethylene in apple fruits, are lost through the cuticle, but very little seems to be known about either the way in which they emerge or what factors affect their loss (see p. 215). Sometimes volatile compounds are distributed and disseminated from the epidermal cells of all the perianth parts; in some plants fragrant oils are secreted only from specialised glands known as 'osmophors' (Esau, 1965). Muller (1966) has shown that *Salvia leucophylla*, *Artemisia californica* and other aromatic shrubs contain phytotoxic terpenes which volatilise and inhibit the establishment of seedlings of a wide variety of plants for some distance from the shrubs. Cineole and camphor are the most toxic terpenes. Muller uses the term 'allelopathy' to describe the unfavourable effect of one plant upon another due to the release of an effective phytotoxin into the environment, and suggests that the lipids in cuticles aid the liberation of the lipid-soluble terpenes from one plant and their absorption by another. The discharge of viscous materials which takes place from stigmas, nectaries and insect trapping-surfaces is better documented, probably because the substances are less volatile.

The loss of nutrients from leaves

The loss of substances from leaves whether by guttation or leaching by rain, dew or mist has not received the general attention it deserves. Mann and Wallace (1925) and Wallace (1930) determined losses of potassium from apple leaves, and associated the spotting of leaves of the *Cox's Orange Pippin* variety with the loss of potassium by leaching. Differences in the susceptibility of leaves of different varieties to the spotting injury was attributed to differences in the control of loss of water-soluble substances by the leaf surface layers. Arens (1934) examined the loss of salts from many species of plant and believed that the loss was due not to a passive leaching effect, but to active secretory processes. The losses of salts may be surprisingly large; Dalbro (1955) reported that orchard apple leaves may lose 25–30 kg of potassium, 9 kg of sodium and 10·5 kg of calcium per hectare per annum. Mecklenburg and Tukey (1963) found that metabolites including carbohydrates, organic acids and amino acids are leached from the leaves of plants by rain and mist; leaching was found in every plant investigated, totalling more than 100 species. The carbohydrates lost include sugars, polysaccharides and sugar alcohols. Tukey and Morgan (1964) showed that leaching is widespread in nature and that large proportions of the nutrients in mature foliage may be lost. Differences occur in the amount of leaching between species and also between cultivars of one species; the differences are apparently related to internal conditions of the plant and to the physical properties of the leaf. Leaves with a waxy surface difficult to wet are usually least subject to leaching, but nutrients are lost even from several species of tropical plants with thick cuticles and waxy surfaces. Tukey, Tukey and Wittwer (1958) showed that greater quantities of nutrient are lost from older leaves than from younger and when the nutrient status of plants is high. Tukey and Mecklenburg (1964) found that radioactive leachates from foliage were reabsorbed by roots and translocated to above-ground parts. They point out that leaching is important in competition between plants and in the development of plant associations and that the recycling of nutrients has important implications in plant nutrition. Tukey and Tukey (1959) suggest that overhead irrigation used for crop production and protection from frost is a potential cause of loss of nutrients which needs investigation.

Hedberg (1964) reports that, like the teazle, an afro-alpine plant *Lobelia keniensis* holds water in its imbricate leaf bases. This water does not primarily come from precipitation since it persists even in prolonged dry weather. This water also contains a considerable quantity of substances apparently, like the water, exuded from the leaf surfaces. This water-soluble material has not yet been identified, but its effect is to prevent all but the surface layer of the liquid from freezing even in the extremely low temperatures endured near the top of Mount Kenya. One effect of this permanently unfrozen water reservoir is to protect the delicate young central part of the rosette from the intense frost; another is to provide a comfortable liquid habitat for Chironomid larvae, probably of the genus *Metriocnemus*. Börner (1960) emphasises the need for more research on organic substances washed from leaves.

The secretion of substances through nectaries and other glands

Nectaries

Nectaries are found on floral parts or occasionally on the vegetative parts of the plant. Floral nectaries can be found, in different species, distributed from the outer perianth to the top of the ovary. Extrafloral nectaries occur on stems, leaves, pedicels and stipules. Their selective value in these positions is not always clear. The nectaries themselves are very variable in form; sometimes the secretory tissue is restricted to the epidermis; sometimes they are elongated like palisade cells or as multicellular hairs. Whatever their general morphology they usually comprise thin-walled cells with a dense cytoplasm and they are covered by a cuticle. Although they vary from species to species, most nectaries secrete a mixture of glucose, fructose and sucrose, in part as sugar phosphates, together with amino acids, organic acids and inorganic salts such as phosphates.

The physiology of secretion, the manner of accumulation of the secreted substances and the nature of the substances in a wide range of species have been examined in detail by Lüttge (1961, 1962a, b) and Frey-Wyssling and Häusermann (1960). Nectaries, like hydathodes, also have a high level of phosphatase activity. The fine structure of the secretory cells has been studied by Schnepf (1964b, 1966a, b) and Mercer and Rathgeber (1962). In spite of considerable variations in their gross morphology, the constituent cells all possess a dense cytoplasm with few vacuoles, large numbers of mitochondria, ribosomes and Golgi bodies, abundant endoplasmic reticulum, large numbers of vesicles possibly derived from the Golgi bodies and frequently an extensively lobed plasmalemma. Perhaps due to the amount of secretion accumulating between the wall and the plasmalemma the wall is sometimes, as in *Gasteria trigona*, extensively lobed into the cytoplasm (Schnepf, 1966a). The prenectar, as it is called by Mercer and Rathgeber (1962), is discharged through the plasmalemma and the cell wall either into the large intercellular spaces which occur between the cells of many nectaries or between the cell wall and the overlying cuticle. In *Abutilon* spp. the nectar is secreted from the tips of multicellular hairs (see Fig. 7.7) (Mercer and Rathgeber, 1962). Prenectar accumulates in the space between the cell wall of the terminal cell and the cuticle. This accumulation causes the overlying cuticle to stretch. Every few seconds or minutes nectar is ejected violently through the cuticle to the outside. The stretched cuticle then snaps back again. An active hair secretes continuously for an hour or more and discharges a volume equivalent to that of the terminal cell every 3–5 min. The cuticle is continuous over the whole hair except at the tip where it is penetrated by small holes. According to Mercer and Rathgeber, when the hydrostatic pressure within the apical cell reaches a certain level, the holes in the cuticle open like a valve and permit the ejection of the nectar. These pores seem to be similar to the pores in the cuticle of the trichome hydathodes seen by Schnepf (1966a) (see Fig. 7.6).

Glandular hairs

Thus most of the secretion of sugars and other substances through the cuticle takes place through prepared pores and does not result in any permanent rupture of the cuticle. However, in the glandular hairs of yarrow (*Achillea millefolium*) Stahl (1953) found that the cuticle bursts open and the whole hair degenerates after a single act of secretion. It is not surprising to discover that glands of various kinds, nectaries and hydathodes, all of which require the temporary or permanent penetration of the cuticle, are preferred ports of entry for certain pathogens, e.g. bacteria (Hildebrand and MacDaniels, 1935).

Systemic insecticides used on plants may contaminate nectar and be carried over into honey. Glynne-Jones and Thomas (1953) found that schradan (octamethylpyrophosphoramide) sprayed on mustard (*Sinapis*) and borage (*Borago officinalis*) plants was translocated into the nectar and concluded that it may appear in an unchanged form in honey obtained from the nectar of plants sprayed less than 4 weeks previously. Lord, May and Stevenson (1968) applied other organo-phosphorus compounds, dimethoate (dimethyl *S*-(*N*-methylcarbamoylmethyl) phosphorothiolothionate) and phorate (diethyl *S*-(ethylthiomethyl)phosphorothiolothionate), to the roots of fuchsia (*Fuchsia*) and nasturtium (*Tropaeolum majus*) plants, chosen because of their copious flow of nectar. The secretion of dimethoate in the nectar was much greater than that of phorate and contamination of the nectaries was at a maximum 4 days after the treatment of the plants.

The secretion of salt from salt glands

A number of species of plants have specialised salt glands on their surfaces from which salt can be discharged. Sometimes salt-laden trichomes are shed from the leaf in a process of auto-amputation.

The more advanced type of salt-secreting gland is found in such species as *Limonium* and *Tamarix*. Thomson and Liu (1967) suggest that the salt accumulates in the very prominent small vacuoles of the secretory cells of *Tamarix*. The vacuoles are then assumed to drift to the periphery of the cells, fuse with the plasmalemma and release the salt into the zone between the plasmalemma and the wall. The salt is then believed to be forced along the wall and out through the cuticular pores at the top of the gland.

Salt secretion is presumably advantageous to plants living in soils with a high salt content. It also appears to have fringe benefits to the forester. According to Waisel and Friedman (1965) the salt-coated leaves of *Tamarix* species are almost nonflammable. Even when set on fire deliberately the leaf litter under *Tamarix* groves extinguishes quickly. The authors therefore suggest that belts of *Tamarix* species should be planted in areas where the fire risk, to susceptible and more valuable trees, is high and where the salt content of the soil makes the *Tamarix* litter safe.

The trapping of pollen grains by secretions

Some sticky exudates are produced, over short periods of time, to entrap pollen grains. A secretion of this type was studied by Konar and Linskens (1966). This secretion brings about a loosening of the cuticle above the epidermal cells of the stigmas of *Petunia hybrida*. The secretion containing fatty acids, amino acids and sugars passes through the epidermal cells and the stigmatic papillae and accumulates between the wall and the cuticle (see Fig. 8.5). The cuticle breaks up into flakes. Some of the epidermal cells become loose at the same time. There is no evidence of action of a 'cutinase'; the cuticle over the stigma is not very thick and probably fragments solely under the pressure exerted from below. The principal function of this secretion, apart from loosening the cuticle and thus easing the passage of the pollen tube, seems to be to act as a 'liquid cuticle' to stop rapid desiccation at the surface and as a liquid trap to retain the pollen grains.

Carbon dioxide exchange through the cuticle

The conventional view of the function of the cuticle is that it limits the loss of water and also the exchange of gas. Had plants developed a cuticle which was impermeable to water but permeable to gases, desiccation would be prevented without impeding photosynthesis. As it is, only plants which evolved stomata parallel with their developing cuticle survived (Walter and Stadelmann, 1968).

The relative impermeability of the cuticle to gases was given experimental support by Blackman (1895). He studied the movement of CO_2, in both directions, through the thick cuticles of *Nerium oleander*, *Prunus laurocerasus* and *Hedera helix* and through the thinner cuticles of *Polygonum sacchalinense* and

Platanus occidentalis. In each case, and regardless of whether young leaves or old leaves were used, he concluded that under normal conditions the stomata are practically the only path of ingress and egress for CO_2. The normal amount of CO_2 in the atmosphere, he believed, was not sufficient to provide the pressure necessary to permit entry through the cuticles of leaves whose stomata were artificially blocked. He conceded that under artificial conditions the cuticle is not entirely impermeable to CO_2. He calculated that a leaf of *Nerium oleander* with its stomata blocked and subjected to an atmosphere of 30% CO_2 in the light, carried out a comparable amount of assimilation to a similar leaf in a normal atmosphere with its stomata open. Impermeability of the cuticle to CO_2 was also found by Brown and Escombe (1905). This opinion was confirmed by the work of Holmgren, Jarvis and Jarvis (1965). They used a range of woody and herbaceous plants, *Betula verrucosa*, *Quercus robur*, *Helianthus annuus*, and *Lamium galeobdolon*. They found that the uptake of CO_2 in light through the adaxial cuticle of leaves with stomata located solely on the abaxial surface, is negligible or indeterminable, regardless of the magnitude of r_c (see p. 200).

On the other hand Mitchell (1936), Freeland (1948) and Dugger (1952) have carried out experiments which suggest that the cuticle may be significantly permeable to CO_2. Mitchell noticed that the amount of CO_2 absorbed by leaves of *Pelargonium*, in which the stomata appeared to be closed, is roughly equal to the amount absorbed by the same leaves when the stomata were open. Starch and other hydrolysable carbohydrates accumulate in the leaves even when the stomata are closed. Freeland used plants selected to obtain a wide variation in the frequency and distribution of the stomata and in the thickness of the cuticle. In some plants the amount of CO_2 exchanged through the epidermal cells alone was significant, sometimes being approximately equal to the amount which diffused through the stomata. In other plants, particularly those with a thick cuticle, little or no exchange through the astomatous cuticle could be detected. Dugger suggested that because of the solubility of CO_2 in cutin it is possible that the gas may penetrate the cutinised wall of the epidermal cell. He used radioactive CO_2 to investigate the permeability of *Hydrangea* and *Coleus* leaves and showed that the non-stomatal cutinised epidermis of *Coleus* in particular is permeable. The permeability of the *Hydrangea* leaf, but not the *Coleus*, was proportional to the partial pressure of the gas above the epidermis. At 1% of CO_2 in the atmosphere the adaxial surface of *Hydrangea* absorbed 3% of that absorbed by the abaxial, whereas the adaxial surface of *Coleus* absorbed 103%. Observed differences in the structure of the epidermal layers did not appear to be sufficient to account for the difference in permeability, and light appeared to have no effect on permeability. Jarvis and Jarvis (1963) found a large amount of cuticular transpiration in leaves of aspen (*Populus tremula*) and suggested that this might indicate a low cuticular resistance to the uptake of CO_2 and hence appreciable photosynthesis when the stomata are closed. Dorokhov (1963) has also claimed that the non-stomatal adaxial surfaces of apple leaves are permeable to CO_2 under normal conditions of growth. Thomas (1965) has reviewed the conflicting literature on this subject.

The conclusion that can be drawn is that some non-stomatal cuticles are significantly permeable to CO_2 and presumably, although little work has been done, to other gases as well. Species in which permeability has been demonstrated have relatively thin cuticles, supported on thin cell walls. Permeability to any significant degree has not been demonstrated in thicker cuticles of plants such as *Prunus laurocerasus* or *Hedera helix*. Permeabilities to gases may be found eventually to be just as variable as to other substances. Dugger (1952) has suggested that pectinaceous pathways through the cuticle, such as those described by Roberts, Southwick and Palmiter (1948) may account for the entry of water-soluble substances such as CO_2. Differences in permeability to CO_2 may be found to be related to differences in the amount and composition of the cuticular wax. The cuticle of the apple fruit is well adapted to gaseous exchange; the cuticular layer is less resistant to the passage of CO_2 than the lenticels and provides the route for most of the emission from the fruit (Marcellin, 1963; Mazliak and Marcellin, 1964).

The reflection of light

Plants growing in exposed areas such as sea shores, alpine slopes or hot or cold deserts are frequently covered either with a white tomentum of hairs, white scales or light-reflecting crystalline waxes or spicules of silica, carbohydrates or terpenoids. A number of authors, including Haberlandt (1914) and Barber and Jackson (1957), have suggested that these surface secretions may reduce insolation. Shade plants rarely if ever possess such superficial characters. Uphof (1962) has reviewed the literature concerning the role of leaf hairs in reducing the light intensity falling upon leaves. Whatever their selective survival value is, these morphological features serve to reduce the amount of visible and infra-red radiation penetrating the leaf. As Shull (1929) in an early review of the subject showed, a white layer can reflect from a leaf up to 50% of the visible radiation and Billings and Morris (1951) found that the surface of the desert peach (*Prunus andersonii*) reflects up to 70% of the infra-red radiation. Pearman (1966) showed that the surface wax increases the reflection of visible radiation from leaves, and Wong and Blevin (1967) found that stripping off the cuticle and epidermal cells of *Agapanthus umbellatus* and a *Gazania* hybrid markedly reduced leaf reflectances.

The general pattern of re-radiation of visible light is as follows. About 5% is reflected at 400 nm rising to a peak of 15% at 550 nm and falling to 5 or 6% at 675 nm. The reflectance then rises to 50% or above in the infra-red region (Billings and Morris, 1951). All hairs or scales on desert plants reflect more visible light, but not necessarily more infra-red. Desert plants, they found, reflect the most, followed by a group of plants from a sub-alpine west-facing pine forest, a north-facing pine forest and finally shaded 'campus' plants. As one might expect the capacity to reflect light falls as the leaves grow older; the decrease may be up to 50% of the original reflected amount.

It is interesting that a number of surface secretions of crystalline material on plants living in highly exposed conditions, e.g. alpine auriculas (*Auricula*) and *Kleinia* spp., are not waxes, but relatively soluble carbohydrates or flavones. The glaucous surface of *Crassula* (*Rochea*) *falcata* is created by microscopic spicules of silica; the surface holds water drops readily. This suggests that the evolutionary advantage of light reflection is a more pressing selective factor, in the Alps or South African deserts, than the shedding of water from the leaves or water loss.

Eveling (1967) in his experiments designed to study the physical effects of spray deposits on leaf surfaces found that the transmission of incident light through leaves coated with dusts was reduced. Inert dusts, such as kaolin, talc, silica and titanium oxide, which are commonly used as carriers for pesticides, reduced the light falling on the leaf surface by up to 20% depending upon the size of the particles and the type of leaf treated.

The perception and concentration of light by leaves

A great many leaves are diaheliotropic, i.e. they move so as to present the greatest area of lamina to the highest light intensity. Examples are common amongst shade plants. Neither the method of perception nor the means whereby leaves are able to respond is understood; little more is known than the fact that the principal perceiving area appears to be the leaf itself and the principal response zone the petiole. Haberlandt (1914) believed that the cells perceiving the stimulus are those of the adaxial epidermis and that they are assisted in this task by thickenings of the outer epidermal walls or cuticles forming a simple 'lens' (Fig. 7.8). These he called papillate cells. When a light beam falls upon the cell, the surface acts as a planoconvex condensing lens (Fig. 7.9). Depending upon the strength of this 'lens' a ray of light is concentrated over a limited region of the inner tangential wall of the epidermal cell, whereas the anticlinal walls remain unlit. An oblique beam of light shifts the area illuminated on the inner tangential wall (Fig. 7.9). Just how such shifts in the illuminated zone can be translated into a response resulting in the reorientation of the leaf as a whole remains within the realm of conjecture. There seems no doubt, however, that such papillate epidermal cells can assist shade plants to gather what light is available even when their surfaces are wet. A smooth leaf covered with a film of water loses much available light by reflection. In a papillate leaf the rounded apices of the epidermal cells still protrude out of the water film and, as Haberlandt has shown, still act as lenses.

FIG. 7.8 Drawings of types of leaf 'lenses'. ***A***, from the adaxial foliar epidermis of *Campanula persicifolia*. ***B***, from the adaxial foliar epidermis of *Vinca major*. ***C***, from the adaxial foliar epidermis of *Petraea volubilis*. ***D***, from the adaxial foliar epidermis of *Lonicera fragrantissima*. (Redrawn from Haberlandt, 1914)

FIG. 7.9 Drawing of the effect of a shift in the light direction on a leaf 'lens'. In vertical illumination there will be a bright central area (**a**–**b**). If the light falls obliquely from the left the illuminated area will shift to **c**–**d**. (Redrawn from Haberlandt, 1914)

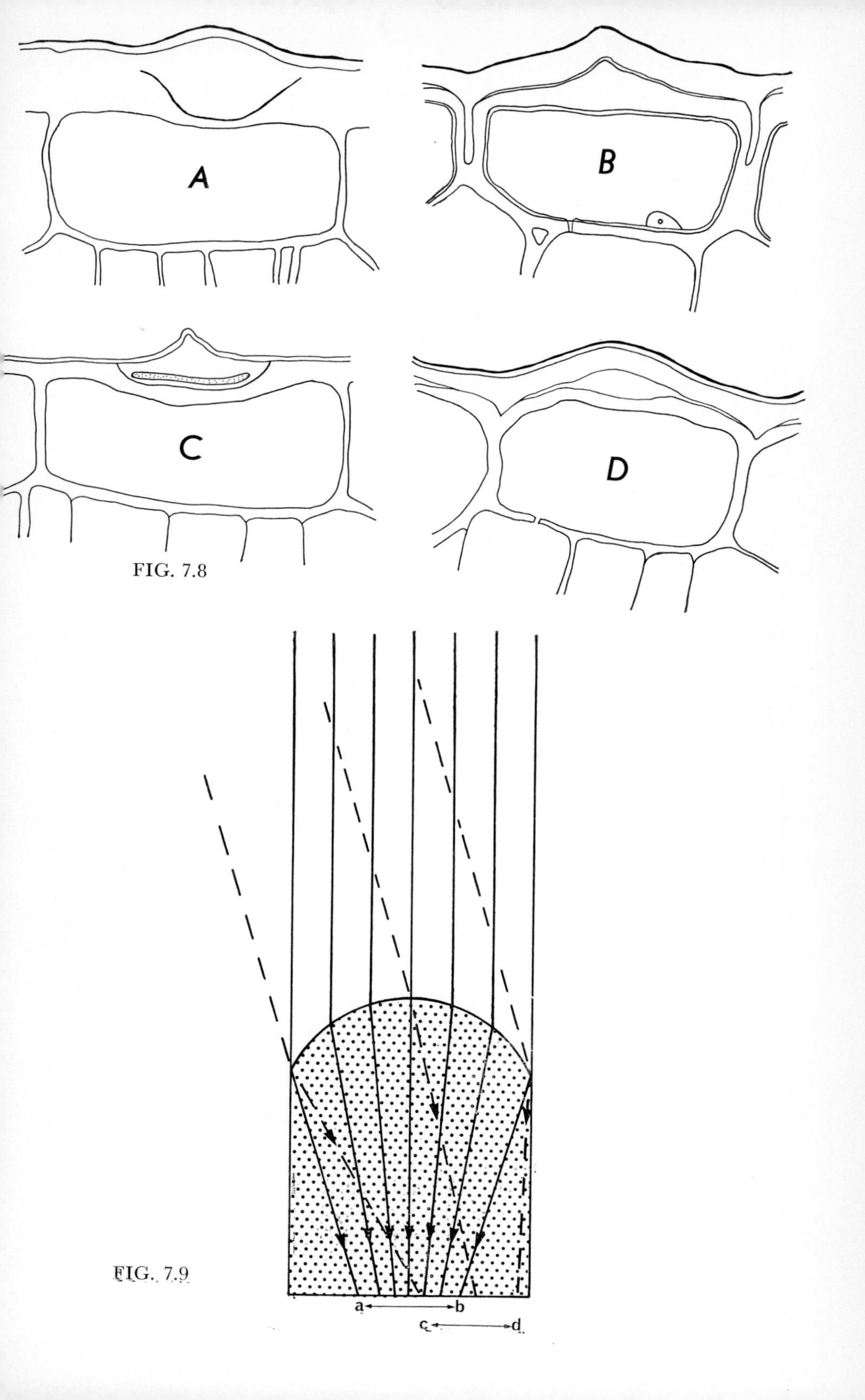

FIG. 7.8

FIG. 7.9

The role of the cuticle in preventing damage by frost

When leaves of frost-susceptible species are examined after they have been exposed to frost, the intercellular spaces are filled with water. These areas injected with water turn brown as they thaw. Grose (1960) noticed that the leaves most severely damaged by frost were those which had remained wet, either through rain or snow, during the period of the frost. Grose also showed that the winter-killing of *Eucalyptus delegatensis* seedlings, grown under controlled conditions, could be simulated at 33 °F (0·5 °C) provided that the leaves were kept in conditions of very high humidity. Moreover the leaves of seedlings grown in the shade were more easily injected with water, and more readily damaged, than those grown in full daylight. Leaf glaucousness tends to prevent this injection by shedding the water from the surface of the leaf and Grose suggests that superficial wax deposits and waxy hairs may, in this way, protect glaucous ecotypes of *Eucalyptus* species against this winter-kill. Since glaucousness is best developed in the full sun, this might explain the greater susceptibility to winter-kill of shade-grown plants.

The cuticle in disorders of fruits

The cuticle is involved in some of the physiological disorders of apple fruits. A thinner cuticle and fewer divisions of the epidermal cells in the shoulder region are at least partly responsible for the more extensive russeting of lateral than terminal fruits (Brown and Koch, 1962). In a study of the cause of Jonathan spot, Tomana (1960) found that the development of the cuticle on the most susceptible variety was much slower than on moderately resistant or resistant varieties.

Shutak and Christopher (1960) suggested that the cuticle appears to play an important part in the development of the serious storage disorder known as scald. Huelin (1964) confirmed that volatile esters are not responsible for the condition. Murray, Huelin and Davenport (1964) detected farnesene, 2,6-dimethyl-10-methylene-2,6,11-dodecatriene, in the cuticle oil of *Granny Smith*, a variety liable to the disorder, and suggested that the compound might play a role in its development. More farnesene was found by Huelin and Murray (1966) in the more susceptible early-picked apples than in late, and in *Granny Smith* than in a resistant variety. They suggested that scald may be caused by products of oxidation of farnesene. Meigh (1967) found that neither the free acids of the fruit wax nor the products of oxidation of the wax acids have any influence on the incidence of scald in cold-stored fruit. The scald-promoting factor is apparently formed in the wax by an oxidative process rather than by natural synthesis. Huelin and Coggiola (1968) found that during the storage of apples at 1 °C the farnesene content in the coating, mainly cuticle, increased to reach a maximum of 15% of the total lipids in the coating. More farnesene was produced in a scald-susceptible variety than in a scald-resistant. Most of the scald appeared after the farnesene reached its maximum level, and symptoms subsequently increased while the farnesene decreased. They concluded that oxidation products of farnesene are probably the causal agents. Meigh and Filmer (1969) found little difference in the relative amounts of farnesene produced by apple varieties of differing susceptibility to scald and concluded that the elucidation of the problem must await the examination of the oxidation products and their effect upon apples in storage. Farnesene occurs also in pear and quince (*Chaenomeles lagenaria*) (Murray, 1969).

The control of germination by the waxiness of seeds and fruits

Some seeds and fruits are prominently waxy (see p. 104). Bonner (1968) studied the uptake of water and germination of the acorns of four species of North American oaks. The acorns of two of the species have a thick waxy coating to the pericarp, but the other two have little wax on their surfaces. Bonner concluded that the presence or absence of wax determines whether or not a particular species can colonise a given habitat.

8

The Cuticle in Action, II Interactions with Chemicals

Chemicals in solution in water or as emulsions or suspensions are sprayed on to plants to control parasites, to regulate growth or to supply nutrients. The amount and distribution of the chemical deposited are greatly affected by the degree of wetting of the plant surface. Wetting (surface active) agents are added to the spray materials to overcome water repellency. To ensure the satisfactory deposition of a chemical on a plant surface, the concentration of wetting agent needed depends upon its kind and efficiency, its compatibility with the chemical and the degree of repellency of the surface. If the surface is inadequately wetted or if too much wetting agent is used, the deposit is reduced. Even under optimum conditions the amount of deposit retained is a function not only of the plant, species but also of different surfaces of a plant (Dormal, 1960). The water-repellent action of some crop plants contributes to the selective effect of herbicides; 2,4-dichlorophenoxyacetic acid (2,4-D), for example, controls the more easily wetted broad-leaved weeds in grassland and dinitro-*sec*-butylphenol (dinoseb) weeds in pea crops (Hartley, 1960). Water repellency is governed by a number of cuticular factors including hairiness, corrugation of the surface and the chemical nature and physical form of the surface wax. The effect of cuticle characteristics on the deposition, distribution and retention of chemicals applied to foliage has been discussed by van Overbeek (1956), Crafts and Foy (1962), Ebeling (1963) and Linskens, Heinen and Stoffers (1965). The chemical and physical nature of plant cuticles in relation to the deposition and penetration of pesticides has been reviewed by Martin (1961, 1966), and the physics of foliar application by Hartley (1966b).

The wetting of plant surfaces

Assessment of wetting

An arbitrary criterion of the wettability of a leaf surface is the advancing contact angle. This is defined as the angle between the surface of a leaf and the tangent plane of a water droplet at the circle of contact between air, liquid and leaf (θ in Fig. 8.1). The angle usually measured is that of a droplet

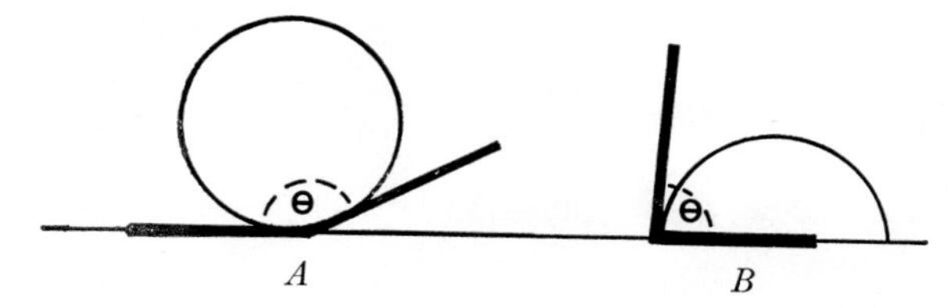

FIG. 8.1 Drawing of large (***A***) and small (***B***) contact angles

at rest on a level surface and immediately after its application. The reason for this will be seen later. A zero angle, although in practice never obtained, would indicate a completely wettable surface and any angle up to 180° demonstrates a degree of unwettability. Any droplet has two angles, a relatively large 'advancing angle' and a smaller 'receding angle'. The former exists when the liquid is at rest on or slowly advancing over a dry solid surface; the latter when the liquid is receding from a previously wetted surface. The difference between these two angles is known as the 'hysteresis' of the contact angle (Adam, 1963). The hysteresis of water on a smooth surface of paraffin wax is 2–3°, but other liquids and surfaces may show hysteresis values up to 60° or more (Linksens, 1952b).

Most contact angle measurements on leaf surfaces have been restricted to advancing angles. Usually an image of the surface and droplet is projected onto a screen or white background (Fogg, 1947; Juniper, 1959b, 1960b; Hall, Matus, Lamberton and Barber, 1965). A converted microscope contact angle goniometer was used by Holloway (1967) to measure the contact angles of water on leaf surfaces, isolated natural waxes and their classes of constituents. Juniper (1960b) used a droplet of 0·001 ml., i.e. about 1 mm in diameter;

1 μ
1 μm

Hall, Matus, Lamberton and Barber (1965) a droplet about 2 mm in diameter; and Fogg (1947) and Holloway (1967) one of about 3 mm in diameter. Linskens (1950, 1952b) dipped the whole surface into a water bath and observed the image formed of the meniscus between the leaf surface and the water through a horizontally mounted microscope. His results for advancing angles agree closely with those obtained by the droplet method.

The magnitude of a contact angle is determined by the characteristics of the solid upon which the liquid rests, and wettability is governed primarily by the nature and arrangement of exposed surface atoms. Fluorinated hydrocarbons are the most hydrophobic compounds known. An advancing contact angle range of 108–114° was recorded on polytetrafluorethylene (Adam, 1958, 1963) and an angle of 120° was given on a fluorinated polymethacrylic ester which exposes $-CF_3$ groups (Hare, Shafrin and Zisman, 1954). The contact angle of water on the smooth surface of any isolated plant wax is unlikely to exceed these values; Holloway (1969a) has measured contact angles on isolated waxes; he reports 94° on ouricury, 98° on carnauba, 102° on *Clarkia elegans*, 103° on *Saccharum officinarum* and *Papaver somniferum*, 104° on *Brassica oleracea* and *Allium porrum*, 106° on *Pisum sativum*, 107° on candelilla and 108° on esparto. Contact angles on the smooth surfaces of extracted plant waxes are usually between 80° and 108°.

Contact angles given by water on leaves can range from 31° (±4°) on the fully grown leaf of *Phragmites communis* to values in excess of 150° on surfaces such as *Triticum* and *Lupinus albus* (Linskens, 1950; Juniper, 1960b; Hall, Matus, Lamberton and Barber, 1965). Contact angles above 100° on leaf surfaces are very common; they are not true but 'apparent' contact angles, which result from the failure of the droplet to make contact with more than a small percentage of the droplet/surface interface. A mathematical treatment of true and apparent contact angles is given by Cassie and Baxter (1945) and Warburton (1963). Factors other than wax affect the magnitude of the contact angle; among these the macro- or micro-roughness of the surface is important. A large number of leaf surfaces show contact angles within the range 80–100°; e.g. the angle on *Sinapis* is 96° (Fogg, 1948a). The importance of wax in water-repellency was demonstrated by Fogg who found that if wax was removed from the cuticle of this plant the contact angle fell to 29°.

Figure 8.2 is drawn from a micrograph of a maize (*Zea*) leaf surface and the dark areas are the surfaces of the acicular crystals of wax, some single and some fused, that project from the surface. A droplet with the surface tension

FIG. 8.2 Drawing of the apparent surface formed by the acicular crystals projecting from the cuticle of a maize (*Zea*) leaf. The black areas (about 28% of the surface area) represent those regions of the cuticle with which a small droplet with a high surface tension would come into contact

FIG. 8.3 Electron micrograph of a carbon replica of the adaxial surface of a white lupin (*Lupinus albus*) leaf. The replica has folded over on the support grid and thus an apparent transverse section of a *Lupinus* leaf has appeared. Compare this micrograph with the true transverse sections in Figs. 1.1 and 4.19. Cr/Au/Pd

of water would be able to wet only the dark areas which comprise 28% of the total area. The smallest droplet that is used in normal commercial practice of low volume spraying is about 50 μm in diameter (Fulton, 1965). A 50 μm droplet would come into contact with about 1000 individual points on the surface of an intact *Zea* or *Lupinus albus* leaf surface (Figs. 4.23 and 8.3). The diagrams which illustrate the review articles of van Overbeek (1956) and Franke (1967a) show a water droplet in contact with a bloomed wax surface of a leaf. In relation to the size of the crystals and the thickness of the cuticular layer the droplet is about 2 μm in diameter.

Sometimes, as in *Pistia* (see Fig. 4.13), contact is made only between the water droplet and the surface of a hydrophobic trichome. Sometimes, as in *Salvinia* (See Fig. 4.12), contact is made with a group of hydrophobic rods. These project perpendicular to the epidermis and fan out at their tips to give a structure like a four-rayed crown. The surface presented to the droplet thus consists of short hydrophobic lengths of trichome with large air spaces in between. This device seems just as effective in preventing wetting as the much smaller acicular crystals which form the surface of most bloomed plants.

Hartley and Brunskill (1958) and Sargent (1966) suggested that high apparent contact angles are due in part to the trapping of air between the acicular projections and to the failure of small droplets of water to displace this air. Hartley and Brunskill attempted to prove this by examining droplets of water on repellent leaf surfaces under a partial vacuum (10 mm of mercury) but the experiment was inconclusive. The repellent hairs on the surfaces of *Salvinia* and *Pistia* have such large air spaces between them that it is difficult to see how significant pressures could be built up by very small water droplets on their surfaces (Fig. 8.4).

Droplets which make contact only with the bloom on a cuticle have contact angles above about 140°. In practice this means that, unless the surface tension of spray droplets is reduced below about 40 dyn/cm, they will be completely reflected from an undamaged bloomed surface. Angles between 105° and 140° indicate that partial contact is being made, parts of the droplet surface being held away from the cuticle by hydrophobic projections. Some retention of small droplets occurs on such a surface. Lowering the surface tension increases the retention on a surface difficult to wet but, as Blackman, Bruce and Holly (1958) have shown, this decreases the retention on certain wettable surfaces.

Receding contact angles have not been measured as much as advancing contact angles. However, in formulating a spray liquid whether for insecticidal, fungicidal or herbicidal purposes, the receding contact angles may be as important as the advancing. The shape of a droplet, once it has arrived on a leaf surface, may be relatively unimportant if penetration through the cuticle is required. However, if a film of material over the surface is needed it is important that the receding contact angle is at or near zero, otherwise the spray, which initially may cover the whole leaf surface, will collect into large discrete droplets and tend to 'pearl off' the leaf. The closer the receding angle approaches zero, the more uniform will be the deposit of material and the less the amount lost by 'run-off'.

Hydrophobic surfaces need not necessarily be composed of hydrophobic materials. The projections from the cuticle of *Kleinia articulata* are hydrophilic and yet create, at least briefly, a strongly hydrophobic surface (Fig. 4.24). Waxy surfaces coated with carbon in the stage of preparation of a standard carbon replica retain the same contact angle as they had uncoated. Carbon as such is not a particularly hydrophobic substance, yet carbon in the form of

FIG. 8.4 Drawing of a droplet of water on the surface of a leaf of *Salvinia*, facsimile reproduction of a micrograph. The droplet was about 1·5 mm in diameter

fine particles on a surface (carbon black) is, as Hartley and Brunskill (1958) demonstrated, highly water repellent. It gives an advancing contact angle approaching 180° and behaves very similarly to the surface of a glaucous pea leaf which shows a contact angle of 140° or above. It seems that any surface is hydrophobic provided that the projections from it are small enough, almost regardless of the nature of the material of which they are made.

Fogg (1947) showed that the contact angle of a droplet of water on a leaf surface varied with the position of the leaf on the plant and that frequently there was a diurnal fluctuation of the contact angle by as much as 30°. However, Fogg used a very much larger size of droplet (3 mm) than used in commercial practice and a smaller droplet (about 1 mm) does not show the same diurnal fluctuation on *Sinapis* and *Triticum* as noted by Fogg. On *Chrysanthemum segetum* using a range of droplet sizes up to 5 mm there appeared to be no diurnal rhythm in angle size (Juniper, 1960b). Nor does the contact angle given by *C. segetum* change with increasing age of the leaf or upon wilting. Some other species show changes in contact angle during the development of successive leaf stages as well as during wilting. The changes seem to be commoner in dicotyledonous species giving contact angles below 100° and in some cereals where turgor changes in the ridged leaves can alter sharply the contact angles of large water droplets.

Linskens (1952a) has shown that differences occur in contact angles on adaxial and abaxial leaf surfaces. He also found that static contact angles on *Narcissus pseudonarcissus* fell from 138° to 74° in 87 days, although the plants were not subjected to weathering. The adaxial surface of *Lupinus albus*, grown under glass, gave contact angles which reached about 120° and suffered no fall, whereas the surface of comparable plants in the open gave angles which reached higher values (up to 140°), but which fell to about 70° during a period of $2\frac{1}{2}$ months.

Contact angles are by no means constant even when the plant is grown under controlled growth cabinet conditions. They may differ appreciably according to altitude or on glaucous and non-glaucous forms as the data of Hall, Matus, Lamberton and Barber (1965) show:

Species	Contact angles	
	Adaxial surface	Abaxial surface
Eucalyptus urnigera		
At 2000 ft (650 m)	99·8° ± 0·9°	108·6° ± 1·0°
At 2300 ft (750 m)	113·6° ± 0·9°	114·5° ± 1·6°
At 3200 ft (1050 m)	141·3° ± 1·1°	143·0° ± 0·4°
Brassica oleracea (cauliflower)		
Non-glaucous	109·9° ± 2·5°	111·3° ± 2·7°
Glaucous	>150°	>150°

Not only the outside surfaces of plants may be strongly hydrophobic. The surfaces of the stomatal chamber, the walls of the palisade cells and the mesophyll cells are often strongly hydrophobic (Ursprung, 1925; Lewis, 1945; van Overbeek, 1956). The only walls within the leaf tissue readily wetted by water are those of the vascular system. Between 40% and 70% of the volume of a mesophytic leaf consists of air space which plays an essential part in gaseous exchange and sometimes buoyancy. If the space were to become waterlogged its role would be entirely impeded.

Because of the difficulty of interpreting the information they convey, contact angles should only be used as a general guide and in confirmation of other evidence. How misleading they can be can be shown as follows. The surfaces of leaves of *Euphorbia helioscopia* and *Kleinia articulata* both give initial contact angles of about 140°. However, if a stream of droplets is aimed at the surfaces, e.g. rain, differences immediately become apparent. The surface of *Kleinia* begins to break up, the advancing contact angle to fall and the droplets begin to be retained on the surface. The *Euphorbia* leaf retains its surface and the droplets are repelled indefinitely. The fine structure of the surface of *Kleinia* (see Fig. 4.24) is composed not of wax but of carbohydrate. The crystals are sufficiently small to ensure that little of the water droplet comes in contact with the carbohydrate and a high contact angle can be maintained at least for a while. If the droplet is kept in position cuticular material breaks free from the bottom of the droplet and floats upwards. The cumulative effect of this damage appears to be responsible for the wettability of *Kleinia* achieved after rain or watering.

Droplet size in relation to wetting

The defence of a crop against insect attack may call for a systemic insecticide or may require an insecticide lodged on or in the cuticle. A contact poison is most efficient if it remains as a crystalline or amorphous deposit on the surface, but here it is most susceptible to weathering. Some insecticides in use are fat soluble and are likely to pass into the wax of the cuticle. Martin (1957) recovered deposits of DDT (dichlorodiphenyltrichloroethane) from the surface of leaves by the use of a thin film of synthetic resin which hardened in contact with the surface. Crystals were picked up from leaves on which droplets of a solution or suspension had dried, but very few crystals were obtained from the dried deposit of an emulsion, the DDT having been absorbed by the wax. Challen (1959) reported a similar effect.

Droplets tend to spread over wettable surfaces. On unwettable surfaces this process is assisted by surfactants. Evans and Martin (1935) studied the area of the surface wetted (A) by a droplet in relation to the contact angle and the surface tension of the spray liquid. The equation

$$A = \frac{k}{\gamma\ (1\text{-}\cos\ \theta_A)}$$

where k is a constant, γ the surface tension of the spray liquid and θ_A the advancing contact angle gives a close agreement to measured areas. Wetting agents reduce both the surface tension and the advancing contact angle and therefore increase A. The equation shows that if the advancing contact angle is zero spreading is unlimited, or limited only by the rate of drying of the carrier whether this be water or some other liquid.

Evans and Martin also found that maximum wetting of a surface is achieved not when 'run-off' from the totally wetted surface occurs, but just before that point. If the spray droplets coalesce into a continuous film most of the liquid may drain off the surface instead of remaining on the surface as more or less discrete areas. This hypothesis can explain the observations of Smith (1946) who showed that 10–20 ml of 2,4-P per yd^2 ($\simeq$ m^2) was the most effective dosage rate in reducing growth in kidney bean (*Phaseolus vulgaris*); quantities below 10 ml or between 40 ml and 100 ml were less effective. The former probably failed because the dose was too low and the latter probably because drainage occurred. Smith was amongst the first to show the influence of droplet size on the efficiency of weed killers. He found that increasing the droplet size enhanced the retention and hence the effect of 2,4-D on the leaf of the kidney bean. The most effective droplets were between 250–560 μm in diameter. This leaf, however, has an easily wetted surface; large droplets are not so readily retained on surfaces difficult to wet.

According to Hartley and Brunskill (1958), who studied the behaviour of water droplets on the repellent pea (*Pisum*) leaf surface, droplets above 243 μm in diameter are totally reflected from the surface and leave no detectable residue. Each droplet flattens as it strikes the surface and, for a brief moment of time, the area that it covers is much greater than that covered by a stationary droplet of the same size. The droplet then retracts violently and comes from the surface with a marked oscillation. The proportion of droplets reflected from the surface decreases as the droplet size is reduced and below 93 μm the droplets are completely retained. The sizes of the droplets that are rejected or retained will obviously vary according to the plant tested, but these results are in good agreement with those of Fogg (1947) on wheat and charlock.

Hartley and Brunskill also tested the effect of decreasing the surface tension of the droplet with methanol and acetic acid. A 250 μm droplet with its surface tension reduced from 70 dyn/cm (0·07 N/m (water)) to below 41 dyn/cm was totally retained. Hartley and Brunskill were unable to obtain bouncing of this sized droplet from smooth water-repellent surfaces such as paraffin wax (contact angle 108°) or silicone-coated slides and suggest that surfaces giving contact angles above 140° are required for bounce to occur.

The work reported suggests that while wettable leaves can be satisfactorily treated by using relatively large droplets, unwettable or partially hydrophobic leaves are best wetted by small droplets. Very small droplets, particularly if they are being blown more or less horizontally, may approach a leaf or stem, but fail to make contact with it due to the boundary layer of still air over the object. The terminal velocities of water droplets also fall away rapidly as they decrease in size below 150 μm. The possibility of drift away from the target is so great with small droplets that the further reduction of droplet size to achieve a higher degree of retention is not usually desirable. In practice, retention can be governed by changing the output of spray, the droplet size and its surface tension. The angle of inclination of the leaf and the angle of incidence of the stream of spray are additional important factors controlling the deposit obtained.

Blackman, Bruce and Holly (1958) showed that the maximum difference in retention between two species is achieved when the spray consists of large drops with a high surface tension and the leaf surface of one species repels the droplets while that of the other does not. It is fortunate that most crops, e.g. brassicas, legumes, grasses and cereals, are difficult to wet whereas most common weeds are easily wetted. Blackman, Bruce and Holly also emphasise the importance of the ages of the crop and its weeds in relation to the selective effect desired. The highest level of retention recorded on *Brassica alba* is in the early seedling stage when the cotyledons still comprise a significant proportion of the total leaf surface. Under the same conditions of high surface tension in the spray, barley (*Hordeum vulgare*) at the same age retains the least. At later developmental stages the difference between these two species narrows. The leaves of barley no longer hold so upright a position and more spray is retained on the leaves, probably because of the reduced angle and the reduced amount of bloom on their surfaces.

Surface roughness in relation to wetting

Cassie and Baxter (1945) attributed the strong water repellency of some leaves to the roughness of their surfaces; the apparent contact angles on their surfaces were much larger than the true angles on their waxes. Linskens (1950, 1952a, b) also found that the ease of wetting of a leaf may be influenced by the corrugation of the surface. He confirmed the conclusion of Fogg (1947) that the extent of corrugation varies according to the turgidity of the underlying tissue. Challen (1960, 1962) further examined the influence of surface characters upon the wetting of leaves and concluded that the macroscopic surface roughness, caused by ridges or trichomes or both, is the chief factor limiting the wetting of some species and that the superficial wax plays a dominant part in limiting the wetting of others. The strong water repellency of *Festuca pratensis* leaves was attributed to macroscopic roughness caused by prominent parallel ridges on the surface. He used *Lycopodium* spores to study the roughness of surfaces in relation to the deposition of standard particles. Hartley and Brunskill (1958) showed that the reflection of water droplets from leaf surfaces was associated with a micro-roughness detectable with the light microscope. Troughton and Hall (1967) have shown that the leaf structure of wheat, in addition to its wax, is an important factor in wetting. The adaxial surface is more repellent than the abaxial; both are waxed but ridges occur, from the tip of the leaf to its base, on the adaxial surface but not on the abaxial.

Holloway (1967, 1969a) found that the differences in the wettability of leaves could not wholly be explained by differences in the physical, chemical or hydrophobic properties of the waxes isolated from them and attributes contact angles greater than 110° on leaves to surface roughness. He distinguishes three types of roughness: macroscopic visible to the eye or with low magnification, microscopic revealed by the light microscope and ultramicroscopic shown by the electron microscope. The surface of the cuticle may be modified by underlying veins; the main veins and their lateral branches may induce a macroscopic roughness and the vein reticulum a microscopic. Trichomes may give macroscopic or microscopic roughness. The topography of the leaf surface is governed in part by the size and shape of the epidermal cells, whose outer surface may be flat, convex or papillose (bearing a small conical structure); when convex or papillose the size of the cells influences the degree of corrugation. The surface of the cuticle itself may be granular, grooved or ridged; most cuticles are smooth or minutely granulated (Stace, 1965). An ultramicroscopic roughness may be imposed by the physical form of the surface wax. Furmidge (1962) made measurements of the surface roughness of leaves and showed that differences in spray retention were closely related to leaf surface irregularities.

Wax structure and composition in relation to wetting

Electron microscopy has proved invaluable in elucidating the role of the superficial wax in preventing wetting. When the superficial wax plays a dominant part in water repellency, its effect depends chiefly upon its physical structure. Water repellency is greatest when the wax has a rough surface in the form of projecting rods or a crystalline or semi-crystalline structure (Mueller, Carr and Loomis, 1954; Schieferstein and Loomis, 1956; Juniper and Bradley, 1958; Juniper, 1959a, b, 1960a; Wortmann, 1965a; Hall, Matus, Lamberton and Barber, 1965; Thrower, Hallam and Thrower, 1965; Silva Fernandes, 1964, 1965a; Troughton and Hall, 1967). The surfaces of leaves of *Lupinus* spp. are notably water repellent; projecting wax rods entrap air and prevent intimate contact between a droplet and the surface (Juniper, 1960a, b). Hall, Matus, Lamberton and Barber combined contact angle determinations with electron microscopy and found that contact angles in excess of 145° are given on leaves when numerous wax rods or plates cover the surfaces. Thrower, Hallam and Thrower showed that the arrangement and distribution of the wax structures can be correlated with the relative ease with which the leaves of *Vicia faba*, *Phaseolus vulgaris* and *Glycine max* can be wetted. Silva Fernandes used a qualitative wetting test in conjunction with electron microscopy and showed that whereas a crystalline or semi-crystalline wax repels water droplets, a non-crystalline or smooth wax permits the acceptance and spreading of droplets. The waxes on the strongly water repellent leaves of *Eucalyptus*, *Exochordia* and *Rhus* spp. are evenly or irregularly distributed crystalline masses, and those on the non-repellent surfaces of *Hydrangea*, apple and broad bean are non-crystalline and smooth. The ultrastructure assumed by a wax, however, is not entirely dependent upon its chemical composition.

Holloway (1967, 1969b) made a detailed examination of the water repellent properties of the classes of constituents isolated from leaf surface waxes. No class of constituents is outstandingly water repellent. Alkanes are the most hydrophobic showing contact angles of 107–108°, but esters, ketones and secondary alcohols are almost as unwettable with contact angles of 103–105°. The primary alcohols, with contact angles of 94–95°, were the least hydrophobic of the natural constituents that he isolated from the waxes; the fatty acids showed contact angles of 101–102°. Individual compounds that are known to occur in natural waxes were also tested; hydroxy-fatty acids gave contact angles of 90–95°, diols angles of 70–71°, triterpenoids angles of 89°–95° and sterols angles of 82–104°. Holloway draws attention to the comparatively small overall range in contact angles shown by the major classes of constituents; the contact angles of the alkanes differ from those of the primary alcohols by only 14°. An ester, ketone or alcohol group substituted within the alkane chain reduces the contact angle by 3–4°, and a terminally substituted carboxylic or alcohol group by 7° and 14° respectively. Because of the small range and of the heterogeneity of natural waxes, their hydrophobic properties usually give little indication of their composition. The contact angle reliably indicates the composition of a wax only when the angle is 106° or larger. Such angles are given only by waxes which contain more than 50% of alkanes, e.g. candelilla, esparto and pea waxes. Holloway also discusses the orientation of the molecules of constituents at the surface of waxes in relation to water repellency. The aliphatic molecules are aligned with their methyl groups exposed on the surface and close packing of the surface methyl groups increases repellency. Functional groups such as hydroxyl or carbonyl in the chains prevent the close packing of the chains and consequently of the methyl groups at the surface. Wettability increases with increasing size and number of substituent groups in the chains, and terminal substitution has a greater effect than substitution within the chains; the greatest reduction in contact angle is shown by the α,ω-diols which contain two terminal hydroxyl groups. Despite their cyclic structure, the triterpenoids and sterols are hydrophobic, which suggests that their surfaces are predominantly hydrocarbon in nature.

Holloway (1967, 1969a) examined the physical properties of the waxes obtained (by washing with chloroform) from the leaves of 40 plant species. They ranged from soft, amorphous waxes melting at about 50 °C to hard, crystalline waxes melting above 200 °C. The high melting waxes probably contained large proportions of triterpenoid compounds. The contact angles of water on the isolated intact waxes varied between 92° and 107°. The chloroform-washed surfaces were then usually less water repellent than the original surfaces, sometimes markedly so; in a few cases, the surfaces devoid of waxes were more repellent than the original. In contrast to the narrow range of contact angles found on the isolated waxes, the contact angles on the leaves themselves varied between 40° (*Rumex obtusifolius*) and 170° (*Triticum aestivum* and *Eucalyptus globulus*). Approximately 60% of the surfaces examined showed contact angles greater than 110°, 15% in the range 90°–110° and the remainder below 90°. Sometimes a large difference was found in the repellency of the two surfaces of a leaf; the contact angle on the adaxial surface of *Acer pseudoplatanus* was 44° and on the abaxial 155°, and on the adaxial surface of *Trifolium repens* 158° and on the abaxial 70°. Holloway classifies leaf surfaces in three categories. A contact angle smaller than 90° suggests that roughness is of little significance and that wax, although playing some part in repellency, is not a predominant feature of the surface. The wax, although of a repellent nature, may incompletely cover the surface thus exposing the less hydrophobic cutin. Alternatively, the surface may carry a wax which, because of its composition, is more hydrophilic. Sometimes the removal of a hydrophilic wax exposes a more hydrophobic surface beneath. A contact angle in the range 90°–110° indicates that a surface is probably covered with a smooth layer of wax; the angles found on such surfaces are in close agreement with those on the isolated waxes. Cuticular roughness may also begin to play a part. A contact angle greater than 110° indicates a surface entirely covered by a wax of ultra-microscopic crystalline form which obscures other cuticular features or a surface whose water repellency is due chiefly to cuticular roughness.

Surface active agents are incorporated in aqueous sprays to assist wetting. Amongst the commonly used wetting agents are the anionic types such as dioctyl and dinonyl sodium sulphosuccinate and non-ionic types such as polyethylene glycol esters and others of which the most commonly used is octyl cresol polyglycol ether (Stanley, 1958). The former are powerful wetting agents, but are unstable in alkaline solution and foam readily, whereas the latter do not foam and are stable at all pH levels. Additives such as hydroxyethylcellulose are used to increase the adhesion of deposits. A full account of spray formulations and adjuvants used is given by H. Martin (1964).

The penetration of substances

Penetration into roots and root hairs takes place through wet fibrillar and pectinaceous material; the so-called 'free space'. There is no water repellent wax covering on roots comparable with that on some leaves and little or no barrier to penetration except perhaps in old suberised roots. Penetration into leaves and stems, on the other hand, involves sooner or later passage through a cuticle. Pores through the cuticle exist in specialised areas; these occur in the trichome hyathodes of *Cicer* and the glandular hairs of *Mentha* see p. 210), at the base of salt glands, below the water-absorbing scales of *Bromeliaceae* (see p. 94) and at the tips of stigmas (Fig. 8.5) (see p. 215). These pores are detectable under the light and electron microscopes.

No general pattern of cuticular pores exists which can account for the observed passage of materials through the cuticle. And yet both polar and non-polar substances can pass through an undamaged cuticle. Crafts (1961) supposes that the cuticle is perforated by micropores which are more or less filled with an aqueous phase depending on the environmental conditions to which the plant is subjected. Through these continuous aqueous phases are thought to pass polar substances moving parallel to non-polar substances which are believed to move through permanent lipoidal pathways. No such pores corresponding to Crafts' hypothetical pathways have been seen in the cuticularised and cutinised layers although possible routes, the ectodesmata, do exist through the epidermal cell walls. Nor are the imbricate cuticles, supposedly interleaved with hydrophilic pectinaceous materials as suggested by Roberts, Southwick and Palmiter (1948), of sufficiently wide occurrence to provide a general purpose route of entry. Many complete and undamaged plant cuticles are freely penetrated by a wide range of both polar and non-polar substances. But the cuticle is a very variable structure; its thickness bears little relationship to its observed properties and, as yet, its properties give little clue to its detailed construction. Silva Fernandes (1964) showed that even the possession of red pigment by an apple cuticle can markedly alter its permeability compared with neighbouring green areas.

The generally accepted structure of the cuticle is of a layer of hydroxy-fatty acids forming a three-dimensional polyester network. Presumably inter-molecular channels exist through which relatively large molecules can move. Dyestuffs, streptomycin and other antibiotics can penetrate quite readily

(Franke, 1967a). The cuticle, in the absence of any added material, is like a thin slice of sponge rubber. If, as is supposed to occur in the cuticles of some older leaves, waxes interleave between the cutin lamellae these lipophilic apolar layers should inhibit the penetration of polar solutions (Meyer, 1938; Hallam, 1964) (see Fig. 4.20). However, no firm conclusions have been reached on either the widespread existence or the continuity of the wax deposits within

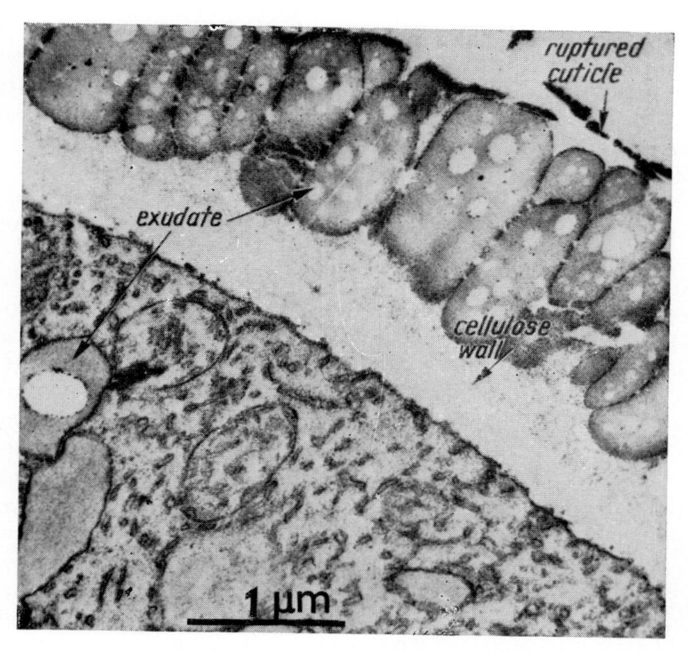

FIG. 8.5 Electron micrograph of a section through the stigmatic surface of a flower of *Petunia hybrida*. A sticky exudate in the form of large vesicles accumulates in the outermost cells of the stigma, passes through the cell wall, accumulates between the wall and cuticle and finally ruptures the cuticle and forces it off in flakes. Glut/Osm. (From Konar and Linskens, 1966, The Morphology and Anatomy of the Stigma of *Petunia hybrida*. *Planta*, **71**, 356–371. Springer, Berlin–Heidelberg–New York)

the cuticle and thus their role in permeability is entirely speculative. Silva Fernandes (1964, 1965b) noticed that a higher temperature (43 °C for 24 and 48 hr) led to an increased permeability of isolated cuticles to phenyl mercuric acetate and this is consistent with the existence of a labile wax layer susceptible to temperature within the cuticle. Skoss (1955) noticed that excised cuticles from the leaves of *Hedera helix* were impermeable to 2,4-D, dinoseb and silver and chloride ions. When the impregnating wax was removed by refluxing the isolated cuticles in ethanol the cuticle became highly permeable to water. Darlington and Barry (1965) found an increased permeability of isolated apricot cuticle to isopropylchloroacetamide after soaking in chloroform. Permeability will obviously also be affected by the charges of polar groups within the macromolecules of cutin.

Apart from entry through the cuticle Shuel (1961) has observed that the nectaries of the flowers of *Streptosolen jamesonii* not only secrete sugars, but reabsorb them.

The role of stomata in uptake

The relative importance of the stomata and cuticle as routes of entry of water-soluble substances into the leaf has been much debated. Stomata are abundant on the abaxial surfaces of leaves of most species, but are often sparse, if not absent, on the adaxial which usually receive by far the greater quantities of spray materials. The stomata at first sight would appear to provide obvious routes of penetration, but Adam (1948) showed that surface tension forces prevent any infiltration of rain. Were this not so, the stomatal cavities of most leaves would be waterlogged for much of the year and gas exchange very much reduced.

Many stomata, e.g. those of the pitchers of *Dischidia*, are so deeply sunken in the cuticle that no water enters even after indefinite immersion. Fogg (1948b) suggested that, although cutin is appreciably permeable to water, wax on the surface of the cuticle may be a considerable barrier to entry. Skoss (1955) concluded that the degree of impregnation of the cuticle with wax is the chief factor controlling the permeability of the stomata-free cuticle to water and that cutin itself may exercise a restraining influence depending upon its thickness. Boynton (1954) reached the tentative conclusion that aqueous solutions of foliar nutrients make an initial entry by way of the stomata, but later also penetrate the cuticle, the cuticular route eventually being as effective as the stomatal. Currier and Dybing (1959) and Dybing and Currier (1961) found that herbicides and a fluorochrome dye penetrated the stomata of leaves of *Zebrina pendula*, *Phaseolus vulgaris*, apple, pear and apricot (*Prunus armeniaca*) provided an efficient wetting agent was incorporated at a suitable concentration. Cuticular penetration of the dye into *Zebrina* leaves was slow, but demonstrated that a relatively large polar molecule can pass through the cuticle. Gustafson (1956, 1957) inclined to the view that solutes enter the leaves of bean (*Phaseolus vulgaris*) and cucumber (*Cucumis sativus*) mainly through the stomata with other routes not excluded; data on an extended range of plants, however, did not support the dominance of stomatal penetration, although the stomata of some plants may have played a part. He suggested that differences in the cuticle of different plants may influence the rate of absorption. Sargent and Blackman (1962), using *Phaseolus* leaf discs and an aqueous solution of 2,4-D, showed that penetration is enhanced in the dark by a wetting agent, but the effect is not proportionately greater in the daylight when the stomata are open. They showed that there is a direct relationship between the ease of penetration of a surface and the number of stomata present on that surface, but that the relationship holds even in darkness when the stomatal pores are closed.

Furmidge (1959) observed that injury by surfactants to apple and plum leaves bore little relation to the distribution of stomata, and concluded that entry is principally other than stomatal. Goodman (1962a) showed that a molecule as large as streptomycin enters the apple leaf as readily through the non-stomatous adaxial surface as through the stomatous abaxial surface and concluded that the cuticle is not the impregnable barrier it was thought to be. Middleton and Sanderson (1965) discount stomatal entry on several grounds. Provided the humidity was high, the greatest uptake of radioactive fission products was from the more concentrated solutions after they had dried on the leaf surface, and the addition of 8-hydroxyquinoline sulphate which caused the stomata to close did not affect uptake. The organo-phosphorus systemic compound schradan is absorbed through the cuticle of *Coleus* and other plants rather than by vapour phase entry through the stomata (Bennett and Thomas, 1954).

One can conclude that some stomata can be penetrated by liquids of low surface tension, especially oils, some of them may be penetrated by some aqueous solutions assisted by surfactants, but they cannot be penetrated by liquid water. Probably, however, the important role of stomata under field conditions is to provide preferential sites of entry through the surfaces of their guard cells and subsidiary cells.

Penetration through the cuticle

The consensus of opinion now is that stomatal penetration of aqueous solutions is relatively unimportant and that the main route of entry of both water- and lipoid-soluble materials is provided by the cuticle. Norman, Minarik and Weintraub (1950) concluded that diffusion through the cuticle is probably the usual means of entry of herbicides; the amount which enters is a function of the period and area of contact, both being related to the degree of wetting of the surface. The nature of the active agent is a factor in entry; Norman, Minarik and Weintraub suggest that because the leaf cuticle is largely lipoidal, non-polar molecules are likely to penetrate more readily than polar. The non-polar molecules, with their greater solubility in lipoids, may use pathways believed to be concerned with the secretion of wax such as those reported by Hallam (1964, 1967) to occur in the leaf membrane of *Eucalyptus cinerea*, by Scott, Schroeder and Turrell (1948) in the outer walls of the epidermal cells of the leaf of *Citrus cinensis* and by Hall (1966, 1967a, b) in the cuticles of apple fruits and *Trifolium* and *Brassica* leaves. Polar compounds may make use of pectinaceous pathways such as those described by Roberts, Southwick and Palmiter (1948) in the cuticle of the apple leaf. The ease with which a compound may penetrate is likely to be greatly influenced by its properties, in particular its water- or lipoid-solubility and by the relative proportions of cutin and wax in the cuticle concerned.

The extent of dissociation of polar compounds is an important factor governing their penetration. The undissociated water-soluble molecule penetrates the cuticle more readily than the dissociated. Acidic arsenical sprays are more effective herbicides than alkaline, and the activity of 2,4-D is greater at a low pH than at a high. Fogg (1948b) found that little 3,5-dinitro-*o*-cresol (DNC) penetrates the stomata of leaves of charlock (*Sinapis arvensis*); a lethal amount enters from aqueous solution by diffusion of the undissociated phenol through the cuticle which behaves as an homogenous lipid membrane. Crowdy (1959) points out that cutin is polar and will absorb water and swell. Its reaction tends to be acidic; below pH 5 a large proportion of the constituent long-chain mono- and di-carboxylic acids are undissociated and the matrix is then more permeable to anions than above this pH when the acids dissociate and the matrix tends to repel anions. He suggests that this effect is illustrated by the results of Orgell and Weintraub (1957) who found that the absorption by leaves of 2,4-D from alkaline solution is aided by the presence of cationic wetters and certain cations which could neutralise the charge on the cuticle. Härtel (1951) reports that cutin shows maximum permeability at a neutral or weakly acidic reaction (pH 5–7).

Some regions of the leaf act as preferential sites of absorption. Trichomes provide an important route of entry (Hull, 1964). This may be a consequence of the collection of spray liquid around them; they may be more easily wetted than the remainder of the surface. The trichome, whether uni- or multicellular, usually carries a cuticle and so absorption is cuticular. Freytag (1957) investigated the physical properties of the cuticle of the glandular hairs of *Matricaria*, and Sifton (1963) the continuous development of cutin on the expanding surface of the hairs of Labrador tea (*Ledum groenlandicum*). Before their abscission from the adaxial surface, the hairs of *Ledum* had a prominent cuticle except at their bases. There is some evidence that the basal portion of a trichome is the most readily penetrated region; Tietz (1954), for example, showed the importance of hairs in the penetration of demeton (diethyl 2-(ethylthio)ethyl phosphorothioate) into *Primula* leaves and concluded that absorption occurred through the basal cells of the hairs. Basal penetration would also occur in *Convolvulus cneorum* (see Fig. 4.11). Wetting of a pubescent surface may be hindered if the hairs are closely matted; the addition of a wetting agent at an appropriate concentration then assists absorption by saturating the mat with the spray fluid and bringing it into close contact with the cuticle.

The cuticles of the guard cells and subsidiary cells of the stomata are probably important sites of absorption. Dybing and Currier (1961) observed that an iron salt in aqueous solution with no wetter entered the guard cells of leaves of *Zebrina*, but not the substomatal cavities; penetration also occurred through the cuticle lying over the veins. Sargent and Blackman (1962) also suggest that the guard and subsidiary cells may be principal sites of penetration. Franke (1967b) points out that these cells often have the highest frequency of ectodesmata and take up sucrose preferentially. But ectodesmata do not penetrate the cuticle; the cuticle over guard cells is often different from that over normal epidermal cells and guard cells have a somewhat different fine structure from that of normal epidermal cells.

The regions immediately adjacent to the mid-vein of leaves offer a preferential site for the absorption of water-soluble materials. Linskens (1950) noted that the maximum wetting of smooth leaves occurred over the veins, and Leonard (1958) found that the base and central part of the midrib of the bean leaf was the most effective site for the absorption of 2,4-D. Thin-walled parenchymatous bundle sheath cells between the epidermis and the vein are believed to provide a facile transport route from the cuticle to the vascular system. This concept, however, is not always a true one. Kamimura and Goodman (1964b) found that a greater absorption of leucine into apple leaves occurred at the interveinal and apical regions; the cells between the epidermis and vein were more compact than the mesophyll cells, with their numerous intercellular spaces, in the interveinal areas. The cuticular membrane of the pear (*Pyrus communis*) leaf is thicker over the veinal than over the mesophyll tissue, and there is no evidence of pectinaceous pathways through the membrane (Norris and Bukovac, 1968). The cuticle lying immediately above the anticlinal walls of the epidermal cells is also believed to be easily penetrated by substances in aqueous solution (Currier and Dybing, 1959). Over the periclinal walls, lamellae of the cuticle run parallel to the surface; over the anticlinal they dip inwards at right angles to the surface (Rudolf, 1925).

Surface active agents usually enhance cuticular penetration, probably due chiefly to the improved wetting of the membrane. Slatyer (1960) points out that a marked increase in cuticle permeability must occur on wetting in order to explain the paradox of the ready cuticular penetration of water into leaves and the cuticular resistance to water transport during transpiration. When wetted or subjected to an atmosphere of high relative humidity, the cuticle is probably in its most open, swollen condition; Sargent (1966) suggests that only under such conditions will a water continuum extend through the cuticle. Such a water path permits the ready absorbtion of polar substances. The positive correlation between humidity and absorption is well documented; Middleton and Sanderson (1965), for example, found that the humidity of the atmosphere appeared to be of greater significance than temperature or light intensity in determining the uptake of radioactive fission products by plants. Under low relative humidity the cuticle probably shrinks, the wax deposits are compressed together, and the passage of water-soluble substances may thereby be impaired. The so-called humectants, e.g. glycerol, have been found in some cases to assist penetration by retarding the drying of the spray film on the surface (Gray, 1955; Holly, 1956; Goodman, 1962a). The formulation of a non-polar substance in oil or as an emulsion may be expected to aid penetration into the cuticle. Better contact with the leaf surface is provided than when formulated as a suspension in water. Hartley (1966a), however, points out that if the oil is fairly volatile it leaves a crystalline deposit of the substance on the surface and if effectively nonvolatile it may be of too high molecular weight to penetrate and then holds back the substance as well. Bukovac (1966) tested classes of compounds isolated from plant waxes for their permeability to water. They differed considerably in permeability, the alkanes being among the least permeable. The significance of this finding will be appreciated when it is realised that the alkane contents of leaf waxes range from virtually nil to 92%, the highest recorded, in *Solandra grandiflora* (Herbin and Robins, 1968b).

Darlington and Barry (1965) suggest that surface active agents can modify the permeability of the cuticle and that this effect could be more fully exploited in the formulation of commercial spray chemicals. The factors which have an important bearing on the design of truly selective herbicides have been discussed by Linser (1964). Furmidge (1959) has considered the influence of surface active agents on the formation of spray droplets and their impaction and retention on surfaces, and on phytotoxicity. He suggests that when the concentration of an agent is above the critical concentration for micelle formation wax may be removed and penetration thereby facilitated, and that the relative safety to plants of the non-ionic agents may be due to their lower reactivity with components of the cuticle.

The penetration of chemicals into foliage is effected by a combination of non-metabolic and metabolic processes. Reversible diffusion through the cuticle is followed by uptake, metabolically controlled, through the cellular membranes. The two phases of uptake of urea by apple leaves were observed by Cook and Boynton (1952). The initial uptake of dalapon by the fronds of duckweed (*Lemna minor*) was predominantly a process of diffusion and the subsequent absorption metabolically controlled (Prasad and Blackman, 1962). That the overall process of uptake is primarily metabolic is indicated by its enhancement due to an increase in light intensity, its dependence upon oxygen and possibly pH, its reduction in the presence of inhibitors of metabolism and its progress and irreversibility against a concentration gradient. The direct correlation between temperature (within a range physiologically compatible with the plant) and absorption is well documented (Fogg, 1948b; Goodman, 1962b; Jackson and Brown, 1963). An increase in light intensity enhanced the absorptive capacity of chrysanthemum and *Coleus* leaves (Bennett and Thomas, 1954). The effect of light intensity is dependent upon humidity; when humidity is adequate, an increase in light intensity enhances uptake (Gustafson, 1956; Barrier and Loomis, 1957; Sargent and Blackman, 1962; Middleton and Sanderson, 1965). Sargent and Blackman (1965) suggest that the light-induced phase of penetration is sensitive to temperature, to sublethal irradiation by ultra-violet light and to metabolic inhibitors, and that it is dependent upon a sufficiently high level of endogenous auxin.

Goodman and Addy (1962a,b) and Kamimura and Goodman (1964a) take the view that the absorption of organic substances by foliage is mediated primarily by the expenditure of respiratory energy by the leaf cells. Sargent (1966) suggests that, among other factors, the rate of penetration by a chemical is governed by the rate at which it is conducted away from the inner surface of the cuticle; illumination stimulates photosynthesis, a rise in the level of transferable carbohydrate promotes flow through the sieve tubes and conduction away from the cuticle is increased. Hartley (1966a) points out that the processes of penetration and translocation are difficult to separate; if the substance has a very low solubility extensive movement away from the local site must take place not only to achieve systemic effect but, on a purely quantitative basis, to accommodate the dose applied. Jyung and Wittwer (1964) showed that the absorption of rubidium and phosphate ions by bean (*Phaseolus vulgaris*) leaves was significantly reduced in the presence of chloramphenicol, suggesting that the absorption process is related to the synthesis of a particular protein and that the carriers of the ions are proteinaceous. Despite extensive investigations the physiology of the uptake of substances by leaves is little understood. Sargent and Blackman (1965) draw attention to the complexity of the process and point out that conclusions reached concerning the penetration of one compound cannot validly be applied to another because of the differences in the nature of chemicals, of formulations used, and of leaf surfaces. The role of the cuticle and the factors involved in the uptake of chemicals by foliage have been reviewed by Hull (1964) and Sargent (1965). The cuticular penetration of chemicals has been further substantiated by Weaver and de Rose (1946), Gustafson and Schlessinger (1956), Teubner, Wittwer, Long and Tukey (1957), Goodman and Goldberg (1960), Herrett and Linck (1961), Brian (1967) and Gabbott and Larmon (1968).

Franke (1961) believes that whatever the path of materials through the cuticle may be they then take advantage of the ectodesmatal channels to enter the epidermal cells. Franke treated whole leaves of *Plantago* and *Helxine* with Gilson fixative. As it penetrates through the cuticle it forms crystals visible under the light microscope and the crystals are concentrated in the ectodesmata. It is possible, however, that the ectodesmata are not preferred paths of entry, but just preferred regions for crystallisation and hence detection. There are, however, other channels in the walls of cells discovered by Gaff, Chambers and Markus (1964). These are about 30 nm in diameter, are obviously not related to ectodesmata since they do not run preferentially in an anticlinal direction, but they are capable of allowing the movement of particles as big as those of colloidal gold. The truth will not emerge until high resolution ***autoradiography*** has been directly applied to observing the passage of labelled substances through intact cuticle.

Many investigations have been made of the penetration of chemicals through cuticular membranes isolated from leaves and fruits by chemical or enzymic treatment. Wittwer and Bukovac and their associates in particular have studied binding sites within membranes in relation to the passage of ions and molecules. Yamada, Wittwer and Bukovac (1965), for example, found that the penetration of urea, maleic hydrazide or N-dimethylamino-succinamic acid through isolated tomato fruit cuticle was greater from the outer to the inner surface than in the opposite direction. Urea enhanced the penetration of rubidium and chlorine ions through the cuticle and also facilitated the absorption of rubidium ions (but not of chlorine) by the palisade and spongy parenchymatous cells separated from the tomato leaf (Yamada, Jyung, Wittwer and Bukovac, 1965). The rate of penetration of ions through the isolated membranes was directly related to the extent of binding on the surface opposite the site of initial entry; the data obtained suggested that uptake dominates the loss of nutrients from the leaf (Yamada, Wittwer and Bukovac, 1964). Ion-binding sites on the leaf membrane were found to be concentrated in areas adjacent to stomata, above the periclinal cell walls and in the continuum of the membrane within the substomatal cavity (Yamada, Rasmussen, Bukovac and Wittwer, 1966). This finding, as Yamada and his associates point out, is of interest in connection with the similar concentrations of ectodesmata. Franke (1969) has studied the binding of radioactive ions and molecules to isolated cuticle. He concludes that the binding sites lie over the ectodesmata in the wall below and that together they form a favoured pathway of penetration of aqueous solutions in both the inward and outward directions.

Silva Fernandes (1965b) showed that wax is an important barrier to the penetration of a water-soluble organo-mercury compound through the isolated cuticle of the apple fruit, and that the effect of the wax depends upon its composition. The sorption of mercury is increased by an increase in the percentage of esters in the wax, by a rise in temperature or in the presence of a wetter.

The information on penetration obtained from the use of excised membranes, however, is of limited value; the properties of the membrane are likely to be changed during isolation, and metabolic processes, which play a part in uptake by the intact leaf, are precluded. Kamimura and Goodman (1964a, c) found that less than 2% of each of a number of organic compounds penetrated the membrane isolated from the apple leaf by chemical treatment; much greater penetration occurred into the intact leaf. Also, the permeability of the membrane to one compound was increased by enzymic treatment. Bukovac and Norris (1966) indicate that the data obtained, while of value in promoting an understanding of the properties of a membrane *per se*, are probably applicable to the isolated system alone.

The cuticle as an ion exchange medium

In a homogenous solution the transfer of material can occur by processes such as diffusion and ionic migration, unrestricted except by the walls of the container. In living cells, transfer by combinations of these processes is restricted or modified by the presence of porous materials which occur within the cells or constitute their walls. Examples of such porous structures are the ground substance between some animal cells, the matrix material of cytoplasm and nucleus, and plant cell walls and cuticular layers. O'Brien (1967) and Clowes and Juniper (1968) have suggested that the epidermal cell wall and the cuticular layers above it may act as a chromatographic column. Even in the short space provided, it may achieve some separation of the mixture of substances that diffuses up into it, and show a partial separation as a different pattern of deposition within the cuticle itself or on the surface.

The uptake of mineral nutrients by roots is associated with an exchange of ions between the root surface and the soil solution or soil particles. Epstein (1953) showed that the absorption of potassium and rubidium involves the formation of an intermediate labile complex of the cation with an ion-binding carrier. Reichenberg and Sutcliffe (1954) proposed a mechanism of uptake of ions based on ion exchange by the root membrane and showed that the energetics of transport by ion exchange account readily for the fact that ions may be absorbed against a concentration gradient. An expenditure of free energy is required.

The cuticle on the aerial parts of plants may absorb and transmit ions in a similar manner. An ion exchanger consists of a polymeric skeleton, insoluble in water and organic solvents, which is held together by cross-linkages from one chain to the next. Ion exchange groups carried on the skeleton exchange readily with other ions in a surrounding solution without any change occurring in the material. When immersed in water, the exchanger takes up water which fills the pores of the network and the polymer swells. The cutin of the cuticle is a polymer composed of chains of hydroxy-carboxylic and carboxylic acids connected chiefly by ester linkages. There is evidence that not all of the carboxylic groups of the cutin acids are esterified (Brauner, 1930; Brieskorn and Böss, 1964); those not involved in the linkages are likely to be ionised, with the hydrogen ions exchangeable with cations. Cutin thus has the attributes of a weak cation exchanger. Seaweeds act as cation exchangers due to the presence of alginic acid, a complex hydroxy-carboxylic acid (Wassermann, 1948). Cellulose, located more deeply in the epidermal layer, is also a weak cation exchanger. The precise chemical nature of the pectinaceous component of the cuticle has not been established. Pectin itself is a linear polymer of D-galacturonic acid residues of pyranose structure, with most of the carboxyl groups of the galacturonic acid residues methylated. Pectic materials function in the plant as carriers of cations and probably act as polyelectrolytes rather than as ion exchangers in the accepted sense because of their tendency to disperse in water. pH has an important influence on the efficiency of a weakly acidic exchanger which cutin may be expected to be. At low pH values the exchanger is in the undissociated acid form and no longer able to accept cations. The process of transmission through the cuticle by ion exchange, if it occurs, is complicated by the presence of wax constituents and of tannin. It is of interest to note that chitin, an acetylated polyaminoglucose derivative present in the cell walls of fungi, is likely to act as an exchanger of anions.

A mechanism of cation exchange through the cuticle would give a preferential absorption of cations over anions; if cations and anions enter at equal rates, a process of diffusion through the water-filled pores of the polymer can be assumed. Yamada, Wittwer and Bukovac (1965) examined the penetration of organic and inorganic substances through the enzymically isolated astomatous cuticles of the onion (*Allium cepa*) leaf and tomato fruit. Urea penetrated more readily than ions, and cations than anions. They suggest that the more rapid penetration of cations than anions may be partially explained by the greater binding of the cations on the negatively charged cuticular surfaces opposite the site of initial entry. The high permeability of urea could not be explained by binding and some factor, other than free diffusion, leading to increased cuticle permeability seemed to be involved. Kuykendall and Wallace (1954) found that the uptake of urea by citrus leaves was directly proportional to the concentration. Yamada, Rasmussen, Bukovac and Wittwer (1966) identified binding sites for calcium and chlorine ions and urea on the tomato cuticular membrane by autoradiography. Urea was bound to a lesser extent than the inorganic ions, but there was no localisation of binding on either the smooth outer or irregular inner surface of the membrane. The binding of naphthaleneacetic acid and naphthaleneacetamide on the isolated and intact cuticular membrane of pear leaves was examined by Bukovac and Norris (1966). The abaxial cuticular membrane bound more than the adaxial, and binding appeared to be due to sorption into the cuticular components with electrostatic forces playing a minor role. Silva Fernandes (1965b) found that the isolated cuticular membrane, 2 cm^2 in size, of the apple fruit absorbed about 35 μg of mercury before diffusion occurred.

Middleton and Sanderson (1965) examined the uptake of ions by barley leaves. Caesium, strontium, iodine, sulphur and phosphorus were applied, as radioactive nuclides, by the immersion of leaves or in droplets under controlled environmental conditions. Evidence was obtained that an ion-exchange mechanism, in addition to metabolic processes, operated in the uptake. Cations were taken in more rapidly than anions, the rate of uptake from the solution in which the leaves were immersed decreased as the concentration was increased, and more cations were removed by washing with salt solution than with water. The effects were more marked when absorption occurred from immersion in solution than from droplets. The fact that the rate of uptake of cations was not directly proportional to concentration indicated that, after the saturation of the cation exchange sites, the rate of uptake was limited by the rate of transport into the plant. The lack of correlation between concentration and uptake of cations and the lower uptake of anions than cations provided evidence of the part played by fixed negative charges in the absorption process.

Keppel (1967) also postulates an ion-exchange mechanism to explain the uptake of cations by leaves. He examined the absorption of sodium, potassium, calcium, caesium, strontium, iron and phosphorus, as radio-nuclides, by the leaves of potato, sugar-beet (*Beta vulgaris*) and barley. The abaxial surface participates only to a small extent in the exchange reactions. The process of uptake involves an initial absorption and the formation of a layer of cations on the 'pectin-like' substances of the cuticle. The ions are exchangeable and their exchange activity depends on their hydration and valency. Ion exchange is followed by a phase of active transport during which the ions diffuse to the wall of the epidermal cells and enter the cytoplasm. Support for an ion-exchange mechanism was obtained from observations that differences occur between cations in their sorption in the surface layer and subsequent transmission into the leaf and that the exchange capacity is unchanged at a relatively low temperature whereas active transport is almost halted. The sugar-beet leaves have a lower sorptive capacity than potato or barley due, it is thought, to a higher degree of polymerisation in the sugar-beet surface layer. The leaves show a relatively low ion-exchange capacity in agreement with their ability to take up mineral nutrients in small quantities. Mecklenburg, Tukey and Morgan (1966) ascribe the leaching of calcium from foliage to a process of diffusion and of ion exchange which involves exchange sites both within the leaf and on the leaf surface. Brian (1967, 1968) concludes that the uptake of a cationic herbicide is consistent with current knowledge on cation absorption into plant tissue. Initial rapid adsorption either on or near the leaf surface is followed by slow absorption. He discusses the loss of efficiency of cationic and anionic herbicides due to adsorption by surface components.

The effect of chemicals on cuticles

Chemicals applied to plants may effect the development of their cuticles. The normal leaf of the pea (*Pisum sativum*) is so water-repellent that droplets are reflected. Dewey, Gregory and Pfeiffer (1956) and Brunskill (1956) observed that when pea plants were grown in soil previously treated with TCA (trichloroacetic acid) the reflection of droplets was eliminated. Pre-emergent treatment of the soil with dalapon (2,2-dichloropropionic acid) also reduced the water-repellency of the leaves. The effects of the soil treatments were examined by Pfeiffer, Dewey and Brunskill (1957) and Dewey, Hartley and MacLauchlan (1962). When cut pea plants were left in the laboratory, those grown in soil treated with TCA wilted first, and the rate of loss of water increased with increasing rates of application of TCA to the soil. At levels of application up to an equivalent of 10 lb/acre there was no obvious sign of damage to the plant, only a reduction in the wax bloom. The effect was almost confined to the fraction of the wax available to solvent in the first few seconds of immersion of the leaves; this fraction was responsible for the reflection of water. Juniper (1959a, b) showed that the contact angles of water on the pea leaves decreased with increasing rates of application of the chemicals. Dalapon at 0·32 lb/acre lowered the angle only from 144° to 135° but altered the shape of the crystals developed on the surface. TCA at 0·63 lb/acre reduced the angle to 129°, at 1·25 lb/acre to 118° and at 2·5 lb/acre to 68°. The progressive reduction of the wax bloom is shown in Fig. 8.6. Gentner (1966) found that N,N-di-*n*-propylthiolcarbamate applied to the soil inhibited wax formation on cabbage seedlings, increasing their susceptibility to dinitrobutylphenol sprays and to the fungus *Rhizopus stolonifer*. The suppression of wax and a consequential increase in the rate of transpiration of the seedlings were proportional to the rate of application of the compound. Several other thiolcarbamates also increased the susceptibility of plants to subsequently applied herbicides.

Wortmann (1965a, b) made electron microscopic studies of the changes in the surfaces of rape (*Brassica napus*), wheat and sugar-beet leaves caused by pesticides. The wax structures were changed and wettability increased by treatment with MCPA (4-chloro-2-methylphenoxyacetic acid) or parathion (diethyl 4-nitrophenyl phosphorothionate); a similar effect was also given by a commercial wetting agent. Different effects were shown by MCPA ester and salt, the effect of the ester being the more severe. The influence on the wax and the increased wetting were greater at the higher concentrations used. Parathion prevented the regeneration of the wax surface structure on the rape leaves after a change in structure had been brought about by brushing or spraying with the wetting agent. There was no regeneration of wax structure on the wheat leaves after any of the treatments. The compounds 2-chloroethyl-trimethylammonium chloride (chlormequat) and N-dimethylaminosuccinamic acid (B 9) have become commercially important as retardants of plant growth. Baker, Dawkins and Smith (1968) examined their effects on the cuticles of blackcurrant leaves. Chlormequat suppressed wax formation during a short period soon after application, whereas B 9 stimulated the production of wax during most of the growing season.

There is some evidence that fungicides may affect leaf epidermal layers. Barner and Röder (1962) examined the effects of fungicides on the leaves of apple, vine and poplar. Copper oxychloride and folpet (N-trichloromethylthiophthalimide) caused thickening of the tissues, the oxychloride increasing the development of the epidermal cell walls and folpet the cuticles. Dithiocarbamate treatment gave leaves with a thin epidermal layer. Mercurial sprays have been widely used for the control of fungi. Batt and Martin (1961) found a lower content of cutin in the cuticles of *Cox's Orange Pippin* apple fruits after treatment of trees with phenylmercuric acetate. The mercury compound may have suppressed enzymic action involved in the formation of cutin. More attention should clearly be given to the effects of chemicals on cuticles.

The cuticle itself may influence the efficiency of pesticides. The protective action of a copper fungicide is augmented by the release of water-soluble copper from the deposit on the plant surface. Wain and Wilkinson (1943), Martin and Somers (1957) and Arman and Wain (1958) obtained evidence that plant secretions promote the solubilisation of the copper. Kovàcs and Cucchi (1964) found that substances on the surface of leaves play a part in the decomposition of zineb (zinc ethylene*bis*dithiocarbamate) as a rule enhancing its activity. DDT deposited from an emulsion passed into the wax of apple leaves, but in this form was no less toxic to insects than an equivalent amount of crystalline material on the surface (Martin, 1957). The addition of a proprietary mixture of amine stearates to DDT wettable powder formulations increases the resistance of deposits on cotton leaves to weathering by rain, and more DDT is absorbed into the leaves from deposits containing the stearates than from deposits of wettable powder alone (Phillips and Gillham, 1968). Burt and Ward (1955) examined the influence of the plant wax upon the toxicity and persistence of DDT crystals, and concluded that absorption by a typical wax is unlikely to influence the toxicity of the deposit provided it does not penetrate more deeply than the wax layer. Dunn, Brown and Montagne (1969) suggest that the efficiency of many fungicides may be reduced by antagonistic materials which occur either on the leaf surface or are available from the plant tissue.

Leaf secretions may promote injury to the plant by an applied chemical. Solubilised copper is likely to be more damaging than the insoluble deposit on

---

FIG. 8.6 Electron micrographs of carbon replicas showing the effect of TCA (trichloracetic acid) and dalapon (2,2-dichloropropionic acid), fed into the soil, prior to germination, on the fine structure of a leaf surface. **A**, control plant of pea (*Pisum sativum* var. *Alaska*) grown at 5000 ft candles (406 lux) light intensity. Advancing contact angle of water 144°. **B**, effect of 0·28 kg/ha of dalapon. Contact angle 135°. *C*, effect of 0·56 kg/ha of TCA. Contact angle 129°. *D*, effect of 1·11 kg/ha of TCA. Contact angle 118°. *E* and *F*, effects of 2·22 kg/ha of TCA. Contact angle 68°. Cr/Au/Pd

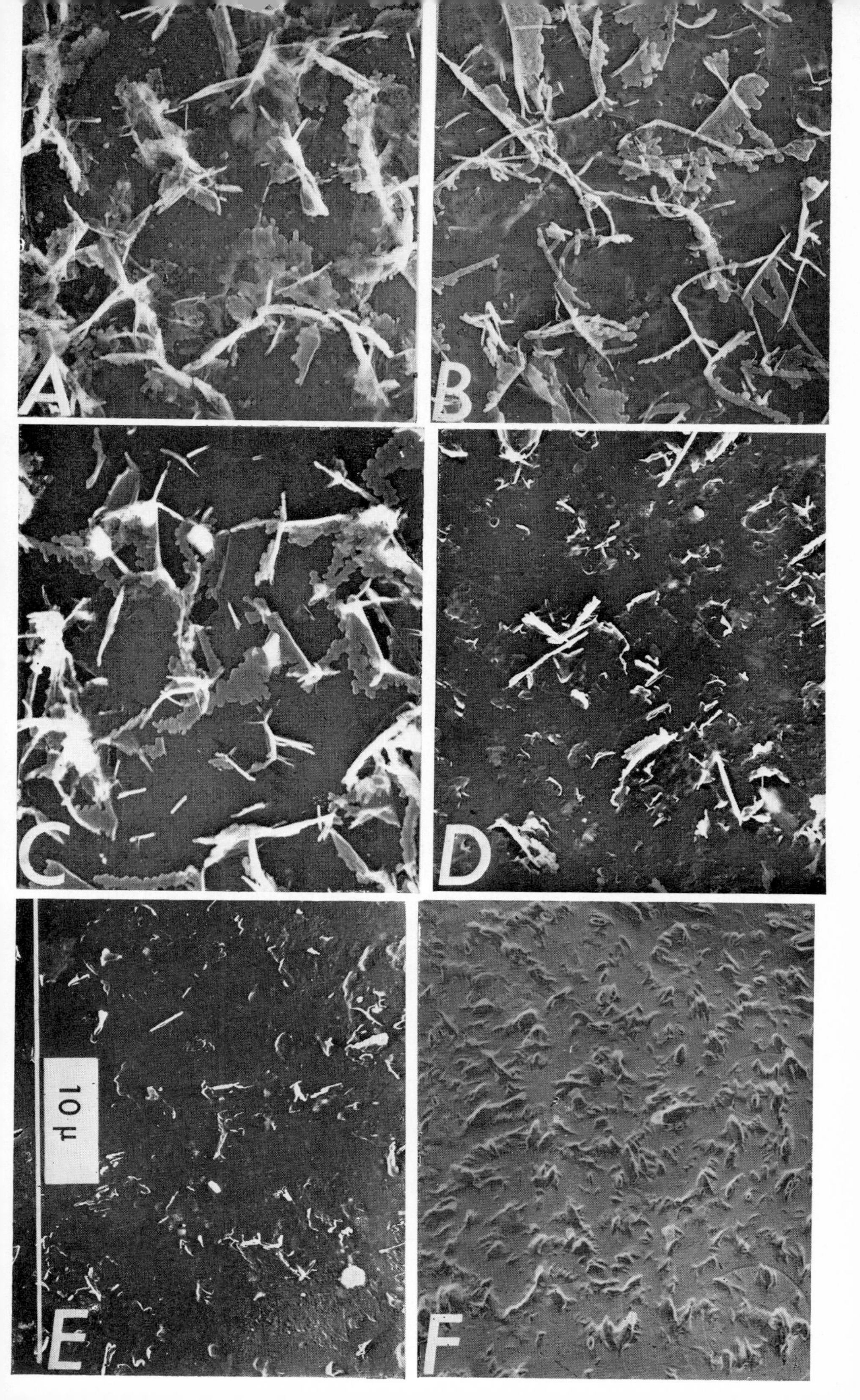
A
B
C
D
10 μ
E
F

the surface. The part played by guttation fluids (Chapter 7) in particular merits examination. Curtis (1944) suggested that guttation fluids increase the solubility of copper deposits, the copper in solution is absorbed into the leaf and injury is induced within the cellular tissue.

The weathering of plant surfaces

All plant surfaces are subject to weathering. Damage is inflicted by the rubbing of leaves on the same or neighbouring plants, by rain, by water splash, by the deposition of foreign bodies such as soot and oils and, probably most important of all, by the scarifying effect of wind-borne sand and dust. The effect of weathering on the properties of the leaf surface can often be dramatic. The copper retained per unit area of banana leaf (*Musa*) increases with leaf age with or without the use of a surfactant (Fulton, 1965). Dewey, Gregory and Pfeiffer (1956) showed that wind, soil blown by the wind and heavy rain damage the surfaces of pea plants. This increases the retention of dinoseb on the pea leaf surface and reduces the normal selective effect between the pea plants and weeds such as *Galium aparine*, *Polygonum aviculare* and *Urtica urens*. Amsden and Lewins (1966) observed that this type of damage was widespread in Lincolnshire and Norfolk, England in 1965. They found that using a standard 300μm drop a high retention of water on pea leaves occurred; in most cases the crops had suffered either from hail or heavy rain storms. Farmers were advised, when it was found that drop retention was 10% or more, to delay spraying for weed control to enable the plants to produce new and better protected growth.

The effect of artificial abrasion of a cabbage leaf surface by a camel-hair brush is shown in Fig. 4.28. Any other surface shows a similar pattern of superficial damage under the same conditions. A brushed pea leaf is shown in Fig. 8.7**A** and a slightly damaged white clover leaf in Fig. 2.6. Amsden and Lewins (1966) devised a simple field method for assessing the wettability of a surface. They dip the whole plant into a 0·5% solution of Crystal Violet with a surface tension of 60 dyn/cm (0·06 N/m) or above. The dye is retained only on the damaged areas. A qualitative assessment can then be made either by photography or by 'contact printing' the freshly dipped leaf onto white paper. Both an adaxial and abaxial estimation, using two pieces of paper simultaneously, can be obtained with the latter method. Amsden and Lewins summarise their observations on field retention as follows. Seed leaves are often highly retentive even when the true leaves are reflective, e.g. kale (*Brassica oleracea* var. *acephala*) and knotgrass (*Polygonum aviculare*). Leaves with smooth

---

FIG. 8.7**A** Electron micrograph of a carbon replica of an adaxial leaf surface of pea (*Pisum sativum* var. *Alaska*) 6 days after brushing with a camel-hair brush. Note the absence of repair of damage and the dense colonies of micro-organisms. Cr/Au/Pd

FIG. 8.7**B** Electron micrograph of a carbon replica of an adaxial leaf surface of pea, as above, after treatment with 2·22 g/m^2 of TCA into the soil prior to germination. Note the absence of wax and the large numbers of micro-organisms. Cr/Au/Pd

FIG. 8,7

wax on their leaves, e.g. broad bean (*Vicia faba*), retain patches of a thin film of the dye. Some plants such as ground ivy (*Glechoma hederacea*) readily absorb the dye. Leaves with a crystalline epicuticular wax, e.g. cereals, grasses and peas, usually show no retention on the young leaves, but an increased retention as they age, due either to weathering, insect damage or handling. The growing points, buds and flower parts of many weeds, e.g. mayweed (*Anthemis* sp.), chickweed (*Stellaria media*) and nettle (*Urtica dioica*), may be relatively non-retentive whereas the rest of the plant is retentive.

Most bloomed plants have some ability to recover from mechanical damage while the leaves are still growing. Some plants, e.g. *Chrysanthemum segetum* (see Fig. 2.10), can recover from mechanical or weathering damage even in old age. Hallam (1967) reports that the leaves of *Eucalyptus* species recover from mechanical damage very quickly indeed. He removed wax by rubbing the leaves with cotton wool; after 3 hr recovery was apparent under the electron microscope and after 48 hr was complete. Peas, however, recover hardly at all (see Figs. 2.7 and 4.21). All of this work concerns the recovery from damage of the superficial wax of the cuticle. De Vries (1968) has also shown that the surface of an apple fruit may respond to damage by inducing excessive cutinisation.

Within a single plant, e.g. *Eucalyptus*, different types of wax have different capacities to resist mechanical damage. Hallam (1967) has shown that the tube waxes of many *Eucalyptus* species are easily removed by rubbing, whereas the plate waxes are difficult to remove. Pielou, Williams and Brinton (1962) found that insecticides tend to be more readily retained on the abaxial than on the adaxial surfaces of the leaves of apple and cherry. They suggest that the differences in retention may be due to some difference in structure between the two surfaces, possibly in the ultrastructure of the wax. Irrigation and weathering tend to accentuate these differences. Weathering of leaf surfaces affects all the surface properties of a leaf, apart from retention. It makes a wide range of plant cuticles more permeable to weedkillers. Sargent (1966) noticed that material collected from the field or even the glasshouse is penetrated much more readily than material grown under constant environmental conditions. Weathering also reduces the ability of cuticles to shed waterborne dust and spores.

The superficial wax structures of many plants are extremely delicate; even a fine spray can destroy the ultrastructure and increase the wettability of a surface such as that of *Hyacinthus orientalis* (see Fig. 4.29). Others such as many *Oxalis* or *Lupinus* species (Fig. 8.3) are highly resistant even to extreme weathering. The soluble secretions on such plants as *Kleinia articulata* are eventually washed away completely. Non-bloomed or damaged surfaces (see Fig. 8.7**A** and **B**) are frequently contaminated by bacteria, spores and dust, whereas surfaces with well-developed blooms (see Fig. 2.7) are often almost completely clean (Juniper, 1960b). The dirt found on non-bloomed surfaces may consist of deposited particles of aerosol size and of dust washed out of the atmosphere by rain or splashed up from the soil. Rain droplets are completely reflected from and fail to deposit dust on surfaces with particulate wax deposits, but are retained even on waxy surfaces if the deposits are not particulate. Günther and Wortmann (1966) suggest that bloomed leaves are more readily cleansed by rain than non-bloomed because large dust particles greater than 1 μm adhere less firmly to the surface of the bloomed leaf than to water. Smaller aerosol particles are occluded between wax projections and not readily removed, presumably because they do not come into contact with water droplets. Günther and Wortmann also note that relatively smooth wax layers, e.g. that of *Crambe maritima*, are cleansed to a greater extent than rougher surfaces, e.g. that of *Chelidonium majus*. On the other hand the adhesion of particles of aerosol size to the surface of non-bloomed leaves is greater than to water. Even large hydrophilic particles may not be washed from them. Moreover rain-carried or rain-splashed dirt adheres to non-bloomed surfaces. Plants accumulate radioactive dusts very readily (Gorham, 1958). Middleton (1958) suggests that waxy surfaces may act as a barrier to the uptake of radioactive sprays falling on plant leaves and Kimber and Booth (1958) have shown that different plants have very different capacities to retain the fall-out of radioactive material.

Some interesting but puzzling observations on the loss of radioactive material from leaf surfaces were made by Moorby and Squire (1963). Radioactive isotopes were applied in a fine spray or in small droplets to the foliage of cabbage, potato and rye grass (*Lolium perenne*). Although the leaves were kept dry the radioactivity was distributed to nearby plants. A greater tendency to loss during summer than winter suggested that some feature of the rapidly growing plants was responsible. Moorby and Squire suggest that the radioactive material may have been distributed by detachment of wax particles, but the potato, unlike cabbage and rye grass, does not have a particulate wax on its surface. A seasonal loss of particulate material can occur from a leaf surface, possibly due to the expansion of the leaf under an inelastic surface deposit. The precise mechanism of loss, however, has not been identified.

The contamination of plant surfaces

Soot is frequently found on leaf surfaces of plants grown in urban areas (Fig. 8.8). It occurs as clumped particles with the constituency of tiny granules of tar. Diesel fumes have an almost identical appearance under the electron microscope (Frey and Corn, 1967). Soot and exhaust particles are hydrophilic and sticky. They are not readily taken up into suspension by rain drops, but stick tenaciously both to wettable and unwettable surfaces. Once in contact with the leaf, although the leaf itself may be undamaged, they modify the properties of its surface. The deposits act as retentive coverings to which dust and spores become attached. Eveling (1967) has shown that inert dusts such as silica or titanium oxide applied to leaf surfaces also modify the properties of the cuticles. Permeability, water loss and gas exchange are affected by processes that are unknown. Dusts deposited by natural agencies may have similar effects.

Darley and Middleton (1966) and Saunders (1968) have reviewed the problem of atmospheric pollution and the damage that such pollution can do

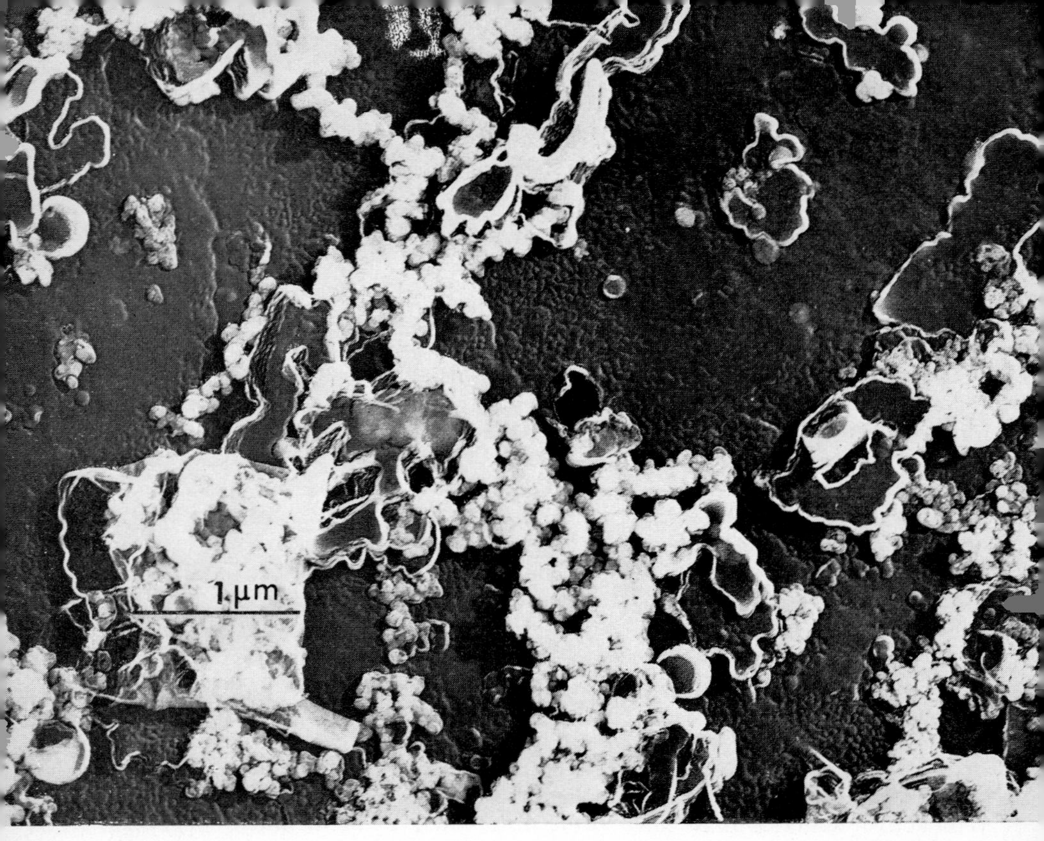

FIG. 8.8 Electron micrograph of a carbon replica of an adaxial leaf surface of cabbage (*Brassica oleracea* var. *capitata*) grown in the open near the chimney of an oil-fired boiler. The dense white granules are soot. Cr/Au/Pd. (Courtesy P. R. Williams)

to plant growth. The problem is causing increasing concern. Nitrogen oxides occur in polluted atmospheres and may exceed 1 part per million (p.p.m.) by volume in the air over densely populated industrialised areas. Exposure of plants to concentrations of 0·3–0·5 p.p.m. of nitrogen dioxide for 20 days may significantly reduce growth without evidence of acute symptoms and at higher concentrations the absorption of carbon dioxide may be suppressed (Taylor, 1968). The stomata of plants are highly sensitive to several atmospheric pollutants; sulphur dioxide caused marked stomatal closure during exposure to polluted air (Mansfield, 1968). Spierings (1968) reports reductions in crops resulting from necrotic injuries to leaves; sulphur dioxide affected fruit trees and hydrogen fluoride bulb plants and cyclamen. Atmospheric fluoride affects the rate of oxygen uptake in plants (Weinstein and McCune, 1968).

A relatively new phenomenon, termed photochemical pollution, results from atmospheric reactions between oxides of nitrogen and certain hydrocarbons present in the exhaust fumes from vehicles. The recognised photochemical reaction products include peroxyacetyl nitrate and to a lesser extent peroxypropionyl nitrate, and ozone. The peroxyacetyl nitrate causes a metallic sheen on leaves typically confined to the lower surface, and ozone a mottling of the upper. Photochemical pollution is now recognised in major metropolitan areas in USA, Europe and Japan (Darley, Taylor and Middleton, 1968). Mudd (1968) has detected biochemical effects of peroxyacetyl nitrate and ozone on certain amino acids and proteins. Wood (1968) states that oxides of sulphur, fluorides, ozone and peroxyacetyl nitrate are the most important pollutants affecting forest trees. Drift from highly industrialised areas provides a serious threat to forests; a chlorotic decline of the ponderosa pine (*Pinus ponderosa*) was found within a range of 60 miles from such an area. Bystrom, Glater, Scott and Bowler (1968) have shown that air pollution can alter the pattern of surface wax development and thus will affect many of the properties of the cuticle as a whole. The effects may be subtle and difficult to detect and may predispose the trees to attack by pathogens. Heck (1968) points out that although the stomata are thought to be the route of entry of gaseous toxins, some plants are known to be sensitive to one pollutant but not to another when stomatal function is not involved. He cites the physiological and environmental factors which influence susceptibility; some of these may well affect the development of the cuticle which merits consideration in relation to uptake.

9

The Cuticle in Action, III Interactions with Pathogens

An important function ascribed to the cuticle is the protection of the plant against pathogens. Even closely related plants differ greatly in their resistance to fungal disease; a rose bush may be devastated by mildew and another alongside it be unscathed. Much work has been done to unravel the basis of resistance to disease and the role of the cuticle has received a due measure of consideration. The cuticle provides a potential barrier to the entry of fungi, bacteria and viruses and may also play a part in protecting plants against insects. Other aspects in which the cuticle becomes involved with organisms include the specialised trapping of insects by insectivorous plants.

Micro-organisms on the plant surface

The surfaces of plants carry a non-parasitic flora apart from pathogens that may be deposited upon them (see Fig. 8.7). Last (1955) coined the name 'phyllosphere' for the environment provided by the leaf surface for the growth of micro-organisms and showed that *Sporobolomyces*, *Tilletiopsis*,

Bullera and *Cladosporium* spp. of fungi are usual inhabitants of the surfaces of cereal leaves. In work on epiphytes in Indonesia, Ruinen (1956) independently proposed the name phyllosphere for the leaf surface as a medium for micro-organisms and showed that the fungal population increases with the age of the leaf, that the ratio of fungi to bacteria is lowest when the leaf is fully grown and that the species of bacteria and yeasts present are characteristic of the host plant. Nitrogen-fixing bacteria (*Beijerinckia*, *Azotobacter*) are present. The phyllosphere is first colonised by bacteria, actinomycetes and fungi and the nutrients supplied by the leaf exudates and by the nitrogen-fixers provide an adequate substrate for successive colonisers. Algae, yeasts and lichens appear and a mixed population of unicellular organisms develops. Arthropods may follow in due course (Ruinen, 1961). Leben (1961) distinguished between 'resident' organisms which are normally present and 'casual' organisms which are deposited; he concluded that most of the inhabitants of the phyllosphere are of the casual kind. Leben (1968) has examined the growth of bacteria on plant surfaces; lactic acid, nitrogen-fixing and cutin-degrading organisms (see Chapter 10) as well as coliforms, erwinias, pseudomonads, xanthomonads and streptococci were isolated from the aerial parts of plants. Last and Deighton (1965) report that bacteria and yeasts on leaves may amount to 10^7 cells/cm^2 of surface.

Interactions occur between the organisms on the surfaces. They compete for nutrients and may secrete toxins harmful to others or substances that are mutually beneficial. A weakening of the surface structure, e.g. by a cutin-degrading bacterium or fungus, may assist attack by another organism. The presence of *Erysiphe graminis tritici*, a normal parasite of wheat, on wheat leaves predisposes the plant to infection by *E. graminis hordei* which normally invades barley (Moseman and Greeley, 1964). Pathogens may have a resident phase on leaves; Crosse (1959) obtained large numbers of cells of *Pseudomonas mors-prunorum*, the causal agent of bacterial canker, from healthy cherry (*Prunus cerasus*) leaves and concluded that the multiplication of the pathogen on the leaves may provide the inoculum for infection.

There is mounting evidence that the phyllosphere microflora may be active in suppressing attack by pathogens. Crosse (1959) found that the non-parasitic flora of cherry leaves was dominated by colonies of at least three species of bacteria, the most common of which reduced the ability of *P. mors-prunorum* to infect when present in the same inoculum. A number of fungi occur normally on the surface of strawberry (*Fragaria chiloensis* var. *ananassa*) fruits, but only a few invade the sub-surface tissue. Among the non-pathogenic residents are found *Aureobasidium pullulans* and *Dendrophoma obscurans*, both of which secrete diffusible materials that suppress the growth of the pathogen *Botrytis cinerea* (Bhatt and Vaughan, 1963). Resident bacteria may inhibit pathogens before penetration occurs, or as secondary invaders they may share in, or limit, tissue destruction (Leben, 1968). Water suspensions of a number of fungi found on sorghum (*Sorghum vulgare*) leaves inhibit the germination of conidia of *Colletotrichum graminicolum*, the anthracnose pathogen, and suspensions of the phyllosphere fungi of pearl millet (*Pennisetum typhoideum*) and chick-pea (*Cicer arietinum*) inhibit the germination of uredospores of the pathogenic rust fungi *Puccinia penniseti* and *Uromyces ciceris-arietini*. When a mixture of the phyllosphere fungi and the uredospores was used to infect host seedlings, rust infections were very much reduced (Sinha, 1968).

Pathogens themselves interact on plant surfaces; the attack by a pathogen on its normal host may be suppressed by the presence of a pathogen of another. *Uromyces phaseoli*, a pathogen of bean (*Phaseolus vulgaris*), does not attack sunflower (*Helianthus annuus*) leaves, but the presence of its uredospores on sunflower leaves prevents infection by *Puccinia helianthi*; similarly the bean leaves are protected from *U. phaseoli* by the presence of *P. helianthi*. Different races of *P. antirrhini* compete on the leaves of antirrhinum (*A. majus*) to the benefit of the host (Yarwood, 1956). Such observations open up the intriguing possibility of a biological approach to the control of fungal diseases. The micro-organisms of the phyllosphere and their interactions in relation to the infection of plants have been described by Leben (1965) and Sinha (1965).

Interactions of the cuticle with fungi

Attachment and penetration by fungi

Fungal pathogens are carried to the plant surface in air currents, rain drops or by other vectors such as insects and mites. Airborne pathogens appear to have little difficulty in achieving contact with the surface although the chances of deposition must be limited by the turbulence and 'side-slip' of air streams; once lodged, they may be washed off by subsequent rain. The deposition of water-borne pathogens is dependent upon a reasonable degree of wetting of the surfaces. The ability of leaves to repel water droplets differs greatly; some leaves are virtually non-wettable, others easily wetted. Davies (1961) found different patterns of deposition of water-borne spores on plant surfaces according to the ease with which the surfaces were wetted; the deposition of spores from droplets was influenced by the degree of wetting of the spores themselves in addition to the ease of wetting of the surfaces. The effect of wettability on the retention of spores is shown by a comparison of Figs. 2.7 and 8.7. The glaucous coating of *Eucalyptus bicostata* affects the deposition of conidia of *Phaeoseptoria eucalypti* (Heather, 1967). The water repellency of a plant surface is clearly an important factor in preventing the establishment of infections from water-borne inocula.

Most spores cannot germinate except in a drop of water or film of mist or dew. This supposition is supported by the fact that unwettable leaf surfaces rarely if ever carry pathogens whereas wettable surfaces are frequently covered (Figs. 8.7**A** and **B**). These generalisations only apply to bacteria and fungi whose propagules are substantially larger than the crystalline wax masses on a leaf surface. The importance of the wetness of the leaf surface for successful attack has been shown by Dickinson (1949a, b, c, d and 1955). Spores of *Botrytis cinerea* deposited onto a cuticle by day are able to take full advantage of the few hours of dew at night to germinate and to be sufficiently far advanced not to be hindered by the desiccating conditions later experienced (Robinson, 1967). Robinson also records that the amount of moisture present is very critical. At 80% humidity *Plasmopora viticola* spores will infect mature vine leaves, but at 70% humidity will only infect young leaves. Some leaf surfaces condense and shed mist much more readily than others; this could be a factor in determining resistance to disease by hindering spore germination at dew fall.

Fungi penetrate into plants either through the intact cuticle, through natural openings such as stomata or through wounds. Many fungi penetrate directly through the cuticle. Some, e.g. *Phytophthora infestans* responsible for the blight disease of potato leaves, were earlier thought to use the stomata, but are now known to penetrate the cuticle as well. Rust-causing fungi as a class prefer the stomatal route, but penetration of the cuticular linings of sub-stomatal cavities is then necessary. It is generally believed that the pathogens that penetrate directly through the main cuticle do so solely by mechanical pressure. Brown and Harvey (1927) suggested that the only satisfactory explanation of direct penetration is that the stimulus to penetration is one of contact and the means of penetration is purely mechanical. The germ tubes of fungi such as *Erysiphe graminis* or *Peronospora*, after emerging from the spores, grow a short way and swell to form an attachment organ called an appressorium. This process appears to be initiated primarily by a contact stimulus from the waxy cuticle. The two fungi parasitise cereals (Gramineae) and crucifers (Cruciferae) respectively, most of which have waxy cuticles. From the appressorium a fine infection thread forces its way through the cuticle into the host tissue. The cellulose cell wall swells as the infection hypha passes through it, suggesting an enzymic action, but no firm evidence of enzymic dissolution of a cuticle by a pathogen has yet been obtained.

Some results by Dickinson (1964) suggest that fungal hyphae have preferred paths over the surface of a cuticle at right angles to certain lines of stress. He also demonstrated that their growth on synthetic membranes follows the same pattern. Dickinson studied the growth of uredospore germ tubes of *Puccinia* species on the surfaces of membranes of polystyrene, polymethyl methacrylate and nitrocellulose. The direction of growth of the germ tubes was at right angles to ridges on the surface of the membranes, i.e. to orientated lines of macromolecules in the membranes. The number of ridges on the surface of the membranes could be changed and the ***thigmotropic*** response was found to vary with their frequency. When the frequency of the ridges was multiplied by their height a constant value K was obtained. Dickinson suggests that any membrane, irrespective of its chemical constitution, will induce the thigmotropic response provided that its K value reaches a certain level. The minimal size of structure that can be detected by a germ tube of these rusts is the size of one cellulose macromolecule 120 nm $\times$ 1·0 nm. Dickinson's hypothesis is that the thigmotropic response results from contact of the germ tube with a repetitive series of changes in thickness in a surface such as a vinyl plastic or with the alternating presence and absence of the ridges in nitrocellulose. The orientated direction of growth is induced by the alignment of one or more of the chitin fibril layers of the germ tube wall parallel to the lines which provide the stimulus. The angle of traverse of the line of stimulus is about 70° connected, it is suggested, with the existence of transversely oblique fibrils in the germ tube wall. The spiral direction of the transversely oblique fibrils changes from clockwise to anti-clockwise with each new growing point, thus explaining the zig-zag growth of a germ tube.

Most monocotyledonous leaves and some dicotyledonous are ridged parallel to the long axis (see Fig. 2.4 and 9.1). Therefore, according to Dickinson, fungi on these surfaces tend to grow across and not along the leaf. The selective advantage of this pattern of growth to the fungus is still obscure. Even when the infection hypha has passed into the stomatal cavity it still has to penetrate through the thinner cuticle of either the stomatal cavity itself or one of the mesophyll cells below. No information except for the limited observations of Arens (1929) seems to be available as to where this penetration actually takes place.

Although views on the existence of a cutinase to assist the penetration of hyphae are conflicting, there is evidence of the secretion of a wax-dissolving enzyme from hyphae. Arens (1929) reported that the conidia of *Plasmopora viticola* secreted some active material when on leaf surfaces. Hall (1968, personal communication) has studied the growth of hyphae of leaf rust (*Puccinia rubigo-vera tritici*) on wheat variety *Hilgendorf*. Carbon replicas of the leaf surface infected with the fungus leave no doubt that the epicuticular wax is dissolved in the vicinity of the hyphae (Fig. 9.1). The wider cleared area around the bulbous tip of the hypha suggests that secretion of the enzymes or solvent takes place predominantly from the apical tip. It is interesting that cuticular ridges appear in the cleared area around the hypha (***R***) whereas no such ridges are visible even in the bare areas farther away from the hypha (***B***). A few platelets of wax lie on the cuticle close to the hypha. Hall believes that this is due to the recrystallisation of the wax already partly dissolved by the hypha. If wax-degrading enzymes are present in some fungi it is not impossible that cutin-dissolving enzymes might be present too, but the minute size of the entry pore (in *Puccinia graminis* less than 1 μm in diameter) and the relative crudity of detection methods prevent their being observed. Lipophilic enzymes are comparatively rare in plants, although a few seeds, e.g. *Simmondsia californica*, have, as food reserves, waxes and not the more usual fats or carbohydrates.

FIG. 9.1 Electron micrograph of a carbon replica of the adaxial leaf surface of wheat (*Triticum vulgare*) on which is growing a hypha of *Puccinia rubigo-vera tritici*. Note the ridges (***R***) visible close to the surface of the hypha. These ridges are not apparent in the bare areas (***B***) amongst the undamaged wax projections away from the hypha. It is possible that the complete dissolution of the epicuticular wax by the hypha exposes extra detail of the cuticle surface. Cr/Au/Pd. (Courtesy D. M. Hall)

FIG. 9.2 Electron micrograph of a carbon replica of the adaxial leaf surface of *Chenopodium album* upon which a mixture of tobacco mosaic virus (TMV) and 'Celite' had been rubbed. The Celite (**C**) and the TMV rods (**V**) are indicated. The undamaged epicuticular wax of this leaf surface is responsible for the lack of wettability as depicted on the dust jacket. Cr/Au/Pd

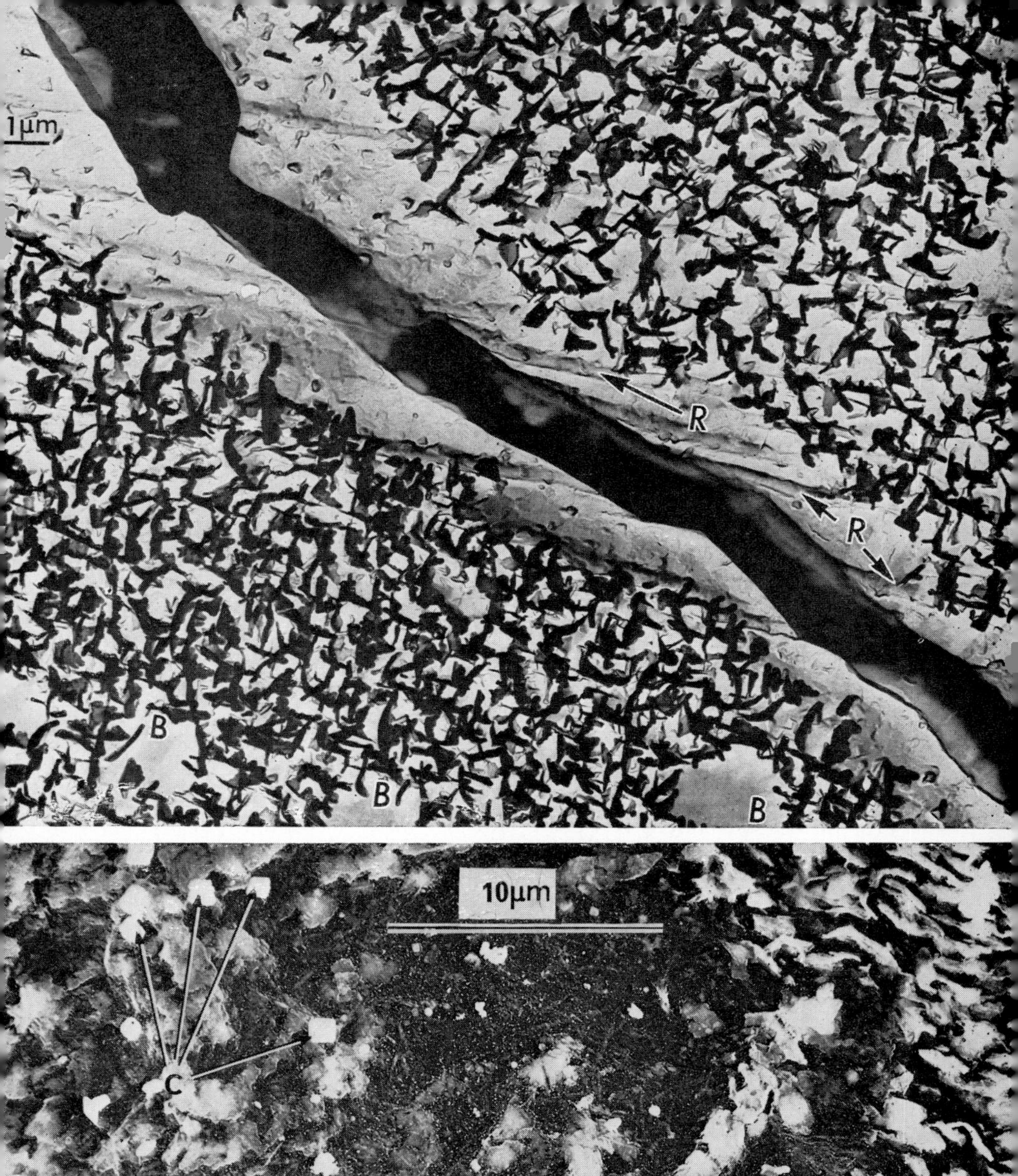
1µm
R
R
R
B
B
B

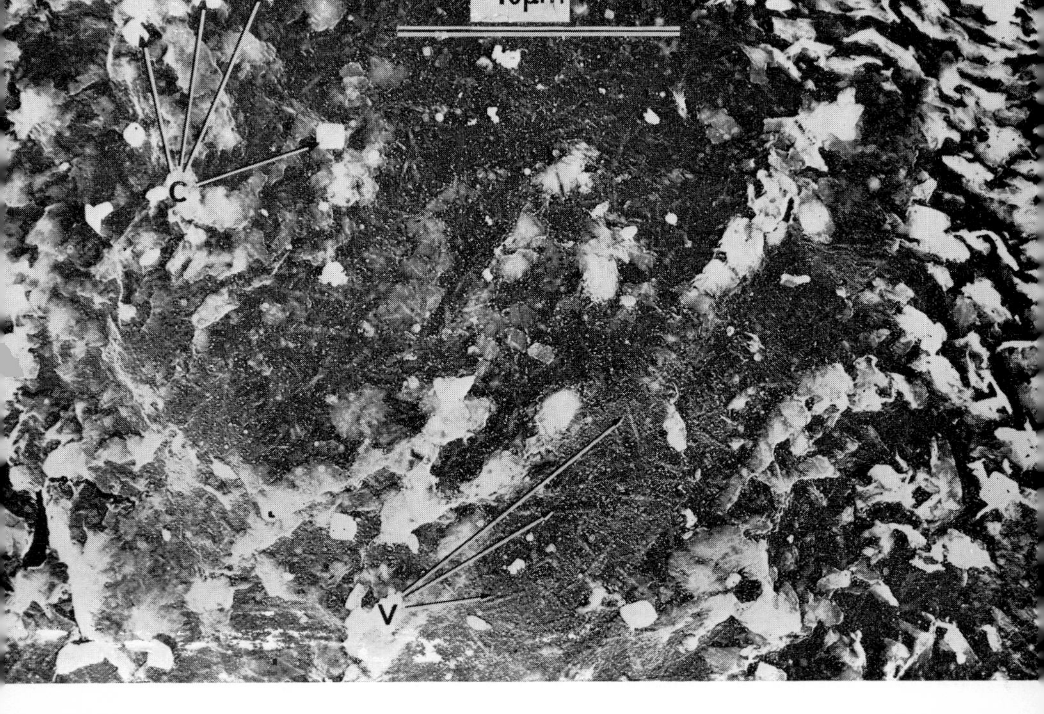
10µm
C
V

Morphology and resistance

Jennings (1962) has produced evidence to suggest that the superficial morphology of the raspberry (*Rubus idaeus*) plays a part in its resistance to a number of fungal diseases. The incidence of spur blight (*Didymella applanata*) was less frequent on seedlings with hairy, spine-free and wax-free canes or non-pigmented canes and also on canes with a moderately dense waxy covering. Grey mould (*Botrytis cinerea*) was similar in its distribution except that it was slightly more abundant on the wax-free canes. Cane spot (*Elsinoe veneta*) was similar to spur blight in its distribution in relation to spininess, wax thickness and pigmentation, but was greater on hairy and wax-free canes. The effectiveness of these cane characters varies greatly from year to year. Cane hairs appear to encourage the repulsion of water from the nodes; this characteristic, as Jennings points out, could enhance the resistance of hairy stems by preventing the moist conditions at the nodes most suitable for fungal infection. The effect of cane hairs on 'run-off' is offset by the waxiness. Droplets of water tend not to run off the surface of waxy raspberry canes and this may be responsible for reducing at the nodes the effectiveness of the water-dispersing property of the hairs. Improved water-shedding from the nodes may explain the finding that the wax-free seedlings have a lower incidence of spur blight, but not of the other diseases. Spur blight is the only disease whose symptoms are frequently confined to the nodes. Thus its incidence may be increased by the waxiness of the stems because the wax impedes 'run-off' from nodes whether or not they are hairy. The wax on the internodes offers resistance not by water-shedding, but by being a fungistatic or mechanical barrier. This suggestion is consistent with the observation that the waxy seedlings are more resistant to cane spot and grey mould, the symptoms of which are usually present between the nodes.

The role of the cuticle as a mechanical barrier to invasion by fungi has been much debated. Some workers support the view that the cuticle acts as a defensive wall, but others reject it. Marks, Berbee and Riker (1965), for example, describe the successful penetration of the young leaves of *Populus tremuloides* by *Colletotrichum gloeosporioides* and the abortive attempts to penetrate the mature, resistant leaves. The resistance of some strawberry (*Fragaria* spp.) plants to the powdery mildew *Sphaerotheca macularis* has been ascribed by Jhooty and McKeen (1965) to the thickness of their cuticles. The penetration of the cuticle, cellulose layer and epidermal cell wall of the susceptible strawberry plants appears to be mechanical, with no apparent dissolution of the cuticular layer around the penetration peg. The upper surfaces of leaves and the stolons of *Fragaria chiloensis* are rarely mildewed, whereas those of *F. ovalis* are often attacked; the cuticle of *F. chiloensis* is consistently up to seven times thicker than that of *F. ovalis*. Similarly Peries (1962) correlates the varietal resistance of strawberry leaves to the fungus with difficulty of penetration of the combined cuticle and epidermal cell wall and with the cutin content of the cuticle.

On the other hand, Mence and Hildebrandt (1966) found no difference in the combined thickness of the cuticle and cell wall of leaves of two varieties of rose which showed a marked difference in resistance to rose powdery mildew. The difference in resistance clearly could not be attributed solely to cuticle thickness, and penetration was not prevented by the cuticle acting as a morphological barrier. The germination of the mildew conidia was apparently stimulated by the leaf surface, but the effect was non-specific and could be exerted by leaves of plants not attacked by the fungus. They suggest that resistance in rose is influenced by factors that operate internally after penetration of the pathogen. A well-developed cuticle does not necessarily prevent infection; in southern England the ornamental shrub *Euonymus japonicus* is severely attacked by the mildew *Oidium euonymi-japonicae* despite an exceptionally heavy cuticular membrane on even the youngest leaves (Roberts, Martin and Peries, 1961). J. T. Martin (1964) has reviewed the role of the cuticle in defence against disease and concludes that in many plants its contribution to protection cannot be great.

Surface stimulation of infection by fungi

It has long been thought that chemotropic factors play an important part in promoting the fungal infection of plants. This view was put forward by Miyoshi (1894, 1895) who showed that various fungi could be induced to penetrate some natural and other membranes only when an attractive chemical substance was placed on the opposite side. Fulton (1906) postulated a negative chemotropism of hyphae to their own staling products. Graves (1916) took a compromise view; while believing that a negative chemotropism was the main effect, he thought that a positive chemotropism could also exist. The passage of the fungal infection thread through the cuticle may be due to a chemotactic response to a nutrient gradient from the cells beneath (Robinson, 1967).

In recent years interest has been taken in substances that, secreted by plants and present on or in the surface layers, stimulate the activities of invading pathogens. Brown (1922) showed that the host may stimulate the parasite before its penetration; nutrients leached from broad bean (*Vicia faba*) leaves into droplets of water or of spore suspensions and detected by conductivity tests assisted the germination of conidia of *Botrytis cinerea*. There have been many reports of the stimulation of this fungus by plant exudates. Sugars in the exudate from grape berries (Kosuge and Hewitt, 1964), leachates from safflower (*Carthamus tinctorius*) blossoms (Barash, Klisiewicz and Kosuge, 1964) and products of exosmosis of carnation petals (Edney, 1967) stimulate the germination of the conidia. Leached materials assisted the infection of castor bean (*Ricinus communis*); high susceptibility to the fungus was associated with large amounts of exuded sugar and low susceptibility with small amounts. The leached material contained glucose, fructose, glycosides, glutamic acid, aspartic acid, histidine, alanine and a compound related to tyrosine. The stimulation of the fungus on the surface appeared to be necessary for initial infection (Orellana and Thomas, 1962). Deverall (1967) demonstrated the presence of sucrose, glucose, fructose, galacturonic acid and amino acids in diffusates which stimulated the fungus. Sol (1966, 1968) examined the effect of pre-treating bean (*Vicia faba*) leaves with sucrose, potassium chloride, lanthanum chloride or decenylsuccinic acid; infection with *B. fabae* was increased due to the stimulation of germination of the conidia, believed to be caused by increased amounts of leached nutrients.

Other fungi are known to be assisted by plant exudates. Blakeman (1968) found that the concentrated leaf washings from five plant species stimulated the infection of chrysanthemum leaf discs by *Mycosphaerella ligulicola*; the infection of some plants resistant to the fungus was promoted by the washings. The leached material of sorghum (*Sorghum vulgare*) influences the germination of conidia of the anthracnose pathogen *Colletotrichum graminicolum*; the material leached from a resistant variety of host inhibits and that of a susceptible variety stimulates germination (Sinha, 1968). Exudates from the buds of sugar beet leaves stimulate the germination of *Peronospora farinosa* conidia more actively than those from the leaves, but the stimulation of germination is probably not an important factor in the epidemiology of sugar beet downy mildew in the field (Russell and Evans, 1968).

Chu Chou and Preece (1968) have made the interesting observation that an exudate of pollen grains stimulates the germination of spores and the development of lesions of *Botrytis cinerea* on the surfaces of strawberry petals and fruits and on bean (*Vicia faba*) leaves. Pollen grains may be expected to be present in the community of the phyllosphere. Another effect of plant exudates is worth recording. Ruinen (1956) suggested that a small epiphytic fern *Drymoglossum piloselloides* attached to the leaves of plants in Indonesia obtains its nutrients from the cuticular secretions of the leaves.

Kerr and Flentje (1957) showed that the presence of a cuticle is important for attachment and penetration by a fungus. The attachment to, organisation on, and penetration of a host by the radish strain of *Pellicularia filamentosa* is controlled by the nature of the cuticular surface and by the presence of diffusible material which is secreted on to the surface from the underlying cells. They suggest that the specificity of the reaction of the different strains of the organism to different hosts is probably due to the variation, from host to host, of the nature of the cuticular surface and of the diffusible secretion. Flentje, Dodman and Kerr (1963) found that exudates from radish (*Raphanus*) stems and cotyledons stimulate the formation of infection cushions during attack by *Thanatephorus cucumeris*. They give the first direct evidence that the chemical stimulant is a natural constituent of the host plant and is not produced by an interaction between the host and the pathogen. When discussing factors concerned in resistance they suggest that stimulatory exudates may be produced by a susceptible host, but not by a resistant; alternatively, stimulatory exudates may be given by both, but on a resistant host cushion development may be suppressed by inhibitory materials. They point out, however, that it is difficult to obtain direct evidence that inhibitors are also exuded. It is of interest that Crosby and Vlitos (1960) and Vlitos and Cutler (1960) found that a plant growth-promoting factor isolated from tobacco had the properties characteristic of long-chain (C_{22}–C_{28}) primary alcohols. These compounds commonly occur in plant surface waxes and it is possible that they contribute to the surface stimulation of fungal growth. Suchorukov (1960) has considered the significance in host-parasite relations of physiologically active substances such as thiamine, biotin and pantothenic acid. These growth factors diffuse out of plant tissues to the surface; infection drops that form on a susceptible host acquire a higher concentration than those on a resistant, thus favouring the development of the pathogen on the susceptible host. Flentje (1959) has discussed the evidence in support of the chemical stimulation of fungi on plant surfaces.

Surface suppression of infection by fungi

While nutrients which diffuse from the host may stimulate a fungus on the surface, other superficial materials derived from the host cells may suppress its growth. Expressed leaf saps are known to inhibit the growth of fungi. Wiltshire (1915) and Johnstone (1931) found a positive correlation between the varietal resistance of apple leaves to scab (*Venturia inaequalis*) and the reduction of conidial germination by the leaf saps. Gilliver (1947) showed that the dried saps of 23% of nearly 2000 plants of many families completely inhibited the germination of the scab conidia. Turner (1956) has reported that the sap of the leaves of seedling oat plants contains an inhibitor of the fungus *Ophiobolus graminis*. Antifungal components of the saps of plants may be carried to the surface layers in exudates. Kovàcs (1955) and Kovàcs and Szeöke (1956) examined the washings from leaves and concluded that inhibitory secretions from poplar (*Populus*), clover and other leaves are probably present in sufficient concentration in water films on the surface to prevent spore germination. Topps and Wain (1957) washed the undamaged leaves of twelve species of woodland trees with water, extracted the washings with ether, and tested the extracts for activity against spores of *Botrytis cinerea*. The extracts of all the species except two (laburnum (*Laburnum anagyroides*) and sycamore (*Acer pseudoplatanus*)) retarded the growth of the conidial germ tubes. The most fungistatic extracts were obtained from elder (*Sambucus*) and privet (*Ligustrum*). Topps and Wain suggest that under natural conditions the compounds on the leaf may well provide a defence against fungal attack. An antifungal substance was detected in exudates of the stem and roots of *Vicia faba* and has been identified as an acetylenic keto-ester (Fawcett, Spencer, Wain, Jones, Le Quan, Page and Thaller, 1965). Free phenolic compounds are present in leaf surface layers; rain drippings from apple leaves gave positive reactions for phenols, especially during the initial phase of periods of continuous steady rain (Richmond and Martin, 1959). It seems probable that the substances migrate to and accumulate in the surface layer of the leaf during dry periods and are intermittently washed away. The germination of the conidia of *Peronospora tabacina* on tobacco leaves is suppressed by water-soluble phenolic compounds which occur in the highest concentration on the upper leaves (Shepherd and Mandryk, 1963). Sharp (1965) showed that spores of *Puccinia striiformis* are inhibited from germinating on wheat leaves, apparently due to secretions on the surface; he suggests that these may originate in the host or in the microflora colonising the leaves.

Cruickshank (1966) makes a distinction between the natural protection of plants due to fungitoxic compounds which occur as normal components of the cells and natural defence called into play after infection by fungi. The differentiation, although useful, is not clear-cut; some naturally occurring antifungal compounds, e.g. phenolic substances, are known to increase in amount following invasion. Much attention has been given to the defensive agents, which are known as phytoalexins (Müller, 1958a, b; Cruickshank, 1963). Examples are pisatin which was obtained from pea (*Pisum sativum*) pods inoculated with conidia of *Sclerotinia* (*Monilinia*) *fructicola* and phaseollin obtained from bean (*Phaseolus vulgaris*) pods similarly inoculated. A detailed discussion of protectants and defensive agents within cellular tissues is not within the scope of this book; they have been reviewed by Martin (1967).

Many of the antifungal agents that are found in leaf exudates are clearly naturally occurring protective substances, but the possibility also arises that substances formed in the cells in response to invasion may also affect the fungus on the surface or in the cuticular layer. Cruickshank (1968) has shown that the formation of the phytoalexin is induced by metabolites of the invading fungus. A metabolite designated monilicolin A was isolated from an *in vitro* culture of *S. fructicola*; it is a sulphur-containing, water-soluble peptide with about 64 amino acid residues and molecular weight of about 8000, and is active as a phytoalexin-inducing agent at very low concentrations. The possibility of phytoalexin activity in exudates has been discussed by Wood (1967). He suggests that it is not improbable that metabolites of the fungus move inwards through the cuticle, particularly when this is highly hydrated and that the underlying cells react to produce the inhibitory substances before they are penetrated by the infection hyphae. The inhibitor may then move out through the cuticle to affect the invader during the early stage of infection.

Various substances described as located in 'surface layers' or 'peels' have been suggested as contributors to defence against disease. Johnson and Schaal (1952) for example reported a higher content of chlorogenic acid in the peel of varieties of potato resistant to common scab caused by *Streptomyces scabies* than in that of susceptible. Gentisic acid (2,5-dihydroxybenzoic acid) was identified, in bound form, in representative species of 73 families of angiosperms; the content of the bound acid in the surface tissues was suggested as a possible factor in resistance to fungal attack (Griffiths, 1958). A fluorescent phenolic compound increased in amount in the peel of apple fruits infected with either *Venturia inaequalis* or *Podosphaera leucotricha*. The compound was considered to be a product of the host and not of the parasite since it occurred only in trace amounts in healthy tissues and was not found in filtrates from cultures or in extracts of mycelium of *V. inaequalis* (Barnes and Williams, 1960). At least ten phenolic compounds, the chief of which are quercetin, various catechins, chlorogenic acid and probably *p*-coumarylquinic acid occur in apple peel consisting of the cuticle with ten layers of attached cells (Hulme and Edney, 1960). The role of the phenolic compounds in the resistance of the fruit to *Gloeosporium perennans* is not clear; Hulme and Edney suggest that if they play a part, some compound or compounds present in relatively small amount must be concerned. Such reports give no indication of the extent to which the substances are located within the cuticle itself; they may well be confined to the flesh attached to the cuticle in tissues such as peels. Fawcett and Spencer (1967, 1968) have shown that when the peel, juice or juice-free pulp of the healthy apple fruit is supplied to a culture of *Sclerotinia fructigena* several antifungal compounds, including 4-hydroxybenzoic acid and 4-hydroxy-3-methoxybenzoic acid, are formed in the juice, but not in the peel or pulp.

Of greater relevance to the subject of this book is the part played by components of the cuticle itself, in particular the wax fraction, in suppressing fungal attack. Martin, Batt and Burchill (1957) examined the leaf wax of apple and other plants since this may be expected to form the first barrier to infection. The wax was obtained by washing leaves with ether, fractionated into acidic, true wax and oil components, and the wax and its fractions were tested for their ability to suppress the germination of conidia of *Podosphaera leucotricha* on apple rootstocks. The wax itself showed some activity, but the most striking fungistatic effect was shown by an ether-soluble acidic fraction extracted from the wax by dilute potash solution. This fraction also suppressed the formation of lesions of *Botrytis cinerea* on broad bean leaves. The acidic fraction in the wax, however, was not thought to be a factor in the varietal susceptibility of apple leaves to mildew. There is mounting evidence that unsaturated fatty acids, which are known to occur in leaf waxes, are involved in mechanisms of resistance to disease. Honkanen and Virtanen (1960) identified an ether-soluble antifungal fraction obtained from rye (*Secale*) plants as a mixture of unsaturated acids chiefly linoleic and linolenic, and Epton and Deverall (1968) found greater lipase activity, with consequent liberation of higher concentrations of linolenic acid, in the leaves of *Phaseolus vulgaris* resistant to *Pseudomonas phaseolicola* than in susceptible leaves. The factors involved in the resistance of lime leaves (*Citrus aurantifolia*) to wither-tip disease caused by *Gloeosporium limetticola* were examined by Roberts and Martin (1963) and Martin, Baker and Byrde (1966a, b). The cuticular membrane develops rapidly and the susceptibility of the leaf to infection is lost when the cutin in the membrane attains a level of about 0·1 mg/cm^2 of surface. The cuticular waxes and also the cutin acids obtained by hydrolysis of the cutin, slightly suppressed mycelial growth when tested at 0·1% in culture medium. The infection thread in the cuticle, however, has to negotiate a barrier of wax in higher concentration which may be more effective; if the thread also enzymically degrades the cutin, it may liberate in its immediate vicinity cutin acids in sufficient concentration to exert a toxic effect. Whether the cuticle is of any great significance in the resistance mechanism was unresolved; a compound of considerable antifungal activity, identified as the furocoumarin isopimpinellin, was obtained from the cellular tissues.

Venkata Ram (1962) showed that the wax of the tea (*Thea sinensis*) leaf contains some fractions that stimulate the germination of the spores and the growth of the germ tubes of *Pestalotia theae* and others that completely inhibit spore germination. Heather (1967) removed the wax layer from young leaves of *Eucalyptus bicostata* and found that infection by *Phaeoseptoria eucalypti* was greatly increased. A fraction of the ether-soluble moiety of the wax was shown by *in vitro* tests to inhibit spore germination, and a negative correlation was found between the degree of susceptibility of the leaf to infection and the amount of the active fraction present in a unit area of the leaf surface. Heather concluded that in this case the hydrophobic nature of the surface, by limiting the deposition of water-borne inoculum, is probably of greater significance in resistance than the chemical inhibition of spore germination. The tree *Ginkgo biloba* is the sole surviving member of the Ginkgoales which date back over 180 million years. The persistence of the species is attributed, at least in part, to its longevity and long reproductive period; an important factor in its longevity is its unusual resistance to pathogens. In a study of its resistance to disease Major, Marchini and Sproston (1960) found an inhibitory substance, 2-hexenal, in the leaves; Major (1967) confirmed that this was derived from linolenic acid. The demonstration by Adams, Sproston, Tietz and Major (1962) that fungal infection pegs do not form in the *G. biloba* leaf cuticle led Johnston and Sproston (1965) to investigate the role of the cuticle in resistance. 2-Hexenal could not be found in the isolated cuticle. The leaf carries an exceptionally heavy adaxial cuticular membrane (0·96 mg/cm^2 just before abscission) containing 60% of cutin. The isolated membrane resisted penetration by the test fungi (*Monilinia fructicola* and *Stemphylium sarcinaeforme*) although they penetrated a collodion membrane and an isolated cherry leaf cuticle. The complete removal of all waxy material from the membrane yielded intact sheets of cutin that were penetrated by *S. sarcinaeforme* but not by *M. fructicola*. A fraction obtained from the wax strongly suppressed spore germination and germ-tube elongation, and the leaf tissue was especially resistant to disintegration by pectin enzyme. Johnston and Sproston show that the factors involved in the resistance of *Ginkgo* to disease are complex, and suggest that the cuticle contributes to resistance both morphologically and chemically.

Whether the material apparently of a condensed tannin type which occurs in appreciable quantities in the cuticles of some leaves and fruits contributes to resistance to fungal attack is unknown. In barks, tannins undoubtedly play an important part. Nienstaedt (1953), for example, found tannin substances to be present in the barks of resistant species of chestnut (*Castanea mollissima* and *C. crenata*) in sufficiently high concentration to retard the development of chestnut blight caused by *Endothia parasitica*. Somers and Harrison (1967) discuss wood tannins in relation to resistance to *Verticillium* wilt disease, and support the view that high molecular weight tannins are widely involved in disease resistance mechanisms. Klöpping and van der Kerk (1951) obtained benzyl gentisate in 0·3% yield from the bark of *Populus candicans*; the compound completely inhibited the growth of *Botrytis cinerea*, *Penicillium italicum*, *Aspergillus niger* and *Rhizopus nigricans* at 100 p.p.m. in the test medium; another product containing a hydrocarbon, probably olefinic, completely inhibited the germination of spores of *P. italicum* at 45 p.p.m. Barks contain a wide variety of other substances (see Chapter 5) whose antifungal activities are largely unknown. Herein lies a potentially fruitful field of research.

The fungus on the surface of the plant may thus find itself under the influence of opposing effects, some stimulatory and others inhibitory. The exudates from plants may contain nutrients that are beneficial to its growth, but also substances that are inimical. The wax on the surface may itself contain inhibitory and possibly stimulatory constituents. In many instances these effects may counteract each other and the overall effect on the pathogen may be slight. Cuticular waxes that contain large proportions of the more hydrophobic constituents (alkanes, ketones, esters and secondary alcohols) are likely not only to limit the deposition of water-borne inoculum but also to prevent the free passage of aqueous exudates to the surface; in such cases, antifungal constituents of the wax may play the dominant role. The work reported by Heather (1967) on the surface factors involved in the resistance of *Eucalyptus bicostata* to *Phaeoseptoria eucalypti* provides a possible example of a situation of this kind.

Waxes that are characterised by significant proportions of the less hydrophobic polar constituents (fatty acids and primary alcohols) probably permit freer exudation of nutrients (and possibly antifungal agents) from the cellular tissues, but are themselves more likely to exert fungistatic action, especially through their unsaturated fatty acids. Wheat plants that are apparently waxless are more susceptible to leaf and stem diseases than those that are waxy (Troughton and Hall, 1967), possibly due to the more ready exudation of nutrients from the less waxy plants. The glossy-leaved mutant of Brussels sprout is more susceptible to white rust (*Albugo candida*) than the normal glaucous plant (Wilson and Jarvis, 1963). The non-glaucous plant carries less wax, of a less hydrophobic nature, than the glaucous. The comparative susceptibility of the non-glaucous plant may be due, at least in part, to the more ready deposition of inoculum and exudation of nutrients permitted by its wax. The balance between the opposing stimulatory and inhibitory effects imposed by the cuticle is likely to be a delicate one. Many factors clearly are involved in the resistance of plants to fungi; even so, those associated with the cuticle are probably of little significance in comparison with others that operate within the cellular tissues.

Abrasion of the leaf cuticle with inert dusts modifies its properties (Chapter 8) and assists the entry of viruses (see p. 206). It may also increase the susceptibility of the leaf to attack by fungi. Berwith (1936) inoculated apple leaves with *Podosphaera leucotricha* at different times after their emergence from buds. Five days after emergence the leaves were resistant to the pathogen, but the resistance could be broken down by abrading the leaves with carborundum powder. Leaves in the field may be abraded by wind-blown particles or other agencies; the effect of this in inducing susceptibility may be significant and has been largely overlooked.

Interactions of the cuticle with viruses

Pathogens such as viruses which are considerably smaller than the crystalline masses of surface wax are able to lodge on an unwettable leaf and often are not easily removed (Fig. 9.2). David and Gardner (1966) examined the possibility of coating the surface of cabbage leaf with the granulosis virus (300 × 120 nm) of the cabbage white butterfly (*Pieris brassicae*) to serve as a non-toxic biological method of control. The virus was not removed to any significant extent by 5 hr of simulated rain or by scrubbing in detergent and rinsing in water. Under natural conditions of weathering the virus persisted on the surface for up to 4 months. The brassica surface (see Figs. 2.5 and 8.8) is peculiarly efficient in retaining such tiny particles and, as the authors point out, such retention and persistence has obvious value in this method of control of insect pests. Figure 9.2 shows tobacco mosaic virus lodged on the surface of a *Chenopodium* leaf. The particles, occluded in crushed wax crystals and protected by projections around them, are very difficult to remove. On the other hand, Brants (1964, 1965) found that tobacco mosaic virus could readily be washed from the wettable leaves of *Nicotiana tabacum* and *Datura stramonium*. Burgerjon and Grison (1965) found that *Smithiavirus pityocampae* was effectively retained on *Pinus nigra*. This pine has a surface structure that would be expected to retain small particles very well. A virus that remains on the surface of a leaf out of contact with a living cell cannot infect its host, but the longer it remains attached the greater is its chance of being carried in by mechanical abrasion of the leaf or by insect or nematode damage. The penetration of viruses into their hosts is discussed by van Nostrand and Goodman (1968). There seems to be no feature of the cuticle that has any relevance to the transmission of viruses by insects although it is not impossible that some features of the cuticle may deter possible vectors.

Mechanical transmission of viruses, using an abrasive such as carborundum powder or diatomaceous earth to abrade the cuticle, has been used for many years. The effect of diatomaceous earth or Celite (**C**) mixed with tobacco mosaic virus on a bloomed leaf surface can be seen in Fig. 9.2. Hirai and Hirai (1964) have shown that viruses can enter a leaf through the broken basal cells of the leaf hairs of tomato. It has also been shown that tobacco mosaic virus accumulates at the base of the hairs of *Nicotiana glutinosa* (Kontaxis and Schlegel, 1962). Trichomes are obviously a vulnerable site of entry for a number of viruses and this is not surprising in view of the relative weakness of the cuticle over hairs and the proximity of the plasmalemma to the surface of most trichomes. However, other work has shown that removal of the trichomes before infection does not alter the number of lesions. Hildebrand (1958) noticed that when trichomes are injured with a micropipette they exude a drop of cytoplasm which is quickly drawn back into the cell. If a virus were present on the surface of the wounded trichome it would also be drawn inside. Wounds artificially inflicted on the surface of a leaf are known to heal rapidly. After abrasion in the manner used to achieve an infection with a virus very few dead epidermal cells are found in the absence of the virus. Moreover, a successful virus infection depends upon the continued existence of the cell into which it gains access. Viruses have only a limited capacity to move from cell to cell and move readily about only in the phloem xylem. Epidermal cells must therefore suffer no substantial loss of fluid even after their plasmalemma has been penetrated. Neither the plasmalemma nor the tonoplast is, incidentally, normally permeable to viruses. Furumoto and Wildman (1963) found that most of the wound sites on a *Nicotiana glutinosa* leaf are closed to subsequent virus infection within 1 min of wounding. Only about 10% of the wounds have a much longer lifetime.

It has been suggested that ectodesmata can act as sites of entry for virus infection. Brants (1964, 1965) found high concentrations of ectodesmata at sites which appear to provide the most ready ports of entry for virus infections. The ectodesmata, however, do not extend through the cuticle (see Fig. 4.6); the cuticle as such is impermeable to viruses and has to be broken above the ectodesmata before they can be (if they are) used as channels of entry. There is no doubt that viruses can pass through plasmodesmata (Esau, Cronshaw and Hoefert, 1967). However, ectodesmata are unlike plasmodesmata in that the electron microscope does not reveal open channels through the ectodesmata. Nor do the ectodesmata penetrate the plasmalemma which is itself impermeable to viruses (Mundry, 1963). The ectodesmata may, however, provide zones of weakness in the epidermal cell walls through which viruses might be thrust.

Tinsley (1953) showed that *Nicotiana glutinosa* and *Nicotiana tabacum* are both more susceptible to virus infection, by artificial means, if they receive unlimited supplies of water. Plants receiving unlimited water produce ten or more times as many local lesions as plants that receive only enough water to prevent wilting. The same effect was noticed in plants grown under reduced light intensity. Plants grown in the shade with just sufficient water to prevent wilting have ten times as many lesions as those under the same water regime, but without shade. A combination of shading and unlimited water supply produces twice as many lesions as unlimited water supply alone. Hence the effects of shading and unlimited watering are thought by Tinsley to be partly but not wholly independent of one another. He suggests that the relevant effects of unlimited watering on enhanced susceptibility are a thinner cuticle and larger and looser cells in the epidermis, spongy parenchyma and palisade tissue. He also discovered that a differential water regime must be imposed on the plants for at least 2 weeks in winter and at least 4 weeks in summer before substantial differences in susceptibility become apparent. The problem is complicated by the fact that at least some of the symptoms of plants grown under shade resemble those of plants grown with unlimited water, e.g. large leaves and thin cuticles, yet at least to a certain extent the regimes are additive and not overlapping in enhancing susceptibility to virus infection. In view of the more recent work on cuticle structure and ectodesmatal distribution this problem needs to be re-examined.

Interactions of the cuticle with bacteria

Stomata are obviously excellent ports of entry for bacteria and presumably viruses too. The relative humidity of a stomatal pore is high; an additional advantage to the establishment of infection. According to work cited by van Nostrand and Goodman (1968) the bacterium responsible for fire blight (*Erwinia amylovora*) of apples and pears will not grow at 97% humidity, grows slowly at 98% and satisfactorily at 99%–99·9%. Rain and dew seem to be particularly favourable for the development of many types of bacterial infection of plants.

Bacteria are capable of entering other natural pores in the cuticle. Hildebrand and MacDaniels (1935) showed that *Erwinia* sprayed onto blossoms caused infection at the nectaries, uncutinised stigmatic surfaces and undehisced anthers. Fire blight is thought to be carried by bees and wasps, hence its predilection for floral parts. Fire blight is more commonly found infecting the nectaries of pears than of apples since in the pear these openings are more accessible. McLean (1921) found that the mandarin (*Citrus nobilis* var. *szinkum*) is resistant to the citrus canker bacterium (*Pseudomonas citri*) whereas *Citrus grandis* is susceptible. They differ, amongst other features, in the size of the cuticular ridge which modifies the size of the opening to each stomatal chamber. He believes that difference in resistance to the movement of liquid water into the stomatal chamber of these two species is sufficient to determine the difference between them in resistance to bacterial infection.

Many bacterial infections take place through wounds in the cuticle and Hildebrand (1942) has shown how minute these wounds may be. He stroked tomato plants with a smooth polished needle previously dipped in a solution of *Agrobacterium tumefaciens*. Within 5 days galls appeared in the vicinity of collapsed trichomes. Thus the normal abrasion of leaf against leaf is quite adequate to create gaps in the cuticle sufficient for bacteria to enter. Trichomes are particularly vulnerable since they are rarely as well protected by cuticle as the normal cells.

The physiological and biochemical aspects of the invasion of plants by fungi and bacteria and the properties of plants that make them resistant to disease have been comprehensively reviewed by Wood (1967).

Interactions of the cuticle with insects

Resistance to insect attack

Anstey and Moore (1954) found that a glossy-leaved mutant of sprouting broccoli (*Brassica oleracea* var. *italica*) is more susceptible to attack by the cabbage flea beetle (*Phyllotreta albionica*) than the normal waxy plant. On the other hand, Thompson (1963) noted that the normal waxy plants in field populations of marrow-stem kale (*B. oleracea* var. *acephela*) had large colonies of the cabbage aphid (*Brevicoryne brassicae*) whereas non-waxy plants were not colonised. Later in the year, the lower surface of the waxy leaves supported whitefly (*Aleuroides brassicae*) at all stages of development, but that of the non-waxy leaves had no larvae or pupae and only a few adults. Thompson concluded that the nature of the leaf surface is a factor in the resistance of the plants to both insects. The difference in resistance of the kale leaves to attack by the aphid was later related to the surface wax. Twice as many winged aphids settled and produced young on the waxy leaves than on the non-waxy. No significant difference was found in the subsequent rate of multiplication on the two surfaces; a difference in the initial infestation was the critical factor (Thompson, 1967). Way and Murdie (1965) also reported that a non-glaucous strain of Brussels sprout was more resistant to the cabbage aphid than the normal glaucous, but relatively more attractive to another aphid, *Myzus persicae*. The surface of the non-glaucous brassica leaf differs considerably, physically and chemically, from that of the glaucous. The glossy brassica leaf is less waxy and less water repellent and its wax has a lower content of paraffins and ketones and a higher content of acidic substances than the wax from the waxy leaf.

The leaves and flower-buds of *Rubus phoenicolasius* are unattractive to the raspberry beetle (*Byturus tomentosus*); the plant is heavily waxed and the wax is richer in acidic substances than other *Rubus* waxes. Lupton (1967) has pointed out that the resistance of plants to insects is based on characters which make the host plant less attractive to the parasite or which in some way limit its reproductive capacity. An aphid placed on a plant repulsive to it walks off without feeding, sometimes in a few minutes. Lupton has exploited this effect in the breeding of raspberry plants that are unattractive to the aphid *Amphorophora rubi*, a vector of virus disease.

Nicotiana gossei exerts a toxic effect upon the aphid, *Myzus persicae* (Thurston and Webster, 1962). Alkaloids, principally nicotine, were identified in secretions from the trichomes of seven species of *Nicotiana*; aphids were killed by contact with the secretions and the resistance of the species to *Myzus persicae* was attributed to this effect (Thurston, Smith and Cooper, 1966).

The factors involved in the specificity of insect attack upon plants are little understood. In this connection, the observation by Way and Murdie (1965) that a glossy form of brassica leaf is resistant to one species of aphid, but more attractive to another is of special interest. The chemical nature of the surface wax may be one factor, hitherto overlooked, in specificity of attack.

The effects on the cuticle of insect attack are often marked. Bystrom, Glater, Scott and Bowler (1968) report than an excess of wax development occurs around the punctures of aphids.

Interesting observations have been made on the composition of the waxes of insects in relation to that of the waxes of host plants. Chibnall, Piper, Pollard, Williams and Sahai (1934) compared the wax of the cochineal insect *Coccus cacti* with that of cactus (*Opuntia* sp., Table 3, 1). Cocceryl alcohol and coccerinic acid occurred in the insect wax but not in the plant wax; cocceryl alcohol was shown by Chibnall, Latner, Williams and Ayre (1934) to be *n*-tetratriacontan-1-ol-15-one and coccerinic acid a mixture of *n*-triacontanoic and *n*-dotriacontan-13-one-1-oic acids. Dixon, Martin-Smith and Subramanian (1965) found that the hydrocarbons produced by the aphid *Megoura viciae* were *n*-alkanes C_{23}–C_{30} and branched-chain alkanes C_{24}–C_{28}; those of the surface wax of the host plant *Vicia faba* were *n*-alkanes C_{26}–C_{33}, with no branched chain compounds. The wax of the scale insect *Ceroplastes destructor* is more complex than that of its orange leaf host; fatty acids and other polar fatty material predominate in the insect wax and hydrocarbons and primary alcohols in the leaf wax (E. A. Baker, unpublished work).

The trapping of insects

A large number of species of plants including the genera *Byblis*, *Drosophyllum*, *Drosera* and *Pinguicula* secrete mixtures of sugars or mucilages whereby an insect can first become trapped and later digested. Schnepf (1963; 1965; 1966a, b) has studied the ***Fangschleim*** (there is no adequate English translation) secretory cells under the electron microscope. He has found that the secretion is associated with enhanced Golgi body activity. The Golgi bodies, it is suggested, accumulate vesicles of galactose, arabinose, xylose, rhamnose and other substances and discharge them towards the cell walls. The vesicles accumulate between the plasmalemma and the cell wall of the gland cells. In *Drosophyllum* the cell wall itself does not appear to hinder the outward passage of the secretion and the cuticle above the secretory glands is penetrated by small pores. Similar pores in specialised areas of the cuticle that release secretions have also been reported by Uphof (1962) and Lloyd (1942). However, according to Schnepf (1965), although the *Fangschleim* in *Pinguicula* seem to be produced in the same way as in *Drosophyllum*, there do not appear to be any pores in the cuticle over the glands.

Many plants trap insects by engulfing them with secreted sugars and some plants repel insects by unpleasant secretions; examples are given by Uphof (1962). No plant yet seems to have adopted the defence by smears used by some aphids (*Aphis fabae*) (Edwards, 1966). Deflection of the aphids' cornicle or touching its body stimulates the release of lipid droplets in water. When this exudate comes in contact with a hard surface or a predator it forms a hard waxy plaque. Many small predators encased in this shell are unable to escape. Aphids have been found with the encapsulated and shrivelled bodies of their predators still stuck to their backs. Edwards has found no evidence of a solvent carrier for this wax and suggests that this rapid conversion from a liquid to a solid state is the conversion of a supercooled liquid wax to a crystalline solid. Such a transfer of supercooled liquid to the surface of plants is not impossible and might explain otherwise puzzling examples of rapid crystallisation of waxes on damaged leaf surfaces as shown by Hallam (1967).

The pitcher plant (genus *Nepenthes*) traps insects by preventing them from escaping from the pitcher and then drowns and digests them in a secreted fluid at the base of the pitcher (Fig. 9.3). *Nepenthes* species, with only a few exceptions, live on the margins of tropical forests and feed on a wide range of arthropods. The plant has a creeping rhizome from which tough climbing stems arise, sometimes of considerable length. The leaves are thick, glossy, often up to a metre in length and the midrib is extended to form a long tendril. At the end of this tendril is developed the pitcher. The pitcher is probably a modified leaf (Lloyd, 1942). The complex details of the peristome and the lid are described by Lloyd (1942) and Juniper and Burras (1962). Just below and between the spines of the peristome are sugar-secreting nectaries and the secretion is highly attractive to insects, but the trapping secretion of most interest is the layer of wax which covers the whole of the inside of the pitcher from just below the peristome to just above the liquid level (Fig. 9.3). The peristome by itself does not seem to be a serious obstacle to the escape of most insects, but no insect seems to be able to climb the wall above the liquid. Flies, earwigs, centipedes, woodlice, and thrips have been found in the pitchers of plants grown in the Botanic Garden of Oxford University and insects as large as wasps and cockroaches in other pitchers. No doubt in a normal habitat an even wider range of arthropods would be found. None of these arthropods with their varied methods of locomotion is able to deal with the unclimbable surface. Only *Misumenops nepenthicola*, a spider found in S.E. Asia, which builds a web inside the pitcher and moves about on its own threads, is able to negotiate the surface. The behaviour of insects attempting to leave the trap is consistent. If they can

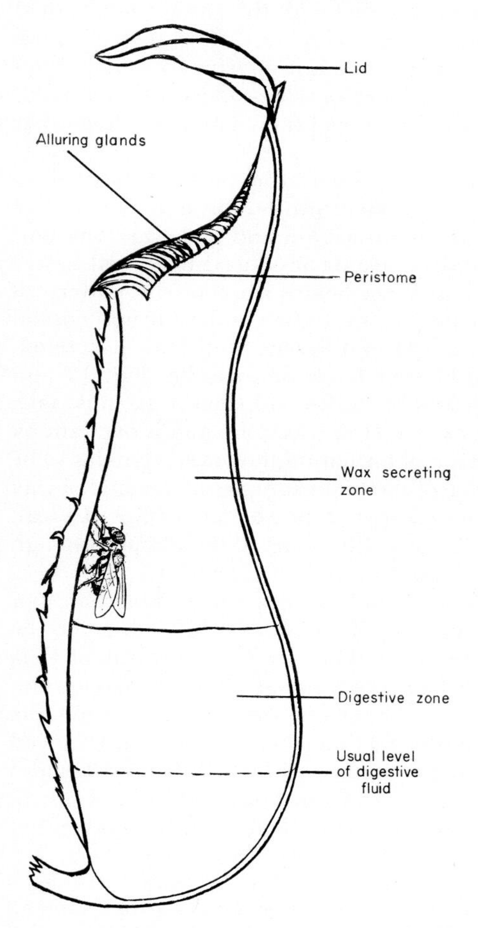

FIG. 9.3 Diagram of the pitcher of *Nepenthes rufescens*. (Redrawn from Juniper and Burras, 1962, by permission from *New Scientist*)

climb at all on the walls of the pitcher they approach the peristome with great difficulty, but its smooth surface prevents them scaling this barrier. They become agitated and wipe their pads against each other and against their mouthparts and abdomen in an effort to clean them. Finally their adhesion fails altogether and they slide to the bottom of the pitcher.

Microscopically the surface is white, faintly iridescent and just slimy to the touch. Electron micrographs show that the intact surface consists of a layer of overlapping scales, each scale being somewhat irregular in shape, but roughly a micron across (Fig. 9.4). When this surface is brushed away it reveals, not the true cuticle of the epidermal cells below, but another layer of wax, this time in the form of projections or ridges sticking out at right angles to the surface of the leaf (Fig. 9.5). The pads of the insect come in contact with the film of scales which, having little contact with or attachment to the projections beneath, detach readily onto any barbed or sticky surface. If the pads of insects which have been in contact with this surface are washed they release large numbers of these scales (Fig. 9.6). The insect, losing adhesion on this submicroscopic scree slope, attempts to wipe the sticky scales from its pads, but even the surface wiped free of scales offers little foothold. More and more scales are detached and become plastered onto feet and other parts until adhesion fails altogether.

Such a structure does, however, present an interesting problem of development. A surface of scales lies above a surface of projections and ridges. The simplest assumption seems to be that the projection and scales are one unit (Juniper and Burras, 1962). The projections are produced first, well before the pitcher lid opens and at some stage just before it opens the pattern of secretion shifts. Either the tops of the projections flatten under the influence of some slight microclimatic change or a wax of different properties is secreted. The second flow of wax spreads out like the blade of a paddle (Fig. 9.7 and Fig. 9.6 insert arrow). Successive pulses of the second stage wax must take place (Fig. 9.7). In the interval the first layer of wax dries and is overlain by successive waves. The exact mechanism of formation, however, remains to be discovered. The genus *Nepenthes* is one of the best examples of a complex wax secretion taking place only over a limited area of the adaxial surface of a leaf. No other crystalline wax is secreted by any other part of the plant, although other areas of the plant secrete sugars and proteolytic enzymes.

This slippery, unclimbable cuticle is not unique in the plant kingdom, but other instances are not so well documented. Haberlandt (1914) observed the behaviour of ants which he transferred from the extrafloral nectaries of *Vicia sepium* to the bloom-covered scape of *Hyacinthus sylvestris*. Their behaviour on this waxy surface was identical to the behaviour of other insects on the unclimbable surface of *Nepenthes*. He noticed that when the wax was removed the ants were able to move about on the surface without difficulty. He also cites other observations suggesting that the blooms on *Salix daphnoides*, *S. pruinosa* and *Fritillaria imperialis* have similar unclimbable surfaces, preventing

FIG. 9.4 (*top, opposite*) Electron micrograph of a carbon replica of the wax-secreting zone of the inside surface of a pitcher of *Nepenthes rufescens*. Cr/Au/Pd. (From Juniper and Burras, 1962, by permission from *New Scientist*)

FIG. 9.5 (*centre, opposite*) As for fig. 9.4, but with the outer scales brushed away. (From Juniper and Burras, 1962, by permission from *New Scientist*)

FIG. 9.6 (*bottom, opposite*) Electron micrograph of the intact scales of a pitcher of *Nepenthes rufescens*, washed from the feet of a fly which had attempted to scale the pitcher wall. Inset, one scale at a higher magnification. Whole scales supported on a carbon film and shadowed with gold/palladium. (From Juniper and Burras, 1962, by permission from *New Scientist*)

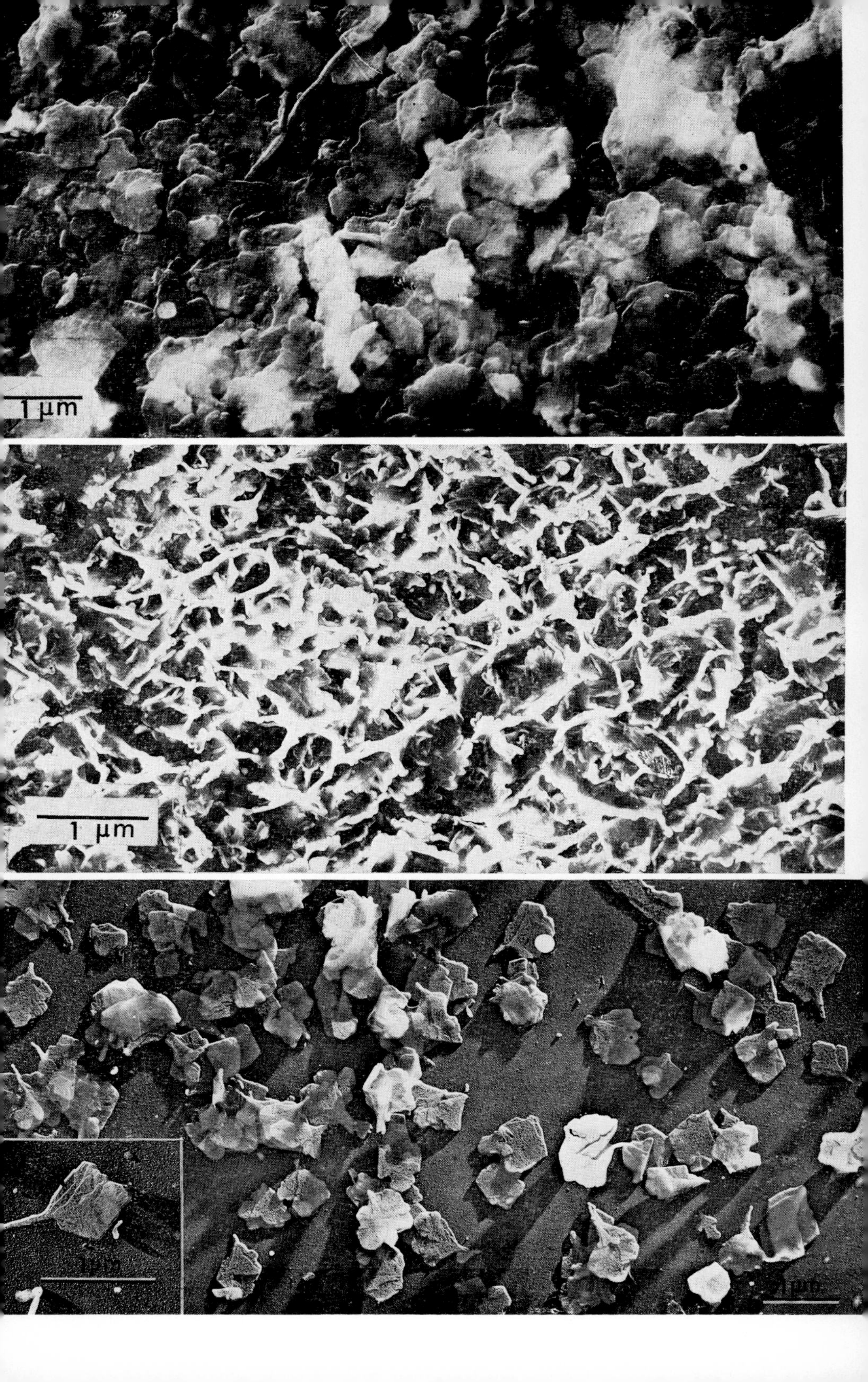
1 µm
1 µm
1 µm
1 µm

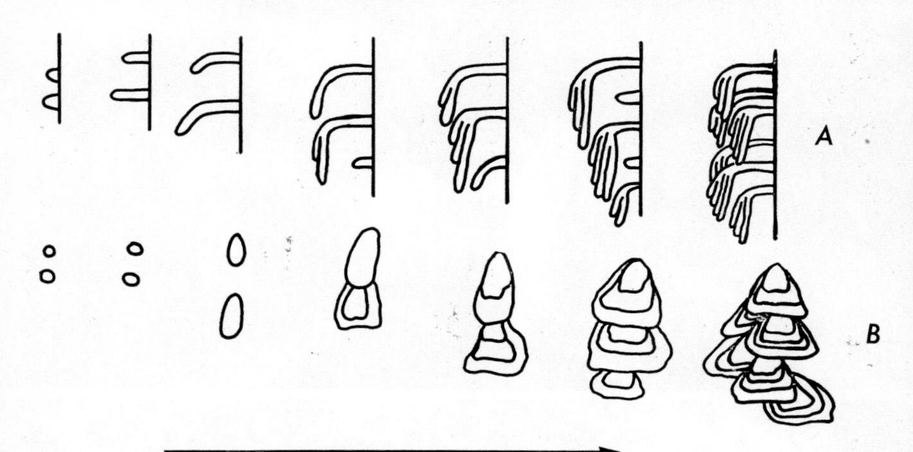

FIG. 9.7 Diagram of the proposed method of development of the scales of a pitcher plant. ***A*** in side view and ***B*** in surface view.

Table 9.1 Arthropod-trapping plants

Kind of trap	Genus	Digestive glands	Mucilage-secreting glands	Sugar-secreting glands or nectaries	Modified stomata	Modified trichomes or emergences	Modifie cuticle
Pitcher plants							
passive	*Heliamphora*			+		+	
	Sarracenia	+		+		+	
	Darlingtonia			+		+	+
	Cephalotus	+		+	+	+	
	Nepenthes	+		+	+	+	+
Lobster pot							
passive	*Genlisea*		+			+	
Fly-paper							
traps	*Byblis*	+	+		+	+	+
passive	*Drosophyllum*	+	+				+
Fly-paper							
traps	*Pinguicula*	+	+				+
active	*Drosera*	+	+			+	
Steel trap	*Dionaea*	+		+		+	+
	Aldrovandra	+	+			+	+
Mousetrap	*Utricularia*		+	+		+	+
	Biovularia		+	+		+	+
	Polypompholyx		+			+	+

access to their floral nectaries. Knoll (1914) reports similar observations on insects attempting to climb on the surfaces of *Iris pallida* and *Cotyledon pulverulenta*. The pitchers of *Dischidia* species frequently contain dead insects and their bodies probably make some contribution to the plant's supply of nitrogen. However, there appears to be no modification of *Dischidia* pitchers either to attract, retain or digest insects.

The forms of cuticular and epidermal modification and secretions which are involved in some way with the trapping of arthropods are listed in Table 9.1. A full description of the morphology and anatomy of these plants is given by Lloyd (1942). A detailed account of their glands and hairs is given by Uphof (1962).

10

The Cuticle in Decay

ANCIENT DEPOSITS

The fossil cuticle

Cuticles and spore and pollen membranes remain almost completely unchanged in the fossil state, although the carbohydrate and other plant constituents may have been completely transformed. In the strict sense the cuticle should not in these circumstances be called a fossil since it has apparently undergone little if any chemical change. It is possible to detach cuticles, with most of their original properties, from fossils of plant fragments as early as the Devonian period (Walton, 1953). Since the cuticle is so resistant to decay it acts, in the fossil state, as a very important indicator of taxonomy, evolution, habitat and possibly climate.

If fossilised plants are found to have a substantial cuticle it can be assumed that they were probably land plants. By its absence the suggestion can be made that the plant belonged to the Thallophyta (algae, mosses (Musci), liverworts (Hepaticae), etc.). In the Australian Silurian deposits algal-like plants possessing a very definite cuticle appear. The suggestion, reinforced by the first traces of an internal vascular structure, is that these plants represent an early stage in the transition from the lower to the higher plant forms. *Zosterophyllum* is one genus from the Australian Silurian with these characteristics. It comprises upright cylindical leafless stems, which branch dichotomously and are crowned with kidney-shaped sporangia. The stems have an epidermis covered by a recognisable cuticle. The even better known genus *Rhynia*, from Rhynie near Aberdeen, Scotland, from the Middle Old Red Sandstone (Mid-Devonian) possesses not only a cuticle, but also recognisable stomata which must have arisen by necessity at about the same time.

The analysis of pollen residues recovered from peats and the beds of former lakes is of great value as a means of dating deposits and has provided much information about the great climatic changes which have occurred in Europe since the last glacial age (Godwin, 1951). Churchill (1968) has used pollen analysis to investigate the prehistoric past of three *Eucalyptus* species which are prominent forest trees in SW. Australia. The extreme resistance of the outer wall of the pollen grain is ascribed to its sporopollenin (see p. 154). Zetzsche and Vicari (1931) and Zetzsche, Vicari and Schärer (1931) obtained preparations of sporopollenin from present-day spores of *Lycopodium clavatum* and pollen of *Picea orientalis*, *Pinus sylvestris* and *Corylus avellana*, and also from fossil lignites. Close chemical similarity was found in the materials from the fossil lignites and from the modern flora.

The remarkable stability of cutin under certain conditions has allowed the plant cuticle to persist unchanged through millions of years. Leaves which become fossils usually do so by falling into water and being buried in mud under anaerobic conditions; under such conditions the cuticle remains intact indefinitely. The cuticle, as preserved, has distinguishing characteristics of its own; its microscopic structure is specific and reveals the structure of the epidermis, and details such as the arrangement of cells around the stomata and the shapes of cells permit the recognition of family characters. The evidence provided by the fossil cuticle is particularly useful in small leaves such as those of the conifers. Until comparatively recently the taxonomic identification of fossilised leaves was very difficult, but a study of their excellently preserved cuticles has completely changed the situation. The fossil conifers are now a well-defined group and may be used as age markers for other groups of plants with which they may be found in physical association (Harris, 1956). It is fortunate that leaves and other parts of plants which develop easily recognisable cuticles are also those which are most readily and frequently fossilised.

Thomas (1925) discovered in the Jurassic deposits of Yorkshire, England, the remains of a group of angiospermous plants, which he named the Caytoniales. On the basis of similarity of cuticle structure he was able to piece together leaves, microsporophylls and seed-bearing organs to recreate the whole original plant.

The technique of linking together scattered fragments of plant remains, on the basis of their cuticular anatomy, has now been widely used. Dilcher (1963), for example, examined fossilised leaves of *Ocotea obtusifolia* collected from the Eocene clays of Tennessee, U.S.A., and gives details of the method of preparing specimens. The distribution and abundance of stomata and of hair bases, the size and shape of cells and characteristics of the guard and accessory cells were used in diagnosis. Isolated tracheids and vessels were found. The fossil cuticle could not be identified with the cuticle of several modern Lauraceae examined.

The characteristics of the fossilised cuticles of two species of *Bothrodendron* (*B. punctatum* and *B. minutifolium*) helped to distinguish between the two species and to confirm that they are actually distinct (Thomas, 1967a). The carboniferous fossil stem genus *Ulodendron* has been known since 1831, but since that time opinions have differed about its generic value and what species should be included in it. From an examination of fossil cuticle characters, Thomas (1967b) concluded that a species called *U. minus* is not distinct from another, *U. majus*, and that another species, known as *U. subdiscophorum*, is most probably also *U. majus*. *Lepidodendron* was shown to be a similar lycopod stem genus to *Ulodendron*.

Fossil leaf cuticles from coal were examined by Sen (1954). The cuticles show no layered structure and probably consist only of cutin. The membrane reacts like modern cutin to staining and to acidic hydrolysing reagents, and responds similarly to optical tests, but is less susceptible than modern cutin to treatment with alkali. The greater resistance to degradation by alkali suggests closer bonding in the fossil cutin or possibly impregnation with resinous or asphaltic material. This probably represents the only significant modification of cutin during and after fossilisation. Sen (1961) has also examined the nature of the cork which persists in ancient buried wood. Chemically identifiable suberin was detected in the cork cell walls of *Betula* spp. and microcrystalline constituents and associated cellulosic and possibly lignified substances were also found. Sen (1956) has reviewed work on the fine structure of organic residues in ancient and buried wood, coal and other carbonaceous materials and peat.

Some cuticles are damaged over relatively brief periods of time much more readily than others. Stewart and Follett (1966) studied under the electron microscope the cuticles of certain Scottish plants preserved in peat. The species they examined were *Phragmites communis*, *Eriophorum vaginatum*, *Calluna vulgaris* and *Sphagnum palustre*. Epicuticular wax similar to that of other Gramineae is present on living *Phragmites* and detectable wax projections can be seen on *Eriophorum*. Details of cell wall outlines and stomata or pores can be seen in the cuticle of *Phragmites*, *Eriophorum* and *Sphagnum*, but not in *Calluna*. *Phragmites* loses its surface wax very quickly and when 5 cm deep in the peat layer virtually all of it has gone. Apart from this loss of wax it suffers little further degradation. *Eriophorum* suffers a loss in surface definition and *Calluna* a progressive etching of the surface. The cuticle of *Sphagnum* is little affected up to the depth in the peat at which it can still be distinguished. As the authors point out, the amount and pattern of damage to different cuticles is very variable. The ability of the different cuticles to withstand erosion would appear to depend upon their inherent chemical composition and physical structure. These findings must modify the overall diagnostic value of more ancient fossil cuticles.

The suggestion has been made that the thick cuticles possessed, for example, by the Bennettitales and some conifers in the Jurassic coal from Brora, Scotland, implies that the climate at the time was hot and presumably desiccating. However, the existence of thick cuticles is not invariably associated with a hostile climate and, as Harris (1956) points out, although inferences based on such evidence are legitimate they are nevertheless speculative. The fossil cuticle has provided much information on early plants and their relationships with living species. Pant and Srivastava (1968), for example, found no significant differences in the external form or cuticular structure of fossil and living scales of *Araucaria* spp. This contribution appears in a collection of reports on fossil plants published as a tribute to the work of Harris; with many others it brings out the value of the fossil cuticle as a diagnostic feature.

The derivation of petroleum from plant waxes

Some components of the cuticular waxes are lost comparatively quickly; others, notably the *n*-alkanes, may persist for long periods. Petroleum is now generally accepted to have been derived, at least in part, from the organic remains of plant and animal life deposited with sediments. *n*-Alkanes are abundant in crude oils. The *n*-alkanes in modern plant waxes show a preponderance of odd carbon-numbered compounds over even. The *n*-alkanes which occur in recent marine sediments also show a predominance of odd carbon-numbered compounds, whereas those in crude oils contain more equal proportions of odd and even carbon-numbered molecules. The *n*-alkanes in recent marine muds therefore resemble plant (and animal) *n*-alkanes more closely than do the petroleum *n*-alkanes (Evans, Kenny, Meinschein and Bray, 1957).

The distribution of *n*-alkanes in crude oils, as a pointer to the origin of petroleum, was examined by Martin, Winters and Williams (1963). The high proportions of *n*-alkanes (sometimes more than 30%) in the oils suggested straight-chain compounds as important starting material, and the evidence obtained supported earlier views that fatty acids were principal precursors. *n*-Alkanes in the range C_{11}–C_{19} predominate only in crude oils from early Palæozoic formations; this, the authors suggest, may be a consequence of the presence of simpler forms of life during that period or of the production of higher proportions of the saturated acids characteristic of living organisms (chiefly C_{16}, C_{18}) due to the uniformly warm climates then prevailing. On the other hand, *n*-alkanes in the C_{20} and C_{30} ranges predominate in most of the crude oils of the Cainozoic age. The authors suggest that waxes of leaves and fruits containing the longer-chain compounds would probably have been supplied in greater quantity by the flowering plants which dominated the Cainozoic era than by the fern and related plants prominent during earlier eras. Algae and mixed phyto- and zoo-plankton contain *n*-alkanes in the overall range C_{15}–C_{32} with C_{15} or C_{17} predominating and with no pronounced alternation between odd and even carbon-numbered compounds such as is found in higher plants. This tends to confirm that the earlier oils, which are characterised by a predominance of the lower alkanes with little or no alternation, originated from the residues of lower organisms. The alkanes of recent marine sediments are predominantly odd carbon-numbered compounds which suggests that they were derived from sources other than algae and plankton (Clark and Blumer, 1967; Stránský, Streibl and Herout, 1967; Stránský, Streibl and Šorm, 1968). Robinson (1963) considers that petroleum hydrocarbons are both biogenic and abiogenic in origin; the evidence for the biogenesis of petroleum is incontrovertible, but may only apply to a part of the material. Indications of biogenesis are clear in young oils, less apparent in those of middle age and all but absent in the older oils.

Wax and cutin in peat and ancient soil

Peat wax contains free fatty acids and their esters, higher alcohols and an unsaponifiable fraction which consists of hydrocarbons, sterols and a small amount of ketones (Bel'kevich, Kaganovich and Trubilko, 1960). The acids liberated on saponification have constants similar to those of arachidic, behenic, lignoceric, montanic and melissic acids (Kaganovich, Bel'kevich and Rakovskii, 1959). The extraction of peat with benzene-ethanol yielded a mixture of resin, asphalt and wax; extraction with petroleum ether or *n*-hexane gave a purer wax which was fractionated by chromatography into hydrocarbons which included the polycyclic compound perylene, esters, acids, hydroxy acids and ketones (Howard and Hamer, 1962). The asphaltic fraction of peat wax is composed of material of relatively high average molecular weight which has a low solubility in polar and non-polar solvents, and the resin of soft, sticky material very soluble in polar solvents. Rauhala (1961) found in peat wax monocarboxylic and monohydroxy acids in the range C_{19}–C_{22}.

On the Somerset levels in England are relics of huge sphagnum bogs, now considerably water logged, which are cut for peat. The horizontal banding of different peat types affords evidence of past climatic change; the bogs had become dry and covered with ling (*Calluna*) and cotton grass (*Eriophorum*), and later with a deep growth of prickly sedge (*Cladium mariscus*) (Godwin, 1951). Peat obtained near Wedmore contained 78% of moisture (air-dry basis, 81% oven-dry) and 6% of ash. Extraction of the air-dried peat with ether yielded 3% of a hard resinous wax which, by thin-layer chromatography, was found to contain hydrocarbons, esters, primary and secondary alcohols and a triterpenoid fraction showing the same R_F value as ursolic acid. The hydrocarbons, esters and the triterpenoid were the major components; the hydrocarbons consisted of *n*-alkanes in the range C_{21}–C_{35} with the C_{29}, C_{31} and C_{33} compounds predominating. Branched hydrocarbons, chiefly C_{33}, were also detected. The wax-free peat was percolated with potassium bicarbonate solution at room temperature to remove pigmented material (17% of the air-dried peat), the residue hydrolysed with 3% ethanolic potash, and the ether-soluble acids liberated were recovered. Among these, 10,16-dihydroxyhexadecanoic acid was identified by thin-layer and gas chromatography. Cutin, assessed from the yield of ether-soluble acids, amounted to 1% of the air-dried peat (Baker, 1968).

A 5000-year old lacustrine sediment obtained from an English fresh-water lake was examined by Eglinton, Hunneman and Douraghi-Zadeh (1968). Hydroxy-fatty acids comprised 0·6% of the dry weight of the sediment and included 10,16-dihydroxyhexadecanoic acid and ω-hydroxy acids in the range C_{16}–C_{24}. These compounds were almost certainly derived from plant cutin and suberin.

CONTEMPORARY PLANTS

Residues in soil

Schreiner and Shorey (1910) identified *n*-hentriacontane, esters, alcohols, tetracosanoic acid and hydroxy-fatty acids in soil organic matter. These compounds are characteristic components of cuticles; Schreiner and Shorey suggested that the hydroxy-fatty acids were probably derived from débris of plant or animal origin. Many soils contain substantial amounts of lipids, up to 20% of the total humus content in some instances. Amongst the lipids found in soils are paraffin hydrocarbons, phospholipids, fats, waxes, fatty acids and terpenoids (Stevenson, 1966). There is some evidence, reviewed by Stevenson, that accumulations of lipids can reduce the fertility of certain soils; sandy soils may become unproductive because waxes coat the surfaces of the sand grains and render them unwettable. According to Stevenson, waxes would seem to be the most stable members of the lipid family in soils. Meinschein and Kenny (1957) extracted soils with benzene-methanol, and analysed the extracts by chromatographic, infra-red and mass spectrometric methods. The principal constituents of the waxes were *n*-aliphatic acids, *n*-primary alcohols and sterols. All the extracts contained cyclic compounds which may have been triterpenes. Butler, Downing and Swaby (1964) found that the wax from an Australian soil contained esterified *n*-alkanoic acids from C_{12}–C_{30} with C_{22}, C_{24} and C_{26} predominating, and also alcohols of similar chain lengths. Morrison and Bick (1966, 1967) examined the wax fractions obtained from a garden soil and a peat. The resin- and asphalt-free wax accounted for 0·6% of the organic matter in the soil (8%) and 1·6% of that in the peat. In addition to long-chain alkanes, alcohols (C_{16}–C_{30}), fatty acids (C_{18}–C_{34}) and hydroxy-fatty acids, methyl ketones (*n*-alkan-2-ones) of chain lengths of C_{17}–C_{37} were found in the wax from both sources. The fractions obtained together amounted to 25–30% of the resin- and asphalt-free waxes. The authors suggest that the ketones may be derived from alkanes and fatty acids of plant cuticular waxes, modified by the action of soil micro-organisms.

The trees of an apple orchard of one acre (0·4 hectare) may shed annually up to 1500 lb (670 kg) of dry leaves (Raw, 1962; Hirst and Stedman, 1962). The cutin content of the litter at leaf-fall is not less than 2% and the wax about 1%. Despite these appreciable annual rates of deposition, cutin and wax fail to accumulate in orchard soil. Baker (1968) examined orchard soil taken from a depth of 3–9 in. (7·5–22·5 cm). Organic matter amounted to 4% of the air-dried soil, and a light-brown wax, extracted by ether, to 0·03%. The wax contained paraffins, esters and acidic materials; no triterpenoid compounds were detected. The paraffins were identified by gas chromatography and compared with those isolated from the waxes of the leaves and fruits of the apple trees and of the dominant grass (*Agrostis stolonifera*) in the orchard (Fig. 10.1). The soil paraffin fraction shows a wider range of

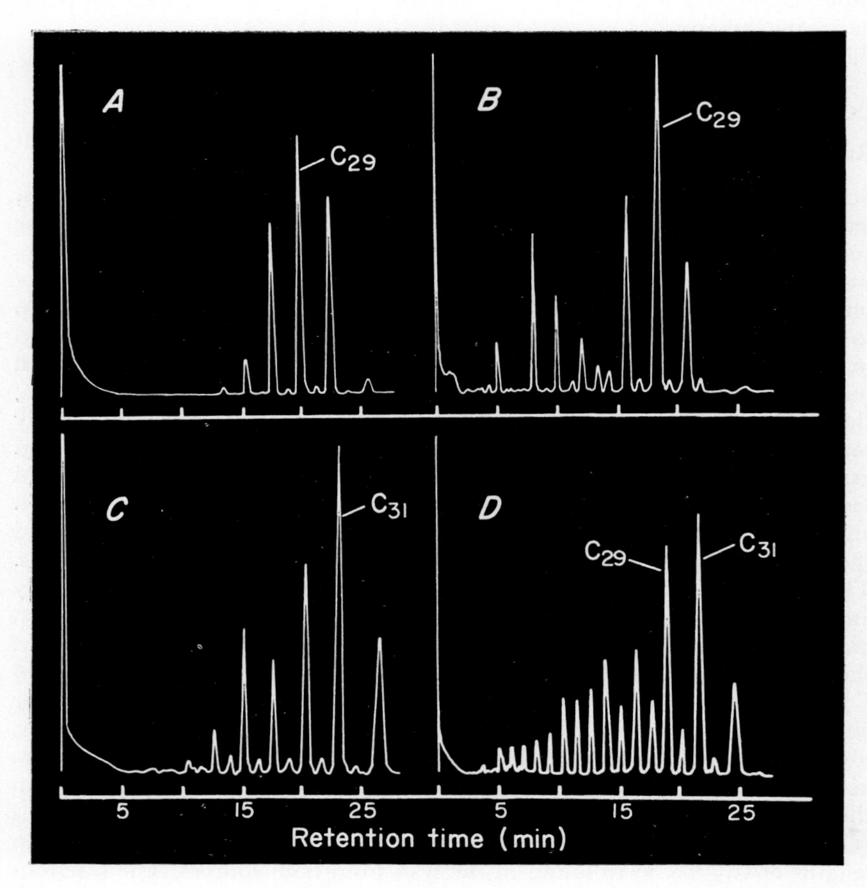

FIG. 10.1 Hydrocarbons of waxes of apple leaf (***A***), apple fruit (***B***), grass (***C***) and soil (***D***) by gas chromatography. Column 6 ft × $\frac{1}{4}$ in, 2% S.E. 30 on 70/80 mesh Chromosorb G, temperature programmed 160–295°C, carrier gas helium, thermal conductivity detector

individual compounds and noticeably larger proportions of even carbon-numbered compounds than the apple and grass fractions. The dominant paraffin of the apple waxes is *n*-nonacosane and that of the grass wax *n*-hentriacontane, and these are both prominent in the soil wax. Cutin was assessed from the yield of ether-soluble acids obtained by hydrolysis of the wax-free soil after extraction with dilute aqueous alkali; it amounted to 0·07% of the soil (air-dry basis). 9,10,18-Trihydroxyoctadecanoic and 10,16-dihydroxyhexadecanoic acids occurred in the ether-soluble acids; the hydroxy acids are prominent constituents of apple cutin.

The breakdown of waxes

As long ago as 1894–1895, Miyoshi recognised the ability of some micro-organisms to use paraffins. This has since been fully confirmed. Ali Khan, Hall and Robinson (1964) showed that a *Pseudomonas* isolated from soil attacks hydrocarbons by terminal or diterminal oxidation followed by β-oxidation, and Fredricks (1967) that *P. aeruginosa* degrades the *n*-decane chain non-specifically whereas *Mycobacterium rhodochrous* does so by successive terminal and β-oxidation. Paraffins are important constituents of some waxes, but the extent to which they are degraded on the plant surface by micro-organisms is unknown. Although paraffins are present in orchard soil, the low content found suggests that breakdown does occur, probably in the soil. Alkanes and alkenes serve as substrates for micro-organisms (McKenna and Kallio, 1965). *n*-Nonacosane is degraded by a soil bacterium (*Micrococcus cerificans*) presumably to shorter chain compounds which are incorporated into esters and other lipids (Hankin and Kolattukudy, 1968). A *Pseudomonas* sp. isolated from orchard soil used ursolic acid as a sole source of carbon; the soil probably contains other organisms able to degrade terpenoid components of the cuticular wax of plants (Hankin and Kolattukudy, 1969).

Some micro-organisms are known to be able to break down waxes. Millman and Yotis (1958) found esterase activity in cell-free extracts of *Aspergillus flavus*, *Candida albicans*, a *Penicillium* sp. and bacteria; the substrates, however, were not plant cuticular waxes. They suggest that de-esterification is an initial step in the use of waxes as sources of energy. Figure 9.1 shows the surface of a wheat leaf on which it appears that the fungus (*Puccinia rubigo-vera tritici*) is carrying out at least a partial degradation of the epicuticular wax. However, no enzymic studies have been carried out on this particular fungus. Ruinen (1963) has shown that yeasts (*Aureobasidium pullulans* and *Cryptococcus* spp.) isolated from the phyllosphere of tropical foliage not only secrete lipids but also possess lipolytic activity. He suggests that lipolysis may be important in the breakdown of the cuticle and in improving the phyllosphere as an environment for the growth of micro-organisms by enhancing the permeability of the outer epidermal wall.

The colonisation and breakdown of leaf litter

Much work on the biological decomposition of tree leaf litter has shown that progressive colonisation occurs by different organisms. *Aureobasidium pullulans* is one of the first to grow on the litter of oak, beech and other trees and is later replaced by other organisms, among which bacteria and a few moulds prevail. *Aureobasidium pullulans* attacks pectin strongly, leading to openings for invasion by others (Smit and Wieringa, 1953). A succession of fungi appears in oakwood litter. Up to one year after leaf-fall, the leaves of oak, ash, birch and hazel each support a characteristic flora of Pyrenomycetes and Fungi Imperfecti; later these are replaced by a flora, common to all, dominated by

Penicillium and *Trichoderma* spp. (Hering, 1965). The progressive colonisation of the leaf litter of *Eucalyptus regnans* was followed by Macauley and Thrower (1966). These workers determined the changes of nutrients available to the fungi in the leaves by extraction of the litter with ethanol, the insoluble fraction being considered to consist of the unavailable structural components. From visual observations, the cuticle appears to persist in decomposing leaf litter as long as any of these, but no critical chemical work on the progressive losses of the waxes and cutin has been done.

Various types of microflora are now known to be involved in the breakdown of the cutin of fallen leaves. Heinen (1960) isolated from rotting leaves a species of *Penicillium* (*P. spinulosum*) which grew in a culture medium with cutin as the sole source of carbon. The cutin showed a change in mechanical properties and was converted into a form easily soluble in warm, dilute alkali. The enzymic liberation of fatty acids was demonstrated, and the enzyme was provisionally termed a cutinase. *Fusarium moniliforme* and the yeast *Rhodotorula* also showed some cutin-degrading activity. The properties of the enzyme were determined by Heinen and van den Brand (1961). The enzyme was enriched by purification of the extract from *P. spinulosum* with protamine sulphate and ammonium sulphate at 70% salt concentration, and the optimum reaction conditions were established. It has a pH optimum of 6·0, is stable and needs no dialysable co-factors and appears to be localised in the cytoplasm of the cell and not associated with any particular organelle. The enzymes present in crude extracts of *P. spinulosum* were then resolved by Heinen (1963a, b). The cutinase yielded several fatty acid-oxidising enzymes at different steps in its purification, and the enzyme solution obtained from a crude extract at 20% saturation with ammonium sulphate contained a peroxidase and fatty acid-oxidising enzymes. Pectinolytic, cellulolytic and catalase activity was also found in the cell-free extract of *P. spinulosum*; the mould is thus able to attack the cellular tissue of leaves (Heinen, 1962). Stadhouders, Heinen and Kraan (1962) studied the breakdown of wax and cutin by *P. spinulosum* with the electron microscope. Discs of leaves of *Gasteria verrucosa* with or without surface wax were used as substrates for both the mycelium and extracts of the fungus. The wax film was destroyed and the surface structure of the exposed cutin destroyed. The results indicated that the fungus secretes not only cutinase but also enzymes, especially wax esterases, capable of breaking down waxes.

At least two enzymes are present in *P. spinulosum* which participate in the breakdown of cutin (Linskens, Heinen and Stoffers, 1965). A cutinesterase catalyses the cleavage of the ester linkages and a carboxycutinperoxidase the peroxide bridges with release of oxygen. Hydroxy-fatty acids are liberated which undergo oxidative breakdown. Heinen and de Vries (1966a) have described a colorimetric method for the determination of long-chain fatty acids from plant cutin. Heinen and de Vries (1966b) found that soil from cultivated land contains a variety of micro-organisms able to bring about the decay of cutin. Seven out of ten isolated organisms, including fungi, the yeasts *Cryptococcus laurentii* and *Rhodotorula* sp. and bacteria, contained cutinolytic enzymes. De Vries, Bredemeijer and Heinen (1967) have studied the natural breakdown of cellulose and cutin by mounting cuticular strips between glass slides for exposure in the soil. They found that cellulose and cutin are normally decomposed in 3–8 months, depending on the season and the type of soil, by existing soil micro-organisms. Rich garden soil acts most rapidly and poor sandy soil most slowly.

The decomposition of litter may be retarded by substances, inhibitory to micro-organisms, that are released on the death of the leaves. Griffiths and Isaac (1963) observed that fallen tomato leaves which had died naturally appeared to have on their surfaces a substance which had an inhibiting effect upon the germination of the conidia of *Verticillium* spp. They suggested that the inhibitor was probably present in a bound state in the living leaf, but on its death passed out through the then permeable cell membranes to the surface in sufficient concentration to prevent the germination of the conidia.

The breakdown of cutin on aerial organs

By pathogens

An aspect of cutin degradation that has intrigued plant pathologists for many years is whether a fungal infection thread dissolves the cutin or otherwise lessens cuticular resistance to its passage. The conflicting views have been discussed by J. T. Martin (1964). The so-called scab diseases of the leaves and fruits of many plants belonging to the Rosaceae are caused by fungi which invade and colonise principally the outer periclinal walls of the epidermal cells. The scab fungi and some other pathogens with this habit have been classified as 'subcuticular fungi' in contrast to the epiphytic forms which develop on the surface and the more usual endophytic which ramify the cellular tissues. Wiltshire (1915) thought it probable that some cuticle-destroying enzyme is produced during the invasion of apple and pear leaves and fruits by the scab fungi *Venturia inaequalis* and *V. pirina*. That this occurs has still not been fully substantiated. The cuticles of apple leaves severely attacked by scab or mildew and of rose leaves attacked by mildew contained appreciably lower amounts of cutin than those of leaves which showed no visible signs of damage (Roberts, Martin and Peries, 1961). This suggests that the fungi degraded the cutin, but the possibility that the presence of the fungi prevented it from being built up in the normal process of cuticle formation cannot be precluded.

Another scab fungus, *Spilocaea oleagina*, invades the olive (*Olea europaea*) leaf. Miller (1949) describes the typical habit of growth in the outer layer of the epidermal cells and, while referring to an earlier belief that the fungus uses cutin as a source of energy, suggests that it probably depends upon food materials in the cells. Graniti (1962) reported that the conidiophores of the fungus which arise from the subcuticular mycelium pierce through the cuticle by a dissolving action, and that the fungal mycelium degrades the materials of the outer cuticular layer and induces a slight decutinisation of the inner cuticular layer. The position was examined in more detail by Graniti (1965a, b). Electron microscope studies showed that the outer wall of the epidermal cells consists of four layers, namely a layer composed of cellulose and pectin, inner and outer cuticular layers and an overlying cuticle proper. The inner cuticular layer is made up of cellulose and pectin with scattered islets of lipoidal material. The outer cuticular layer is heavily cutinised and contains wax and partially polymerised, unsaturated lipoids; ingrowths of this layer form the cuticular pegs between the epidermal cells. The overlying cuticle appears as a continuous sheet of cutin, sometimes layered. The colonies of *S. oleagina* appear to establish themselves in the outer cuticular layer, the constituents of which are enzymically degraded. The process of infection is visualised as follows. The fungus pierces the cuticle proper and meets the outer cuticular layer where cutin, wax, pectin and cellulose become available to it, together with pH conditions favourable to its extracellular enzymes (Graniti, de Leo and Bagordo, 1962). The colonies grow parallel to the leaf surface, sheltered by the thick cuticle proper overhead and prevented from penetrating deeper by a defensive response by the host in which a phenolic aglucone, liberated from the glycoside oleuropin, plays an important part (Graniti and de Leo, 1966). Defensive measures against scab fungi in some plants include the formation of corky layers which separate the colonies from the cellular tissues. The work of Graniti suggests that cutin is degraded by the pathogen; materials other than cutin in the outer cuticular layer may, however, be used in preference to cutin.

The general view, still widely held, that pathogenic fungi do not actively degrade cutin may have to be modified in the light of the report by Linskens and Haage (1963) that *Rhizoctonia solani*, *Botrytis cinerea*, *Cladosporium cucumerinum* and *Pyrenophora graminea*, all pathogens, display cutin-degrading (cutinase) activity. A sample of cutin was obtained from a normal host of each fungus by isolating the cuticular membrane and extracting it with alcohol, and a 'standard' cutin was prepared similarly from *Gasteria verrucosa*, which is not susceptible to any one of them. A purified enzyme extract from each fungus was tested on the cutin of its host and on the standard cutin; the activity of each enzyme preparation was determined by titration of the fatty acids liberated. *Rhizoctonia* and *Pyrenophora* freed more acids from their host cutins (*Solanum tuberosum* tuber and *Triticum aestivum* leaf) than from the standard, whereas *Botrytis*, a relatively less specific pathogen, was less active against a host cutin (*Solanum tuberosum* leaf) than the standard. *Cladosporium* showed only slight activity on the cutin of the fruit of a variety of *Cucumis sativus* highly resistant to it. Another fungus, *Penicillium spinulosum*, attacked the standard cutin more readily than that of the potato tuber. Linskens and Haage concluded that the fungi are capable of breaking down the plant cuticle enzymically. The results further suggest that selectivity in the choice of host is associated with activity against the host cutin.

Suberin is also susceptible to fungal attack. Swift (1965) found a 60% loss of suberin in the bark tissue of roots of *Brachystegia spiciformis* after ten months rotting by *Armillaria mellea*.

By cutinase in the pollination process

Heinen and Linskens (1961) showed that the cutinase of *Penicillium spinulosum* and other fungi attacks the cutin of the stigma of *Brassica nigra*. This work was followed by interesting observations on the role of cutinase in the pollination process. The germinated pollen of a number of plants was found to contain an enzyme believed to be identical with the cutinase detected in the moulds (Linskens and Heinen, 1962). Active preparations of cutinase were obtained from the pollen of *B. nigra*, *B. campestris*, *Tropaeolum majus* and *Sinapis alba*, all of which have a cutin layer at the stigma surface. The enzyme, however, was absent from the pollen of *Petunia hybrida* and *Cannabis sativa*, which is not required to penetrate a cutin layer at the stigma surface (see p. 215). The enzyme was thus associated with the pollen of plants whose stigmata are covered with a cuticle. Linskens and Heinen suggest that the function of the enzyme is to assist the pollen tubes to penetrate the stigmatic cuticular layers.

Christ (1959) had shown that the cuticle of the stigma is involved in the self-sterility of *Cardamine pratensis*. After self-pollination, the growth of the pollen tube is inhibited, the effect being localised at the stigma surface. Christ suggested that foreign pollen possesses an enzyme that makes the cuticle permeable, permitting the pollen tubes to penetrate the walls of the stigmatic papillae, and that pollen from the same flower also possesses the cutin-dissolving enzyme but in a form incapable of breaking down the cutin. Possible explanations of the inactivity of the enzyme in self-pollination were put forward. The pollen enzyme may occur in an active form but is inactivated by a substance on the stigma; this occurs only when the enzyme and inhibitor have been formed under the influence of the same sterility allelomorphs. Alternatively, the pollen enzyme may be in an inactive form and is activated by a compound secreted by the stigma; this occurs only if the pro-enzyme and the activator have been formed under the influence of different sterility allelomorphs. The penetration of the cuticle by the pollen tube differs from fungal penetration in that no appressorium is formed; this was regarded by Christ as evidence that the pollen tube breaks down the cuticle enzymically. Kroh (1963) thought that the penetration of pollen tubes into the stigmatic papilla wall requires both cutin- and pectin-dissolving enzyme systems. He suggested that after self-pollination, the formation of the activator-enzyme complex is probably inhibited by the stigma, with the consequent blocking of pollen germination and pollen tube penetration, and that after cross-pollination the initiation of the processes which occur in the stigma of self-incompatible Cruciferae probably involves an irreversible activation of the pollen cutinase system by the stigma.

The breakdown of waxes and cutin by animals

Grazing animals consume considerable quantities of cuticular waxes and cutin in the herbage, but the extent to which they are broken down during digestion of the food is largely unknown. The fate of the waxes and cutin in leaves and fruits consumed by man, so far as is known, has not been investigated; the cuticular membrane at least appears not to be digested. The paraffins of the waxes are known to pass unchanged through the gut of cattle. Oró, Nooner and Wikström (1965) investigated the origin of the predominating C_{29}, C_{31} and C_{33} *n*-alkanes found in the manure of cattle. The hydrocarbons of *Medicago arabica*, which formed part of the pasture diet, were chiefly C_{29}, C_{31} and C_{33} with C_{31} the major component; the relatively greater abundance of the C_{33} hydrocarbon in the manure was attributed to the grass and other pasture plants rich in this compound. Martin (1955) has used the cuticle as an aid to the analysis of the natural diet of hill sheep; the rumen contents were examined to ascertain qualitatively the species of plants consumed. The contribution made by large animals to cuticle degradation is well worthy of further investigation.

The contribution to wax and cutin breakdown by animals in the soil is likely to be at least as important as that of the soil microflora. Witkamp and van der Drift (1961) and van der Drift (1963) examined the influence of environmental conditions upon the activity of the soil macrofauna in reducing forest litter. Rainfall that gave adequate moisture conditions seemed more important than temperature, and much larger numbers of saprophagous macrofauna were collected from a calcareous mull than from an acid mull or mor. Bacteria, actinomycetes and the larger litter-feeding animals predominated in a calcareous mull litter, containing relatively large amounts of protein and easily decomposable carbohydrates. Fungi and micro-arthropods, on the other hand, were most numerous in a mor litter which contained more lignin and cellulose and less protein.

The importance of the activities of earthworms in maintaining aeration and fertility of the soil is recognised. Large quantities of fallen leaves are pulled into the soil and consumed. The amount of orchard leaf litter removed during winter is proportional to the population of *Lumbricus terrestris* which may amount to 3/4–1 ton of worms per acre (1875–2500 kg per hectare) (Raw, 1962) and over 90% of the litter is often removed before spring (Hirst and Stedman, 1962). The efficiency of the earthworm in disposing of surface leaf litter is shown by the observations of Hirst, Le Riche and Bascomb (1961) and van Rhee (1963). When copper from sprays used in orchards accumulated in the soil earthworms were destroyed and mats of fallen leaves, resistant to decomposition, built up on the ground beneath the trees. Edwards and Heath (1963) examined the role of soil animals in the breakdown of leaf litter. Discs of leaves were placed in bags of nylon of differing mesh, which were buried in the soil, and the extent of breakdown was estimated visually. Earthworms disposed of the discs much more efficiently than smaller invertebrates, the most important of which were springtails (*Collembola* spp.), enchytraeids and dipterous larvae. Edwards and Heath noticed that some discs of beech (*Fagus sylvatica*) leaves which were tanned remained uneaten, whereas those not tanned were heavily attacked. An interaction between polyphenolic materials and protein complexes in leaves may affect their digestibility in litter breakdown (Handley, 1961).

The importance of the small soil animals in comminuting plant material is beyond dispute, but whether they break down waxes and cutin in the process is unknown. The loss of wax and cutin during the digestion of apple leaves by the earthworm has, however, been investigated by Baker, Martin and Stringer (1968). Earthworms (*L. terrestris*) were obtained by treating the ground beneath orchard trees with formalin, well washed and kept at 13 °C in water, 2 mm deep, in lidded containers. The apple leaves, which had received no spray treatment of any kind, were taken just before leaf-fall. Before being used experimentally, the worms were starved for 3–4 days when the gut was seen by transmitted light to be empty. The leaves as gathered were accepted by the earthworms with great reluctance, but after leaching the leaves in water for not less than 24 hr they were readily consumed. During leaching for 24 hr, the water-soluble phenolic constituents of the leaves were almost completely removed. This accords with the observation of Edwards and Heath and suggests that, under natural conditions in the soil, leaves must be leached before they become acceptable as diet. The tests included feeding worms with whole leached apple leaves, discs cut from the leaves and cuticular membranes isolated from the leaves by treatment with pectinase solution. Analyses for waxes and their constituents and cutin were made on each material, the amounts consumed were recorded and solid and liquid faecal material was collected daily, bulked and analysed separately. Cutin assessed from the yield of ether-soluble acids obtained on saponification, of which 75% consisted of hydroxy-fatty acids, amounted to 2·9% (dry basis) of the leached leaves and 24% of the cuticular membranes. The test with the cuticular membranes lasted for 2 weeks during which 1230 membranes, each 2 cm^2 in size, were consumed by 14 earthworms. In the early stage, each earthworm was supplied with and could only ingest 2 membranes each day; after a few days 10 membranes were given to each worm daily and all were consumed. The earthworms showed visible signs of loss of weight but remained active; they quickly regained weight when fed on leached leaves. On a dry weight basis, 78% of the membrane diet was recovered as faecal solids. The test with the cuticular membranes gave the more dependable results; 45% of the wax (25–55% of various classes of constituents) and 70% of the cutin fed to the worms were recovered in the faeces. A similar loss of cutin (*c.* 30%) during passage through the gut was recorded in the test with leaf discs. It seems probable that during the normal digestion of leaf litter some degradation of the cuticle is effected by the earthworm; its more important role is probably to present leaf material, as faeces, in a form more accessible to attack by micro-organisms.

11

Commercial Uses of Cuticular Materials

A book on the cuticles and protective coverings of plants would not be complete without a brief reference to their commercial exploitation.

Waxes

Cuticular waxes have been widely used in the manufacture of polishes, varnishes, candles, cable coverings, electrical insulators, sealing wax, dental wax, gramophone records, waxed and carbon papers, waxed cartons, photographic plates and pharmaceutical preparations (Howes, 1936). The most important vegetable waxes are carnauba wax, candelilla wax, ouricury wax, esparto or fibre wax, sugar cane wax, bayberry or candleberry wax, Japan wax and China tallow tree wax. Some of the waxes used in the polish industry are mentioned by Law (1965).

Carnauba wax is obtained from the leaves of *Copernicia cerifera* the wax palm of Brazil and Venezuela. The palm, 20–40 ft (7–13 m) high, grows in groves flanking rivers; the leaves are deeply lobed, flabelliform and densely arranged.

The upper surface of young leaves is the most waxy; the young leaves are collected and heated with a little water to melt the wax which is run into moulds. Each leaf yields about 5 g of wax. Carnauba is harder and more brittle than other waxes and is the best of the polishing waxes; it gives a high lustre on leather and furniture and is used as a hardener in the manufacture of gramophone records and typing paper. The Cuban palm (*C. hospita*) matures more quickly than the Brazilian palm and its larger leaves give a better yield of wax which resembles that from *C. cerifera.*

The source of candelilla wax is *Euphorbia cerifera* and *E. antisyphilitica* which grow in profusion under desert conditions in N. Mexico and southern parts of the U.S.A. *Euphorbia cerifera* produces many stems, 2–5 ft (60–150 cm) high, from a woody rootstock. The stems are heavily waxed and the leaves small and inconspicuous. The wax is obtained by boiling the plant in water with added sulphuric acid or more recently by extraction with organic solvent at a wax collection centre. The wax is brown or reddish but can be bleached to make it resemble carnauba; it is not so hard or brittle as carnauba. Its uses are similar to those of carnauba. *Pedilanthus pavonis* of Mexico is also used as a source of the wax.

Ouricury wax is obtained from the ouricury wax palm *Syagrus coronata* (or *Cocos coronata*), a native especially of the Amazon basin. Another source is *Attalea* spp. e.g. *A. excelsa.* The wax palm is 18–30 ft (6–10 m) high; the wax is isolated from the leaf by melting or scraping and is dense and hard. Esparto wax is derived from esparto grass, *Stipa tenacissima*, a plant 3–4 ft (90–120 cm) high, of the Mediterranean basin and N. Africa. The wax is isolated by solvent extraction, 1 ton of the grass yielding about 3 lb of wax. It is a by-product of the esparto paper industry and is used in the manufacture of fine paper. The Australian grass *Glyceria ramigera* yields a colourless wax which resembles carnauba. Sugar cane wax, from *Saccharum officinarum* is extracted by solvent from filter press cake obtained in the process of manufacture of sugar. The wax is refined by treatment with solvent and also resembles carnauba.

The fruits of various species of *Myrica* of northern, central and southern America and southern Africa yield a low-melting triglyceride material, the so-called bayberry wax, which has been used for many centuries for illumination purposes. The fruits are boiled in water and the wax collected and moulded. The early settlers in N. America used the incrustations of the bayberry or wax myrtle (*M. cerifera*) and of the small waxberry (*M. carolinensis*) for candles. Myrtle wax (a misnomer) is obtained from the berries of *M. arguta* and related species of central and southern America, and has a composition similar to that of bayberry wax. The capeberry wax of S. Africa is derived from the fruits of *M. cordifolia.* The chief uses of the triglycerides from *Myrica* fruits are for candles for the Christmas trade, soaps, ointments and leather polishes and, mixed with paraffin wax, for embedding material for use with the microtome (Williams, 1958). Japan wax, a vegetable 'tallow', occurs between the kernel and husk of berries of *Rhus succedanea* of the Himalayas, Indo China and Japan. The wax is obtained by solvent extraction or steam treatment. It has been used in Japan from remote times, especially for candles and polishes.

The China tallow tree (*Sapium sebiferum*) is widely grown in China for the wax which covers the seed. This wax has been used since antiquity for candles and soap making (Dimbleby, 1967).

Waxes are isolated, sometimes for local use, from a large number of other plants. The wax palm of the Andes, *Ceroxylon andicola*, has a waxy coating 0·5 cm thick on its bole. The wax is scraped off, melted and clarified, and used locally for candles. Flax wax, from *Linum usitatissimum*, is obtained as a by-product in the preparation of flax fibre. Sisal wax is isolated, by extraction with solvent, from species of *Agave*, and hemp wax, also by solvent extraction, from *Cannabis sativa*. The leaves of wild banana plants, *Musa* spp., are used in Java and Malaya as sources of wax. Raffia wax is obtained from the leaflets, 5–6 ft long, of the raffia palm (*Raphia pedunculata*) of the Malagasi Republic. The wax is especially prominent on the under surface of the leaflet. The upper epidermis is stripped off to give, when dried, the fibrous 'raffia' used in horticulture. The wax is then removed from the dried leaflets by melting; it is hard and brittle and resembles carnauba wax. The Chinese insect wax is produced by *Coccus cerifera* of the order Hemiptera on the twigs of *Ligustrum lucidum* or *Fraxinus chinensis*. The bark with its coating of wax is removed and the wax obtained by treatment with hot water. A full account of commercial vegetable waxes and their properties is given by Warth (1956).

Plants with conspicuous wax deposits other than commercially used plants are the wax gourd (*Benincasa hispida*), the castor-oil plant (*Ricinus communis*) and species of *Eucalyptus*, *Buxus* and *Mesembryanthemum*. The stems of some species of *Rubus*, notably *R. biflorus* of the Himalayas and *R. giraldianus* of China, are copiously waxed (Howes, 1936). The commercial exploitation of peat wax has been considered from time to time, but has met with little success.

Oak cork and barks

The cork of commerce is manufactured from the bark layer of the cork-oak tree *Quercus suber* which grows in Portugal, Spain, some districts bordering the Mediterranean and in N. America. The tree, 20–40 ft (7–13 m) high, yields supplies about every 10 years from an age of 20 years, and may live for 150 years. The cork layer is stripped from the tree by longitudinal and transverse incisions, soaked in water, pressed and dried. The compressibility and resilience and its resistance to attack by micro-organisms and many chemicals, including acids, make it invaluable as a stopper for containers. Strong alkali solutions are detrimental, causing a breakdown of the suberin which holds the cork cells together. Cork is also susceptible to attack by halogens.

Other bark layers are used for a variety of purposes. They are removed by making incisions and peeling; sometimes the tree is felled or uprooted. Douglas fir (*Pseudotsuga menziesii*) contains up to 11% of wax which has been extracted for commercial use. Some tree barks are rich in tannin (30–60%) and are used in the manufacture of leather. Oak bark is the traditional British tanning material. The bark of the mimosa or wattle tree (*Acacia mollissima*) of S. and E. Africa and other regions is an important source of tanning material, imports of the extract in recent years exceeding those of quebracho (the wood of *Schinopsis* spp. of S. America) or myrobalan (the fruit of *Terminalia chebula* of India). Other barks used for tanning are those of *Eucalyptus*, *Tsuga*, *Rhizophora* and *Larix* spp.

Barks are valuable sources of medicinal preparations. Species of *Cinchona*, chiefly *C. ledgeriana*, *C. officinalis*, *C. calisaya*, *C. succirubra* and *C. robusta*, some of which are hybrids, are cultivated in S. America, India and Central Africa to yield cinchona bark. Well-grown trees may reach a height of 100 ft (33 m) with a trunk circumference of 7 ft. The dry bark may contain 10% or more of total alkaloids, the chief of which are quinine, quinidine, cinchonine and cinchonidine. Cascara sagrada is the bark of *Rhamnus purshiana*, a small tree cultivated in North America and central Africa. The bark contains about 2% of a mixture of purgative glycosides of emodin, aloe-emodin and chrysophanol. Alder buckthorn (*Rhamnus frangula*) bark contains a related glycoside, frangulin, and is less objectionable in taste than cascara sagrada. The emodins are derivatives of anthraquinone. Oil of cinnamon is obtained, in about 1% yield, by distillation from the bark of *Cinnamomum zeylanicum*, a small evergreen tree cultivated in Ceylon. Cassia (*Cinnamomum cassia*) bark yields an oil similar to that of *C. zeylanicum*. The oils contain 55–75% of cinnamic aldehyde. The bark of the slippery elm, *Ulmus fulva*, a small tree indigenous to central and northern U.S.A., has demulcent and emollient properties because of its content of mucilage. Witch-hazel bark is obtained from *Hamamelis virginiana*, a small shrub common in north America. Species of *Quillaia*, chiefly *Q. saponaria* which are tall trees indigenous to Chile and Peru, yield quillaia or soap bark, a source of commercial saponin. Galls of *Quercus infectoria* contain 50–70% of tannin and have been used in the tanning and dyeing industries. An account of the characteristics and medicinal uses of barks is given by Wallis (1955).

Glossary

Abaxial surface Of a leaf, the one farthest away from the axis, usually the lower surface.

Adaxial surface Of a leaf, the one closest to the axis, usually the upper surface.

Adduct A crystalline mixture, not a true compound, in which the molecules of one of the components are contained within the crystal lattice framework of the other component.

n-*Aldehyde* An alkane with an aldehyde group substituted in a terminal position in the chain. Formula $CH_3(CH_2)_nCHO$.

Aliphatic An open-chain (alkyl) compound in which the carbon atoms are linked in such a manner that no closed rings containing carbon are formed.

Alkane A straight-chain saturated (normal, *n*-) hydrocarbon. Also named paraffin. Formula $CH_3(CH_2)_nCH_3$ The systematic names of alkanes are given on p. xix.

Alkene A straight-chain hydrocarbon which has two adjacent carbon atoms united by a double bond. Also named olefin.

Alkyl See ***Aliphatic***

Allele (allelomorph) In Mendelian inheritance, one of a pair of contrasted characters inherited alternatively with its partner and thought to be based on genes situated in homologous chromosomes.

Anisotropic Showing different refractive indices when viewed in different directions. See ***birefringence***.

Anteiso-*Alkane* An aliphatic saturated hydrocarbon branched by the substitution of a methyl group on the third carbon atom of the chain. A 3-methylalkane.

Anthelminthic A drug toxic to intestinal or nematode worms.

Autoradiography A photographic technique which indicates the location of radioactive substances.

Birefringence (double refraction) The formation of two light rays with different refractive indices when polarised light passes through an anisotropic substance.

Cambium The sheet of meristematic tissue between the xylem and phloem. Its cells divide tangentially to produce more xylem and phloem. Also applied to the meristem that produces the cork.

β-Carotene A carotenoid of structure

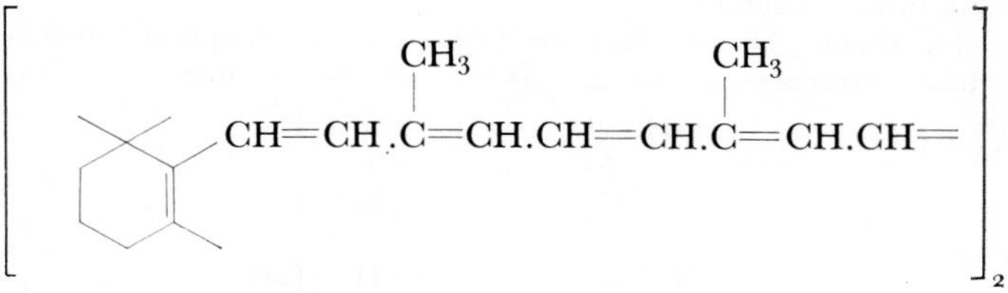

$$\left[\text{(trimethylcyclohexenyl)}\text{–}CH{=}CH.\overset{CH_3}{\overset{|}{C}}{=}CH.CH{=}CH.\overset{CH_3}{\overset{|}{C}}{=}CH.CH{=}\right]_2$$

Carotenoid See ***Terpenoid***

Chelation The equilibrium reaction between a metal ion and a complexing agent involving the formation of more than one bond between the metal and the agent and resulting in a ring structure incorporating the metal ion.

Cline A gradient within a continuous population in the frequencies of different genotypes (a genocline) or phenotypes (a phenocline) in different localities.

Co-enzyme A Co-enzymes serve as donors or acceptors of atoms or groups of atoms and mediate the transfer of these fragments. Co-enzyme A is synthesised from pantothenic acid, a water-soluble vitamin. It activates acyl groups (e.g. acetyl, propionyl) through the formation of a thiol ester and participates in the transfer of the groups.

Coleoptile The tubular leaf-like sheath that covers the developing leaves of embryos and seedlings.

Condensed tannin A tannin usually formed from two or more molecules of catechin or leucocyanidin or mixtures of the two. A non-hydrolysable tannin which tends to polymerise especially in acid solution to give an insoluble, amorphous coloured product known as a phlobaphene.

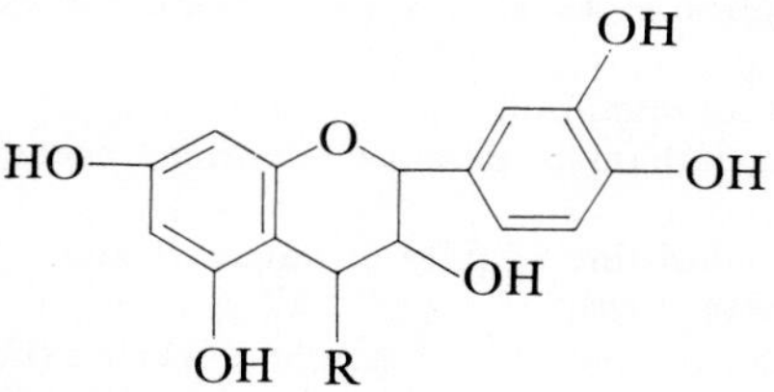

[Structure: flavan ring with HO, O, OH, OH, OH, OH, R substituents]

Catechin R = H

Leucocyanidin R = OH

Covalent bond The union of two atoms by the sharing of a pair of electrons.

Cystolith A crystal or concretion of calcium carbonate in a cell.

n-*Dicarboxylic acid* An alkane with carboxyl groups substituted in the terminal positions in the chain. An alkanedioic acid. Formula $COOH(CH_2)_nCOOH$

n-*β-Diketone* An alkane with two carbonyl groups separated by CH_2 substituted within the chain. Formula $CH_3(CH_2)_nCOCH_2CO(CH_2)_mCH_3$

n-*α, ω-Diol* An alkane with hydroxyl groups substituted in the terminal positions in the chain. Formula $OHCH_2(CH_2)_nCH_2OH$

Diterpene See ***Terpenoid***

Ectodesma A thread-like structure or region in the outer epidermal wall and terminating at the cuticle.

Electrophoresis The phenomenon of migration of suspended or colloidal particles in a liquid due to the effect of an electromotive force or potential difference across immersed electrodes.

Endoplasmic reticulum (ER) The system of double membranes that permeates the cytoplasm forming flat sacs or tubes.

Epithem A tissue of thin-walled, more or less isodiametric cells with large air spaces found inside hydathodes.

Erythro and ***threo*** Denote diastereo-isomers of a compound which contains hydroxyl groups on adjacent carbon atoms. See ***Vicinal.***

```
 OH OH          OH H
 |   |          |  |
—C — C—       —C — C—
 |   |          |  |
 H   H          H  OH
erythro        threo
```

Estolide (1) Acid estolide. A compound formed by inter-esterification from ω-hydroxy-fatty acids only. (2) Neutral estolide. A compound similarly formed from ω-hydroxy-fatty acids and an α,ω-diol.

Etholide See ***Estolide***

Exine Outer layer of the wall of pollen consisting mainly of sporopollenin.

Fangschleim The sticky exudation from the cuticle that traps insects in certain insectivorous plants.

Flavonoid A compound with the basic structural unit of flavone. C_6–C_3–C_6.

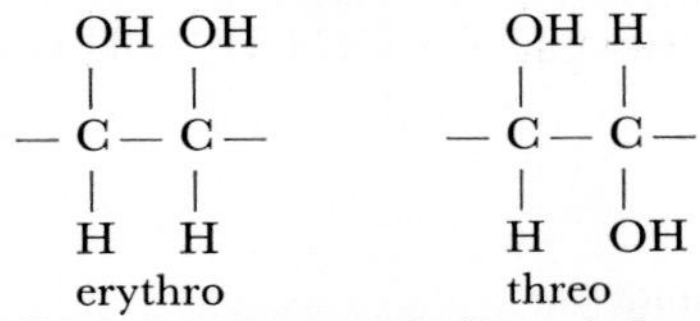

Freeze-etching Technique in which a frozen specimen is splintered, its surface etched by sublimation and replicated to give a preparation for electron microscopy.

Genotype The genetic constitution of an organism.

Gilson fixative A mixture containing ethanol, mercuric chloride, acetic acid, oxalic acid and formaldehyde.

Golgi bodies Cytoplasmic organelles consisting of piles of flattened sacs. Named after the discoverer, C. Golgi (1844–1926).

Head-to-head condensation The head end of a long-chain carboxylic acid refers to the end which bears the carboxyl group. The term implies the formation of a carbon-to-carbon bond by the linkage of the head ends of two molecules.

Homologue A compound which differs in molecular formula from the next member in the series by CH_2.

n-*ω-Hydroxy acid* An alkane with an hydroxy group and a carboxyl group substituted in the terminal positions in the chain. Formula $OHCH_2(CH_2)_n$ COOH

n-*Hydroxy-β-diketone* An alkane with an hydroxyl group and two carbonyl groups separated by CH_2 substituted within the chain. Formula $CH_3(CH_2)_nCOCH_2$ $CO(CH_2)_xCHOH(CH_2)_mCH_3$

Intine Inner layer of the wall of pollen consisting mainly of carbohydrates.

Iso-*alkane* An aliphatic saturated hydrocarbon branched by the substitution of a methyl group on the second carbon atom of the chain. A 2-methylalkane.

Isotropic Showing the same refractive index when viewed in different directions.

n-*Ketol* An alkane with an hydroxyl group and a carbonyl group substituted within the chain. Formula $CH_3(CH_2)_nCO(CH_2)_xCHOH(CH_2)_mCH_3$

n-*Ketone* An alkane with a carbonyl group substituted within the chain. Formula $CH_3(CH_2)_nCO(CH_2)_mCH_3$

Lignan A compound which has as its basic skeleton two units of the phenylpropane C_6–C–C–C structure joined by carbon-carbon bonds between the middle carbon atoms of the side chains.

Lignin A complex polymer built up of phenylpropane units which incrusts the cell walls of tracheids, fibres and other parts of vascular plants.

Meristem A tissue of dividing cells.

Microfibril A thin thread of cellulose which forms the basis of the skeleton of the plant cell wall.

Middle lamella The first formed portion of the cell wall. Said to consist mainly of pectin.

Mitochondrion A cytoplasmic organelle concerned with respiration.

n-*Monocarboxylic acid* An alkane with a carboxyl group substituted in a terminal position in the chain. An alkanoic acid. Formula $CH_3(CH_2)_nCOOH$.

Ochrea or ***Ocrea*** A tubular sheath-like expansion at the base of the petiole.

Olefin See ***Alkene***

Organelle Term used to describe various separate parts of a cell having different functions.

Paraffin See ***Alkane***

Periderm The protective tissue of organs whose epidermis is replaced after growth. Formed from the cork cambium (phellogen) which produces phellem (cork) cells to the outside and phelloderm to the inside.

Phenotype The type of organism produced by the interaction of its genotypic characters with the environment.

Phlobaphene See ***Condensed tannin***

Plasmalemma The membrane which encloses the cytoplasm of living cells.

Plasmodesma A fine protoplasmic connection between plant cells. It passes through the cell wall and is bounded by an extension of the plasmalemma.

Plastid A cytoplasmic organelle. There are several kinds, of which the best known are the chloroplasts.

Polarised light Light whose vibration is confined to a single plane (plane polarised).

n-*Primary alcohol* An alkane with an hydroxyl group substituted in a terminal position in the chain. An alkan-1-ol. Formula $(CH_3(CH_2)_nCH_2OH$.

Primary wall The cell wall formed before cell extension ceases.

Ribosome A submicroscopic particle, containing ribonucleic acid and protein, concerned with protein synthesis. Found in the cytoplasm, nucleus, mitochondria and plastids.

Saturated hydrocarbon A hydrocarbon in which each carbon atom is linked to one or more other carbon atoms by single bonds. See ***Alkane***.

n-*Secondary alcohol* An alkane with an hydroxyl group substituted within the chain. Formula $CH_3(CH_2)_nCHOH(CH_2)_mCH_3$.

Secondary wall The cell wall formed after cell extension ceases.

Spherosome A cytoplasmic organelle with a high fat content and bounded by a single unit membrane.

Steroid A compound which possesses the cyclic carbon skeleton of cyclopentenophenanthrene. A sterol is a steroid alcohol.

Tannin An amorphous high molecular weight polymer of phenolic compounds.

Terpenoid A compound which possesses a carbon skeleton which can be regarded as built up by the fusion of two or more isoprene units (see p. 171). Diterpenoids (C_{20}) are built from four, triterpenoids (C_{30}) from six and carotenoids (C_{40}) from eight. Carotenoids are tetraterpenoids (see ***β-Carotene***).

Thigmotropism The response to mechanical contact expressed by clinging or curving, as in tendrils.

Threo See ***Erythro***

Triterpene See ***Terpenoid***

Unsaturated hydrocarbon A hydrocarbon in which carbon atoms are linked by one or more double or triple bonds. See ***Alkene***.

Vicinal (*vic*) Denotes neighbouring or adjoining positions on a carbon chain or ring. Used to name derivatives with substituting groups in such positions e.g. *vic*-dihydroxy acids, formed by the addition of hydroxyl groups to the double bond of an unsaturated acid.

Vitamin A palmitate Vitamin A is a primary alcohol derived from half the molecule of β-carotene. Esterification with palmitic acid gives the palmitate.

Xeromorphic Structurally modified to resist drought.

Xerophyte A plant capable of living in dry conditions.

References/Author Index

The *italic* figures in parentheses are the page numbers upon which authors are cited.

A

ADAM, N. K. (1948). *Discuss. Faraday Soc.* No. **3**, 5. *(236)*
— (1958). *Endeavour* **17**, 37. *(225)*
— (1963). In *Waterproofing and Water-repellency*, ed. J. L. Moilliet, Elsevier, Amsterdam, London and New York, p. 1. *(223, 225)*
ADAMS, P. B., SPROSTON, T., TIETZ, H. and MAJOR, R. T. (1962). *Phytopathology* **52**, 233. *(266)*
AGETA, H. (1959). *J. pharm. Soc. Japan* **79**, 58. *(132)*
ALBERSHEIM, P. (1965a). In *Plant Biochemistry*, ed. J. Bonner and J. E. Varner, Academic Press, New York and London, p. 151. *(77)*
— (1965b). In *Plant Biochemistry*, ed. J. Bonner and J. E. Varner, Academic Press, New York and London, p. 298. *(77)*
ALI KHAN, M. Y., HALL, A. N. and ROBINSON, D. S. (1964). *Antonie van Leeuwenhoek* **30**, 417. *(285)*
ALVAREZ-VAZQUEZ, R. and RIBAS-MARQUÉS, I (1968). *An. R. Soc. esp. Fis. Quim.* **64 B**, 783, 1001. *(153)*
AMBRONN, H. (1888). *Ber. dt. bot. Ges.* **6**, 226. *(58)*
AMELUNXEN, F., MORGENROTH, K. and PICKSAK, T. (1967). *Ztschr. Pflanzenphysiol.* **57**, 79. *(32, 110)*
AMSDEN, R. C. and LEWINS, C. P. (1966). *Wld Rev. Pest Control* **5**, 187. *(10, 248)*
ANDERSON, D. B. (1928). *Jb. wiss. Bot.* **69**, 501. *(4, 60)*
— (1934). *Ohio J. Sci*, **34**, 9. *(58)*
— (1935). *Bot. Rev.* **1**, 52. *(4, 60)*
ANSTEY, T. H. and MOORE, J. F. (1954). *J. Hered.* **45**, 39. *(191, 270)*
APLIN, R. T., CAMBIE, R. C. and RUTLEDGE, P. S. (1963). *Phytochemistry* **2**, 205. *(137, 140)*

ARBER, A. (1920). *Water Plants.* Cambridge University Press, Cambridge. (*96*)
ARENS, K. (1929). *Jb. wiss. Bot.* **70**, 93. (*258*)
— (1934). *Jb. wiss. Bot.* **80**, 248. (*212*)
ARENS, T. (1968). *Protoplasma* **66**, 403. (*84, 201*)
ARMAN, P. and WAIN, R. L. (1958). *Ann. appl. Biol.* **46**, 366. (*246*)
ARNOLD, L. K. and HSIA, P. R. (1957). *Ind. Engng. Chem. ind. Edn* **49**, 360. (*136*)
ARTZ, T. (1933). *Ber. dt. bot. Ges.* **51**, 470. (*60, 96*)
AWASTHI, Y. C. and MITRA, C. R. (1968). *Phytochemistry* **7**, 1433. (*150*)

B

BAILEY, L. H. (1949). *Manual of Cultivated Plants.* Macmillan, New York. (*192*)
BAIN, J. M. and McBEAN, D. McG. (1967). *Aust. J. biol. Sci.* **20**, 895. (*204*)
— (1969). *Aust. J. biol. Sci.* **22**, 101. (*204*)
BAKER, A. J., EGLINTON, G., GONZALEZ, A. G., HAMILTON, R. J. and RAPHAEL, R. A. (1962). *J. chem. Soc.* 4705. (*130, 140*)
BAKER, E. A. (1968). Personal communication. (*282, 283*)
BAKER, E. A., BATT, R. F. and MARTIN, J. T. (1964). *Ann. appl. Biol.* **53**, 59. (*47, 52, 145*)
BAKER, E. A., BATT, R. F., ROBERTS, M. F. and MARTIN, J. T. (1962). *Rep. agric. hort. Res. Stn Univ. Bristol for 1961*, 114. (*147, 181, 185*)
BAKER, E. A., BATT, R. F., SILVA FERNANDES, A. M. S. and MARTIN, J. T. (1963). *Rep. agric. hort. Res. Stn Univ. Bristol for 1962*, 69. (*185, 187*)
— (1964). *Rep. agric. hort. Res. Stn Univ. Bristol for 1963*, 106. (*46, 129, 136, 185*)
BAKER, E. A., DAWKINS, D. J. and SMITH, B. D. (1968). *Rep. agric. hort. Res. Stn Univ. Bristol for 1967*, 116. (*183, 245*)
BAKER, E. A. and MARTIN, J. T. (1963). *Nature, Lond.* **199**, 1268. (*146*)
— (1967). *Ann. appl. Biol.* **60**, 313. (*52, 147, 148, 184, 185, 186*)
BAKER, E. A., MARTIN, J. T. and STRINGER, A. (1968). Unpublished work. (*290*)
BAKER, W. (1949). *Nature, Lond.* **164**, 1093. (*138*)
BANDULSKA, H. (1926). *J. Linn. Soc. (Bot.)* **47**, 383. (*14*)
— (1931). *J. Linn. Soc. (Bot.)* **48**, 657. (*14*)
BARASH, I., KLISIEWICZ, J. M. and KOSUGE, T. (1964). *Phytopathology* **54**, 923. (*261*)
BARBER, H. N. (1955). *Evolution* **9**, 1. (*11, 101, 180, 191*)
BARBER, H. N. and JACKSON, W. D. (1957). *Nature, Lond.* **179**, 1267. (*180, 190, 217*)
BARBER, H. N. and NETTING, A. G. (1968). *Phytochemistry* **7**, 2089. (*171*)
BARNER, J. and RÖDER, K. (1962). *Proc. 10th International Orthocide Conference*, Belgrade. (*246*)
BARNES, C. S., GALBRAITH, M. N., RITCHIE, E. and TAYLOR, W. C. (1965). *Aust. J. Chem.* **18**, 1411. (*125*)
BARNES, E. H. and WILLIAMS, E. B. (1960). *Phytopathology* **50**, 844. (*264*)
BARRETT, A. J. and NORTHCOTE, D. H. (1965). *Biochem. J.* **94**, 617. (*76*)
BARRIER, G. E. and LOOMIS, W. E. (1957). *Pl. Physiol., Lancaster* **32**, 225. (*240*)
BARTHÉLEMY, A. (1868). *Ann. Sci. nat. (Bot.)* **9**, 287. (*54, 55*)
BARY, A. DE (1871). *Bot. Ztg.* **29**, 129, 145, 161, 566, 573, 589, 605. (*56, 103, 109*)
— (1884). *Comparative anatomy of the vegetative organs of the phanerogams and ferns*, transl. Bower and Scott, Oxford University Press, London. (*57*)
BATT, R. F. and MARTIN, J. T. (1961). *Rep. agric. hort. Res. Stn Univ. Bristol for 1960*, 111. (*46, 246*)
— (1966). *J. hort. Sci.* **41**, 271. (*46*)
BECKER, E. S. and KURTH, E. F. (1958). *TAPPI* **41**, 380. (*148*)

BEL'KEVICH, P. I., KAGANOVICH, F. L. and TRUBILKO, E. V. (1960). *Trudy Inst. Torfa, Minsk* **9**, 274. (*282*)

BENDORAITIS, J. G., RUSANIWSKY, W., STEDMAN, R. L. and SWAIN, A. P. (1960). *Chemy Ind.* 838. (133)

BENNETT, S. H. and THOMAS, W. D. E. (1954). *Ann. appl. Biol.* **41**, 484. (*237, 240*)

BERG, R. (1914). *Chemikerzeitung* **38**, 1162. (*62*)

BERMEJO-BARRERA, J., ESTEVEZ-REYES, R. and GONZALEZ-GONZALEZ, A. (1964). *An. R. Soc. esp. Fis. Quim.* **60 B**, 601. (*134*)

BERWITH, C. E. (1936). *Phytopathology* **26**, 1071. (*267*)

BESCANSA-LÓPEZ, J. L., GIL CURBERA, G. and RIBAS-MARQUÉS, I. (1966). *An. R. Soc. esp. Fis. Quim.* **62 B**, 865. (*149*)

BESCANSA-LÓPEZ, J. L. and RIBAS-MARQUÉS, I. (1966). *An. R. Soc. esp. Fis. Quim.* **62 B**, 871. (*149*)

BHATT, D. D. and VAUGHAN, E. K. (1963). *Phytopathology* **53**, 217. (*255*)

BIANCHI, A. and MARCHESI, G. (1960). *Z. Vererb. Lehre* **91**, 214. (*103, 188, 189*)

— (1961). *Atti. Accad. gioenia Sci. nat.* **6**, 395. (*103*)

BIANCHI, A. and SALAMINI, F. (1964). *Genetica* **18**, 183. (*188, 189*)

BILLINGS, W. D. and MORRIS, R. J. (1951). *Am. J. Bot.* **38**, 327. (*217*)

BISHOP, C. T. (1955). *Can. J. Chem.* **33**, 1521. (*76*)

BISHOP, C. T., BAYLEY, S. T. and SETTERFIELD, G. (1958). *Pl. Physiol., Lancaster* **33**, 283. (*76*)

BISHOP, C. T., HARWOOD, V. D. and PURVES, C. B. (1950). *Pulp Pap. Can.* **51**, 90. (*148*)

BLACKMAN, F. F. (1895). *Phil. Trans. R. Soc. Ser. B* **186**, 503. (*56, 215*)

BLACKMAN, G. E., BRUCE, R. S. and HOLLY, K. (1958). *J. exp. Bot.* **9**, 175. (*226, 230*)

BLAIR, E. H., MITCHELL, H. L. and SILKER, R. E. (1953). *Ind. Engng. Chem. ind. Edn.* **45**, 1104. (*66*)

BLAKEMAN, J. P. (1968). *Ann. appl. Biol.* **61**, 77. (*262*)

BLASDALE, W. C. (1945). *J. Am. chem. Soc.* **67**, 491. (*138, 139*)

— (1947). *Jl. R. hort. Soc.* **72**, 240. (*138*)

BOLLIGER, R. (1959). *J. Ultrastruct. Res.* **3**, 105. (*113, 158, 202*)

BONNER, F. T. (1968). *Bot. Gaz.* **129**, 83. (*106, 221*)

BONNETT, H. T. and NEWCOMB, E. H. (1966). *Protoplasma* **62**, 59. (*107*)

BORGES DEL CASTILLO, J., BROOKS, C. J. W., CAMBIE, R. C., EGLINTON, G., HAMILTON, R. J. and PELLITT, R. (1967). *Phytochemistry* **6**, 391. (*114, 140*)

BÖRNER, H. (1960). *Bot. Rev.* **26**, 393. (*213*)

BOSE, S. and GUPTA, K. C. (1961). *Proc. annu. Conv. Sug. Technol. Ass. India* **29**, 70 (*126*)

BOUGAULT, J. and BOURDIER, L. (1908). *C.r. hebd. Séanc. Acad. Sci., Paris* **147**, 1311. (*61, 62*)

— (1909). *J. Pharm. Chim., Paris* **29**, 561; **30**, 10. (*62*)

BOUSSINGAULT, J. B. (1864). *Annls. Sci. nat. (Bot.)* **1**, 314 (*55*)

— (1865). *C.r. hebd. Séanc. Acad. Sci., Paris* **61**, 493, 605, 657 (*55*)

BOYNTON, D. (1954). *A Rev. Pl. Physiol.* **5**, 31. (*236*)

BRADLEY, D. E. (1965). In *Techniques for Electron Microscopy* (2nd Ed.) ed. D. Kay. Blackwell Scientific Publications, Oxford, p. 96. (*20, 23*)

BRANTS, D. H. (1964). *Virology* **23**, 588. (*84, 268, 269*)

— (1965). *Virology* **26**, 554. (*268, 269*)

BRAUNER, L. (1930). *Jb. wiss. Bot.* **73**, 513. (*61, 243*)

BREDEMEIJER, G. and HEINEN, W. (1968). *Acta bot. neerl.* **17**, 15. (*174*)

BRENNER, R. R. and FIORA, J. (1960). *Industria Quim., B. Aires* **20**, 531. (*131, 160*)

BRIAN, R. C. (1967). *Ann. appl. Biol.* **59**, 91 (*241, 244*)

— (1968). In *Physico-chemical and biophysical factors affecting the activity of pesticides.* Society of Chemical Industry Monograph No. 29, p. 303. (*244*)

BRIAN, R. C. and CATTLIN, N. D. (1968). *Ann. Bot.* **32**, 609. (*89*)
BRIESKORN, C. H. (1959). *Pharm. Zentralhalle Dtl.* **98**, 638. (*147*)
BRIESKORN, C. H. and BÖSS, J. (1964). *Fette Seifen Anstr-Mittel* **66**, 925. (*145, 152, 154, 243*)
BRIESKORN, C. H. and FEILNER, K. (1968). *Phytochemistry* **7**, 485. (*133, 170*)
BRIESKORN, C. H. and HOFMANN, H. (1962). *Arch. Pharm., Berl.* **295**, 505. (*51*)
BRIESKORN, C. H. and KLINGER, H. (1963). *Z. Lebensmittelunters. u.-Forsch.* **120**, 269. (*124*)
BRIESKORN, C. H. and MUSTAFA KESKIN. (1954). *Pharm. Acta Helv.* **29**, 338. (*136*)
BRIESKORN, C. H. and REINARTZ, H. (1967a). *Z. Lebensmittelunters. u.-Forsch.* **133**, 137. (*50, 129*)
— (1967b). *Z. Lebensmittelunters. u.-Forsch.* **135**, 55. (*50, 53, 146, 154*)
BRIESKORN, C. H. and SCHNEIDER, W. (1961). *Z. Lebensmittelunters. u.-Forsch.* **115**, 513. (*51, 124, 145*)
BRIGGS, L. H. and LOCKER, R. H. (1951). *J. chem. Soc.* 3131. (*151*)
BRODIE, P. B. (1842). *Proc. geol. Soc.* **3**, 592. (*55*)
BRONGNIART, A. (1830). *Annls. Sci. nat.* (*Bot.*) 1st Ser. **21**, 420. (*55, 67*)
— (1834). *Annls. Sci. nat.* (*Bot.*) 2nd Ser. **1**, 65. (*55*)
BROOKS, J. and SHAW, G. (1968). *Nature, Lond.* **219**, 532. (*155*)
BROWN, D. S. and KOCH, E. C. (1962). *Proc. Am. Soc. hort. Sci.* **81**, 35. (*220*)
BROWN, H. T. and ESCOMBE, F. (1905). *Proc. R. Soc. Ser. B* **76**, 29. (*216*)
BROWN, W. (1922). *Ann. Bot.* **36**, 101. (*13, 261*)
— (1936). *Bot. Rev.* **2**, 236. (*12*)
BROWN, W. and HARVEY, C. C. (1927). *Ann. Bot.* **41**, 643. (*257*)
BROWN, W. V. and JOHNSON, C. (1962). *Am. J. Bot.* **49**, 110. (*89*)
BRUNSKILL, R. T. (1956). *Proc. 3rd Br. Weed Control Conference* **1**, 593. (*245*)
BUCHNER, G. (1918). *Chemikerzeitung* **42**, 373. (*62*)
BUKOVAC, M. J. (1966). Personal communication. (*239*)
BUKOVAC M. J. and NORRIS, R. F. (1966). *Atti Simp. Int. Agrochim.* **6**, 296. (*242, 243*)
BURGERJON, A. and GRISON, P. (1965). *J. invert. Path.* **7**, 281. (*268*)
BURT, P. E. and WARD, J. (1955). *Bull. ent. Res.* **46**, 39. (*246*)
BUTLER, J. H. A., DOWNING, D. T. AND SWABY, R. J. (1964). *Aust. J. Chem.* **17**, 817. (*283*)
BYSTROM, B. G., GLATER, R. B., SCOTT, F. M. and BOWLER, E. S. C. (1968). *Bot. Gaz.* **129**, 133. (*253, 271*)

C

CALAM, D. H. (1968). *Phytochemistry* **7**, 1419. (*136*)
CANDY, H. A., McGARRY, E. J. and PEGEL, K. H. (1968). *Phytochemistry*, **7**, 889. (*149*)
CARLQUIST, S. (1956). *Am. J. Bot.* **43**, 425. (*81*)
CAROTHERS, W. H. and HILL, J. W. (1932). *J. Am. chem. Soc.* **54**, 1559. (*147*)
CAROTHERS, W. H. and VAN NATTA, F. J. (1933). *J. Am. chem. Soc.* **55**, 4714. (*147*)
CARRUTHERS, W. and JOHNSTONE, R. A. W. (1959). *Nature, Lond.* **184**, 1131. (*48, 121, 133*)
CASSIE, A. B. D. and BAXTER, S. (1945). *Nature, Lond.* **155**, 21. (*225, 231*)
CHALLEN, S. B. (1959). *Chemist Drugg.* **101**, 18. (*229*)
— (1960). *J. Pharm. Pharmac.* **12**, 307. (*231*)
— (1962). *J. Pharm. Pharmac.* **14**, 707. (*231*)
CHAMBERS, T. C. and POSSINGHAM, J. V. (1963). *Aust. J. biol. Sci.* **16**, 818. (*46, 204*)
CHANNON, H. J. and CHIBNALL, A. C. (1929). *Biochem. J.* **23**, 168. (*64, 66, 158, 163, 164, 169*)

CHATTERJEE, A. and MITRA, S. S. (1949). *J. Am. chem. Soc.* **71**, 606. (*151*)
CHEVREUL, M. (1807). *Annls. Chim. Phys.* **62**, 323. (*69*)
— (1815). *Annls. Chim. Phys.* **96**, 141. (*69*)
CHIBNALL, A. C., EL MANGOURI, H. A. and PIPER, S. H. (1954). *Biochem. J.* **58**, 506. (*65*)
CHIBNALL, A. C., LATNER, A. L., WILLIAMS, E. F. and AYRE, C. A. (1934). *Biochem. J.* **28**, 313. (*271*)
CHIBNALL, A. C. and PIPER, S. H. (1934). *Biochem. J.* **28**, 2209. (*66, 159, 163*)
CHIBNALL, A. C., PIPER, S. H., EL MANGOURI, H. A., WILLIAMS, E. F. and IYENGAR, A. V. V. (1937). *Biochem. J.* **31**, 1981. (*65*)
CHIBNALL, A. C., PIPER, S. H., POLLARD, A., SMITH, J. A. B. and WILLIAMS, E. F. (1931). *Biochem. J.* **25**, 2095. (*63, 158*)
CHIBNALL, A. C., PIPER, S. H., POLLARD, A., WILLIAMS, E. F. and SAHAI, P. N. (1934). *Biochem. J.* **28**, 2189. (*64, 65, 271*)
CHIBNALL, A. C., WILLIAMS, E. F., LATNER, A. L. and PIPER, S. H. (1933). *Biochem. J.* **27**, 1885. (*64*)
CHICOISNE, A., DUPONT, G. and DULOU, R. (1957). *Bull. Soc. chim. Fr.* 1232. (*152*)
CHRIST, B. (1959). *Z. Bot.* **47**, 88. (289)
CHU CHOU, M. and PREECE, T. F. (1968). *Ann. appl. Biol.* **62**, 11. (262)
CHURCHILL, D. M. (1968). *Aust. J. Bot.* **16**, 125. (279)
CLARK, R. C. Jr. and BLUMER, M. (1967). *Limnol. Oceanogr.* **12**, 79. (282)
CLENSHAW, E. and SMEDLEY-MACLEAN, I. (1929). *Biochem. J.* **23**, 107. (*63, 158*)
CLOWES, F. A. L. and JUNIPER, B. E. (1968). *Plant Cells.* Blackwell Scientific Publications, Oxford. (*20, 23, 73, 75, 82, 242*)
COLE, L. J. N. (1956). *Diss. Abstr.* **16**, 1570. (*48, 134, 137*)
COLE, L. J. N. and BROWN, J. B. (1960). *J. Am. Oil chem. Soc.* **37**, 359. (*48, 134, 137*)
COLLISON, D. L. and SMEDLEY-MACLEAN, I. (1931). *Biochem. J.* **25**, 606. (*63*)
COMMONWEALTH BUREAU OF SOILS, HARPENDEN, ENGLAND. *Dew as an Ecological Factor and its Measurement 1947–1961*, Bibliography No. 691. (*199*)
CONSDEN, R., GORDON, A. H. and MARTIN, A. J. P. (1944). *Biochem. J.* **38**, 224. (*37*)
COOK, J. A. and BOYNTON, D. (1952). *Proc. Am. Soc. hort. Sci.* **59**, 82. (*240*)
CORBETT, R. E. and McCRAW, E. H. (1959). *J. Sci. Fd Agric.* **10**, 29. (*149*)
CORBETT, R. E. and McDOWALL, M. A. (1958). *J. chem. Soc.* 3715. (*149*)
CORBETT, R. E., McDOWALL, M. A. and WYLLIE, S. G. (1964). *J. chem. Soc.* 1283. (*149*)
CORBETT, R. E., YOUNG, H. and WILSON, R. S. (1964). *Aust. J. Chem.* **17**, 712. (*149*)
COREY, E. J. and URSPRUNG, J. J. (1956). *J. Am. chem. Soc.* **78**, 5041. (*69*)
CORMACK, R. G. H. (1937). *New Phytol.* **36**, 19. (*109*)
CORNER, E. J. H. (1965). *The Life of Plants.* Weidenfeld and Nicolson (Nat. Hist. Series), London. (*90*)
CRAFTS, A. S. (1961). *The Chemistry and Mode of Action of Herbicides.* Interscience, New York and London. (*82, 234*)
CRAFTS, A. S. and FOY, C. L. (1962). In *Residue Reviews*, ed. Francis A. Gunther. Springer-Verlag, Berlin. Vol. 1, p. 112. (*223*)
CRAWFORD, M. A. and MENEZES, F. A. (1963). *Biochem. J.* **89**, 72P. (*132*)
CRISP, C. E. (1965). *The biopolymer cutin.* Ph.D thesis, University of California, Davis, California. (*4, 8, 11, 53, 145, 153, 176*)
CRISP, D. J. (1963). In *Waterproofing and Water-repellency*, ed. J. L. Moilliet. Elsevier, Amsterdam, London and New York, p. 416. (*2*)
CROSBY, D. G. and VLITOS, A. J. (1960). *Contr. Boyce Thompson Inst. Pl. Res.* **20**, 283. (*262*)
CROSSE, J. E. (1959). *Ann. appl. Biol.* **47**, 306. (*255*)
CROWDY, S. H. (1959). In *Plant Pathology: Problems and Progress 1908–1958*, ed. C. S. Holton, G. W. Fischer, R. W. Fulton, H. Hart and S. E. A. McCallan. University of Wisconsin Press, Madison, p. 231. (*238*)

CRUICKSHANK, I. A. M. (1963). *A. Rev. Phytopath.* **1**, 351. (*264*)
— (1966). *Wld Rev. Pest Control* **5**, 161. (*263*)
— (1968). *Abstrs 1st Int. Congr. Pl. Path.*, London, p. 39. (*264*)
CUNZE, R. (1926). *Beih. bot. Zbl.* **42**, Abt. 1, 160. (*179*)
CURRIER, H. B. and DYBING, C. D. (1959). *Weed Science* **7**, 195. (*236, 239*)
CURTIS, L. C. (1944). *Phytopathology* **34**, 196. (*248*)

D

DALBRO, S. (1955). *Proc. 14th Int. hort. Congr.* **1**, 770. (*212*)
DALE, H. M. (1951). *Science, N.Y.* **114**, 438. (*109*)
DALY, G. T. (1964). *J. exp. Bot.* **15**, 160. (*46, 180*)
DAMM, O. (1901). *Beih. bot. Zbl.* **11**, 219. (*59*)
DARLEY, E. F. and MIDDLETON, J. T. (1966). *A. Rev. Phytopath.* **4**, 103. (*251*)
DARLEY, E. F., TAYLOR, O. C. and MIDDLETON, J. T. (1968). *Abstrs 1st Int. Congr. Pl. Path.*, London, p. 41. (*253*)
DARLINGTON, W. A. and BARRY, J. B. (1965). *J. agric. Fd Chem.* **13**, 76. (*235, 239*)
DASHEK, W. V. and ROSEN, W. G. (1966). *Protoplasma* **61**, 192. (*107*)
DAVENPORT, J. B. (1956). *Aust. J. Chem.* **9**, 416. (*123, 139*)
— (1960). *Aust. J. Chem.* **13**, 411. (*123, 162*)
DAVID, W. A. L. and GARDNER, B. O. C. (1966). *J. invert. Path.* **8**, 180. (*267*)
DAVIES, R. R. (1961). *Nature, Lond.* **191**, 616. (*256*)
DEVERALL, B. J. (1967). *Ann. appl. Biol.* **59**, 375. (*261*)
DE VRIES, H. A. M. A. (1968). *Acta bot. neerl.* **17**, 229. (*250*)
DE VRIES, H. A. M. A., BREDEMEIJER, G. and HEINEN, W. (1967). *Acta bot. neerl.* **16**, 102. (*286*)
DEWEY, O. R., GREGORY, P. and PFEIFFER, R. K. (1956). *Proc. 3rd Br. Weed Control Conference* **1**, 313. (*245, 248*)
DEWEY, O. R., HARTLEY, G. S. and MACLAUCHLAN, J. W. G. (1962). *Proc. R. Soc. Ser. B* **155**, 532. (*45, 245*)
DICKINSON, S. (1949a). *Ann. Bot.* **13**, 89. (*256*)
— (1949b). *Ann. Bot.* **13**, 219. (*256*)
— (1949c). *Ann. Bot.* **13**, 337. (*256*)
— (1949d). *Ann. Bot.* **13**, 345. (*256*)
— (1955). *Ann. Bot.* **19**, 161. (*256*)
— (1964). *Trans. Br. mycol. Soc.* **47**, 300. (*257*)
DILCHER, D. (1963). *Am. J. Bot.* **50**, 1. (*279*)
DIMBLEBY, G. W. (1967). *Plants and Archeology*. Baker, London. (*117, 294*)
DIXON, A. F. G., MARTIN-SMITH, M. and SUBRAMANIAN, G. (1965). *J. chem. Soc.* 1562. (*271*)
DOLZMAN, P. (1964). *Planta* **60**, 461. (*94, 95, 195*)
— (1965). *Planta* **64**, 76. (*94, 95, 195*)
DORMAL, S. (1960). *Bull. Inst. agron. Stns Rech. Gembloux* **3**, 1169. (*222*)
DOROKHOV, B. L. (1963). *Bot. Zh. Kÿyiv* **48**, 893. (*10, 216*)
DORSCHNER, K. P. and BUCHHOLTZ, K. P. (1956). *Agron. J.* **48**, 59. (*180*)
DOUS, E. (1927). *Bot. Arch.* **19**, 461. (*58*)
DOWNING, D. T. (1961). *Rev. pure appl. Chem.* **11**, 196. (*151*)
DOWNING, D. T., KRANZ, Z. H., LAMBERTON, J. A., MURRAY, K. E. and REDCLIFFE, A. H. (1961). *Aust. J. Chem.* **14**, 253. (*51*)
DOWNING, D. T., KRANZ, Z. H. and MURRAY, K. E. (1960). *Aust. J. Chem.* **13**, 80. (*51*)

DOWNING, D. T., KRANZ, Z. H. and MURRAY, K. E. (1961). *Aust. J. Chem.* **14**, 619. (*48, 49, 125*)
DRAKE, N. L. and JACOBSEN, R. P. (1935). *J. Am. chem. Soc.* **57**, 1570. (*69, 149*)
DRAWERT, H. and MIX, M. (1963). *Protoplasma* **57**, 270. (*73*)
DUCHARTRE, P. (1856). *C.r. hebd. Séanc. Acad. Sci., Paris* **42**, 37. (*55*)
DUDMAN, W. F. and GRNCAREVIC, M. (1962). *J. Sci. Fd Agric.*, **13**, 221. (*46*)
DUGGER, W. M. (1952). *Pl. Physiol., Lancaster* **27**, 489. (*216, 217*)
DUHAMEL, L. (1963). *Annls. Chim.* **8**, 315. (*149*)
— (1965). *Bull. Soc. chim. Fr.* 399. (*152*)
DUNCAN, I. J. and DUSTMAN, R. B. (1936). *J. Am. chem. Soc.* **58**, 1511. (*147*)
DUNN, C. L., BROWN, K. F. and MONTAGNE, J. T. W. (1969). *Phytopath. Z.* **64**, 112. (*246*)
DUPERON, P., VETTER, W. and BARBIER, M. (1964). *Phytochemistry* **3**, 89. (*50*)
DUPONT, G., DULOU, R. and CHICOISNE, A. (1956). *Bull. Soc. chim. Fr.* 1413. (*152*)
DUPONT, G., DULOU, R. and COHEN, J. (1955). *C.r. hebd. Séanc. Acad. Sci., Paris* **240**, 875. (*152*)
— (1956). *Bull. Soc. chim. Fr.* 819. (*152*)
DUTROCHET, R. J. H. (1832). *Annls. Sci. nat. (Bot.)* **25**, 242. (*55*)
DYBING, C. D. and CURRIER, H. B. (1961). *Pl. Physiol., Lancaster* **36**, 169. (*236, 238*)
DYSON, W. G. and HERBIN, G. A. (1968). *Phytochemistry* **7**, 1339. (*142*)

E

EBELING, W. (1961). *Hilgardia* **30**, 531. (*209*)
— (1963). In *Residue Reviews*, ed. F. A. Gunther. Springer-Verlag, Berlin. Vol. 3, p. 35. (*223*)
EBELING, W. and WAGNER, R. E. (1961). *Hilgardia* **30**, 565. (*209*)
ECHLIN, P. (1968). *Scient. Am.* **218**, 81. (*119*)
EDNEY, K. L. (1956). *Ann. appl. Biol.* **44**, 113. (*118*)
— (1967). *Ann. appl. Biol.* **60**, 367. (*261*)
EDWARDS, C. A. and HEATH, G. W. (1963). In *Soil Organisms*, ed. J. Doeksen and J. van der Drift. North-Holland Publ. Co., Amsterdam, p. 76. (*290*)
EDWARDS, J. S. (1966). *Nature, Lond.* **211**, 73. (*272*)
EDWARDS, W. N. (1935). *Biol. Rev.* **10**, 442. (*114*)
EGLINTON, G., GONZALEZ, A. G., HAMILTON, R. J. and RAPHAEL, R. A. (1962). *Phytochemistry* **1**, 89. (*45, 114, 130, 131, 132, 133, 134, 140*)
EGLINTON, G. and HAMILTON, R. J. (1967). *Science, N.Y.* **156**, 1322. (*114, 170*)
EGLINTON, G., HAMILTON, R. J., KELLY, W. B. and REED, R. I. (1966). *Phytochemistry* **5**, 1349. (*130*)
EGLINTON, G., HAMILTON, R. J. and MARTIN-SMITH, M. (1962). *Phytochemistry* **1**, 137. (*122, 140*)
EGLINTON, G., HAMILTON, R. J., MARTIN-SMITH, M., SMITH, S. J. and SUBRAMANIAN, G. (1964). *Tetrahedron Lett.* No. 34, 2323. (*131*)
EGLINTON, G., HAMILTON, R. J., RAPHAEL, R. A. and GONZALEZ, A. G. (1962). *Nature, Lond.* **193**, 739. (*140*)
EGLINTON, G. and HUNNEMAN, D. H. (1968). *Phytochemistry* **7**, 313. (*40, 44, 50, 53, 146, 153*)
EGLINTON, G., HUNNEMAN, D. H. and DOURAGHI-ZADEH, K. (1968). *Tetrahedron* **24**, 5929. (*283*)
EGLINTON, G., HUNNEMAN, D. H. and McCORMICK, A. (1968). *Organic Mass Spectrometry* **1**, 593. (*53*)

EGLINTON, G., SCOTT, P. M., BELSKY, T., BURLINGAME, A. L. and CALVIN, M. (1964). *Science, N.Y.* **145**, 263. (*14*)
EPSTEIN, E. (1953). *Nature, Lond.* **171**, 83. (*242*)
EPTON, H. A. S. and DEVERALL, B. J. (1968). *Ann. appl. Biol.* **61**, 255. (*265*)
ESAU, K. (1953). *Plant Anatomy.* John Wiley, New York, London, Sydney. (*4*)
— (1965). *Plant Anatomy* 2nd ed., John Wiley, New York, London, Sydney. (*73, 75, 82, 104, 212*)
ESAU, K., CRONSHAW, J. and HOEFERT, L. L. (1967). *J. Cell Biol.* **32**, 71. (*269*)
EVANS, A. C. and MARTIN, H. (1935). *J. Pomol.* **13**, 261. (*229*)
EVANS, E. (1968). *Plant Diseases and their Chemical Control.* Blackwell Scientific Publications, Oxford. (*206*)
EVANS, E. D., KENNY, G. S., MEINSCHEIN, W. G. and BRAY, E. E. (1957). *Analyt. Chem.* **29**, 1858. (*281*)
EVELING, D. W. (1967). *Physical effects of particulate sprays on leaf physiology.* Ph.D. thesis, University of London. (*206, 218, 251*)

F

FAHN, A. (1967). *Plant Anatomy.* Pergamon Press, Oxford. (*89, 90, 92, 100, 107, 109*)
FARGHER, R. G. and PROBERT, M. E. (1923). *J. Text. Inst.* **14**, 49T. (*62*)
— (1924). *J. Text. Inst.* **15**, 337T. (*62*)
FAWCETT, C. H. and SPENCER, D. M. (1967). *Ann. appl. Biol.* **60**, 87. (*265*)
— (1968). *Ann. appl. Biol.* **61**, 245. (*265*)
FAWCETT, C. H., SPENCER, D. M., WAIN, R. L., JONES, SIR EWART, R. H., LEQUAN, M., PAGE, C. B. and THALLER, V. (1965). *Chem. Commun.* 422. (*263*)
FINDLEY, T. W. and BROWN, J. B. (1953). *J. Am. Oil Chem. Soc.* **30**, 291. (*51, 132, 134*)
FISHER, D. J. (1966). *Rep. agric. hort. Res. Stn Univ. Bristol for 1965*, 255. (*148*)
FLENTJE, N. T. (1959). In *Plant Pathology, Problems and Progress 1908–1958*, ed. C. S. Holton, G. W. Fischer, R. W. Fulton, H. Hart and S. E. A. McCallan. University of Wisconsin Press, Madison, p. 76. (*263*)
FLENTJE, N. T., DODMAN, R. L. and KERR, A. (1963). *Aust. J. biol. Sci.* **16**, 784. (*262*)
FOGG, G. E. (1947). *Proc. R. Soc. Ser. B* **134**, 503. (*223, 225, 227, 230, 231*)
—(1948a). *Discuss. Faraday Soc.* No. 3, 162. (*225*)
— (1948b). *Ann. appl. Biol.* **35**, 315. (*236, 237, 240*)
FRANKE, W. (1960). *Planta* **55**, 390, 533. (*12, 85*)
— (1961). *Am. J. Bot.* **48**, 683. (*84, 241*)
— (1962). *Umschau* **62**, 501. (*75*)
— (1964). *Planta* **63**, 279. (*84*)
— (1967a). *A. Rev. Pl. Physiol.* **18**, 281. (*226, 235*)
— (1967b). *Planta* **73**, 138. (*83, 85, 201, 238*)
— (1969). *Am. J. Bot.* **56**, 432. (*241*)
FRAPS, G. S. and RATHER, J. R. (1910). *Ind. Engng. Chem. ind. Edn.* **2**, 454. (*62*)
FREDRICKS, K. M. (1967). *Antonie van Leeuwenhoek* **33**, 41. (*285*)
FREDERIKSEN, P. S. (1957). *Linnaea* **11**, 33. (*179*)
FREELAND, R. O. (1948). *Pl. Physiol., Lancaster* **23**, 595. (*216*)
FREMY, E. and URBAIN, V. (1882). *Annls. Sci. nat.* (*Bot.*) **13**, 360. (*69*)
— (1885). *C.r. hebd. Séanc. Acad. Sci., Paris* **100**, 19. (*67, 69*)
FREY, A. (1926). *Jb. wiss. Bot.* **65**, 195. (*4, 60*)

FREY, J. W. and CORN, M. (1967). *Nature, Lond.* **216**, 615. (*251*)
FREYTAG, K. (1957). *Planta* **50**, 41. (*238*)
FREY-WYSSLING, A. (1930). *Z. wiss. Mikrosk.* **47**, 1. (*60*)
— (1948). *Submicroscopic morphology of protoplasm and its derivatives.* Elsevier, Amsterdam, London and New York, p. 183. (*61*)
FREY-WYSSLING, A. and HÄUSERMANN, E. (1941). *Ber. schweiz. bot. Ges.* **51**, 430. (*60*)
— (1960). *Ber. schweiz. bot. Ges.* **70**, 150. (*213*)
FREY-WYSSLING, A. and MÜHLETHALER, K. (1959). *Vjschr. naturf. Ges. Zurich* **104**, 294. (*96, 158*)
— (1965). *Ultrastructural Plant Cytology.* Elsevier, Amsterdam, London and New York. (*158*)
FRITZ, F. (1937). *Planta* **26**, 693. (*82, 178*)
FUJII, M. and KURTH, E. F. (1966). *TAPPI* **49**, 92. (*150*)
FUJITA, A. and YOSHIKAWA, T. (1951). *J. pharm. Soc. Japan* **71**, 913. (*137*)
FUKUI, Y. and ARIYOSHI, H. (1963). *J. pharm. Soc. Japan* **83**, 1106. (*131*)
FULTON, H. R. (1906). *Bot. Gaz.* **41**, 81. (*261*)
FULTON, R. H. (1965). *A. Rev. Phytopath.* **3**, 175. (*226, 248*)
FURMIDGE, C. G. L. (1959). *J. Sci. Fd Agric.* **10**, 267, 274, 419. (*236, 240*)
— (1962). *J. Sci. Fd Agric.* **13**, 127. (*231*)
FURUMOTO, W. A. and WILDMAN, S. G. (1963). *Virology* **20**, 45. (*268*)

G

GABBOTT, P. A. and LARMON, V. N. (1968). *In Physico-chemical and biophysical factors affecting the activity of pesticides.* Society of Chemical Industry Monograph No. 29, p. 268. (*241*)
GAFF, D. E., CHAMBERS, T. C. and MARKUS, K. (1964). *Aust. J. biol. Sci.* **17**, 581. (*82, 113, 241*)
GALE, J. and HAGAN, R. M. (1966). *A Rev. Pl. Physiol.* **17**, 269. (*209*)
GANE, R. (1931). *Rep. Fd Invest. Bd for 1931.* H.M.S.O. London, p. 241. (*65*)
GASTAMBIDE-ODIER, M. and LEDERER, E. (1959). *Nature, Lond.* **184**, 1563. (*164, 169*)
GATES, D. M. (1968). *A Rev. Pl. Physiol.* **19**, 211. (*200*)
GÄUMANN, E. and JAAG, O. (1935). *Ber. schweiz. bot. Ges.* **45**, 411. (*96, 98, 200*)
GÉNEAU DE LAMARLIÈRE, L. (1906). *Revue gén. Bot.* **18**, 289, 372. (*58*)
GENSLER, W. J. and SCHLEIN, H. N. (1955). *J. Am. chem. Soc.* **77**, 4846. (*70*)
GENTNER, W. A. (1966). *Weed Science* **14**, 27. (*245*)
GERARDE, J. (1597). *The Herball or General History of Plants.* John Norton, London, p. 1005. (*194*)
GIDDINGS, J. C. and KELLER, R. A. (1965). *Advances in Chromatography.* Edward Arnold, London. Vol. 1. (*44*)
GILLIVER, K. (1947). *Ann. appl. Biol.* **34**, 136. (*263*)
GILSON, E. (1890). *Cellule* **6**, 63. (*69*)
GINDEL, I. (1965). *Nature, Lond.* **207**, 1173. (*199*)
GLADDING, R. N. and WRIGHT, H. E. (1959). *Tobacco Sci.* **3**, 81. (*133*)
GLAUERT, A. M. (1965). In *Techniques for Electron Microscopy* (2nd Ed.), ed. D. Kay. Blackwell Scientific Publications, Oxford, p. 166. (*27*)
GLEN, A. T., LAWRIE, W., McLEAN, J. and YOUNES, M. EL-G. (1965). *Chemy Ind.*, 1908. (*131*)
GLOVER, J. and GWYNNE, M. D. (1962). *J. Ecol.* **50**, 111. (*195*)
GLYNNE-JONES, G. D. and THOMAS, W. D. E. (1953). *Ann. appl. Biol.* **40**, 546. (*214*)
GODDU, R. F., LEBLANC, N. F. and WRIGHT, C. M. (1955). *Analyt. Chem.* **27**, 1251. (*51*)

GODWIN, H. (1951). *Endeavour* **10**, 5. (*279, 282*)
— (1968). *New Phytol.* **67**, 667. (*154*)
GOHLKE, R. S. (1959). *Analyt. Chem.* **31**, 535. (*44*)
GONZALEZ-GONZALEZ, A., SOLA, J., IGLESIAS-MARTIN, F., RIVAS-PARIS, V. and RIBAS-MARQUÉS, I. (1968). *Quim. Ind.* **15**, 139. (*152*)
GOODMAN, R. N. (1962a). In *Antibiotics in Agriculture*, ed. M. Woodbine, Butterworths, London, p. 165. (*4, 237, 239*)
— (1962b). In *Advances in Pest Control Research*, ed. R. L. Metcalf. Interscience Publishers, New York and London. Vol. 5, p. 1. (*240*)
GOODMAN, R. N. and ADDY, S. K. (1962a). *Phytopathology* **52**, 11. (*240*)
— (1962b). *Phytopath. Z.* **46**, 1. (*47, 98, 240*)
GOODMAN, R. N. and GOLDBERG, H. S. (1960). *Phytopathology* **50**, 851. (*241*)
GORHAM, E. (1958). *Nature, Lond.* **181**, 1523. (*251*)
GOTTFRIED, S. and ULZER, F. (1926). *Chem. Umsch. Geb. Fette* **33**, 141. (*62*)
GRANITI, A. (1962). *Phytopathologia Medit.* **1**, 157. (*287*)
— (1965a). *Phytopathologia Medit.* **4**, 38. (*287*)
— (1965b). *Atti. Semin. Studi Biol.* **2**, 217. (*287*)
GRANITI, A. and LEO, P. DE (1966). *Phytopathologia Medit.* **5**, 65. (*288*)
GRANITI, A., LEO, P. DE and BAGORDO, F. (1962). *Phytopathologia Medit.* **2**, 20. (*288*)
GRAVES, A. H. (1916). *Bot. Gaz.* **62**, 337. (*261*)
GRAY, R. A. (1955). *Pl. Dis. Reptr* **39**, 567. (*239*)
— (1956). *Phytopathology* **46**, 105. (*206*)
GRESHOFF, M. and SACK, J. (1901). *Recl. Trav. chim. Pays-Bas Belg.* **20**, 65. (*62*)
GRIFFITHS, D. A. and ISAAC, I. (1963). *Ann. appl. Biol.* **51**, 231. (*287*)
GRIFFITHS, L. A. (1958). *Nature, Lond.* **182**, 733. (*264*)
GRNCAREVIC, M. and RADLER, F. (1967). *Planta* **75**, 23. (*203*)
GROOM, P. (1893). *Ann. Bot.* **7**, 223. (*194*)
GROSE, R. J. (1960). *Silviculture of Eucalyptus delegatensis.* Ph.D. thesis, University of Melbourne. (*220*)
GUILLEMONAT, A. and CESAIRE, G. (1949). *Bull. Soc. chim. Fr.* 792. (*152*)
GUILLEMONAT, A. and STRICH, A. (1950). *Bull. Soc. chim. Fr.*, 860. (*151, 152*)
GUILLEMONAT, A. and TRAYNARD, J.-C. (1963). *Bull. Soc. chim. Fr.*, 142. (*150*)
GUILLEMONAT, A. and TRIACA, M. (1968). *Bull. Soc. chim. Fr.*, 950. (*153*)
GÜNTHER, I. and WORTMANN, G. B. (1966). *J. Ultrastruct. Res.* **15**, 522. (*251*)
GUSTAFSON, F. G. (1956). *Am. J. Bot.* **43**, 157. (*236, 240*)
— (1957). *Pl. Physiol., Lancaster* **32**, 141. (*236*)
GUSTAFSON, F. G. and SCHLESSINGER, M. J. (1956). *Pl. Physiol., Lancaster* **31**, 316 (*241*)

H

HABERLANDT, G. (1914). *Physiological Plant Anatomy.* Macmillan, London. (*75, 86, 90, 94, 95, 103, 119, 195, 217, 218, 274*)
HACKNEY, F. M. V. (1943). *Proc. Linn. Soc. N.S.W.* **68**, 33. (*187*)
HAFIZ, A. (1952). *Phytopathology* **42**, 422. (*210*)
HALL, D. M. (1966). *Aust. J. biol. Sci.* **19**, 1017. (*201, 237*)
— (1967a). *Science, N.Y.* **158**, 505. (*112, 181, 237*)
— (1967b). *J. Ultrastruct. Res.* **17**, 34. (*28, 112, 181, 237*)
HALL, D. M. and JONES, R. L. (1961). *Nature, Lond.* **191**, 95. (*11, 202*)
HALL, D. M., MATUS, A. I., LAMBERTON, J. A. and BARBER, H. N. (1965). *Aust. J. biol. Sci.* **18**, 323. (*6, 11, 129, 130, 189, 192, 223, 225, 228, 232*)

HALLAM, N. D. (1964). *Aust. J. biol. Sci.* **17**, 587. (*113, 235, 237,*
HALLAM, N. D. (1967). *An electron microscope study of the leaf waxes of the genus Eucalyptus, L'Heritier.* Ph.D. thesis, University of Melbourne. (*46, 73, 77, 81, 82, 100*) *101, 102, 103, 104, 114, 115, 141, 181, 204, 237, 250, 272*)
HALLER, A. (1907). *C.r. hebd. Séanc. Acad. Sci., Paris* **144**, 594. (*62*)
HANDLEY, W. R. C. (1961). *Pl. Soil* **15**, 37. (*290*)
HANKIN, L. and KOLATTUKUDY, P. E. (1968). *J. gen. Microbiol.* **51**, 457. (*285*)
— (1969). *J. gen. Microbiol.* **56**, 151. (*285*)
HARE, E. F., SHAFRIN, E. G. and ZISMAN, W. A. (1954). *J. phys. Chem., Ithaca* **58**, 236. (*225*)
HARLAND, S. C. (1947). *Heredity, Lond.* **1**, 121. (*189, 190*)
HARMS, U. and WURZIGER, J. (1968). *Z. Lebensmittelunters. u.–Forsch.* **138**, 75. (*139*)
HARRIS, T. M. (1956). *Endeavour* **15**, 210. (*279, 281*)
HART, N. K. and LAMBERTON, J. A. (1965). *Aust. J. Chem.* **18**, 115. (*149*)
HÄRTEL, O. (1951). *Protoplasma* **40**, 107. (*238*)
HARTLEY, G. S. (1960). *Chemy Ind.*, 448. (*223*)
— (1966a). *Formulations and availability of pesticides.* In Society of Chemical Industry Monograph No. 21, p. 122. (*239, 240*)
— (1966b). *Proc. 8th Br. Weed Control Conference* **3**, 794. (*223*)
HARTLEY, G. S. and BRUNSKILL, R. T. (1958). In *Surface Phenomena in Chemistry and Biology*, ed. J. F. Danielli, K.G.A. Pankhurst and A. C. Riddiford. Pergamon Press, Oxford, p. 214. (*226, 227, 230, 231*)
HATHWAY, D. E. (1958). *Biochem. J.* **70**, 34. (*150*)
— (1959). *Biochem. J.* **71**, 533. (*150*)
HATT, H. H. and LAMBERTON, J. A. (1956). *Research, Lond.* **9**, 138. (*6, 137*)
HATTORI, S. and NAGAI, W. (1930). *J. chem. Soc. Japan* **51**, 162. (*139*)
HÄUSERMANN, E. (1944). *Ber. schweiz. bot. Ges.* **54**, 541. (*96*)
HÄUSERMANN, E. and FREY-WYSSLING, A. (1963). *Protoplasma* **57**, 370. (*210*)
HAYES, H. K. and BREWBAKER, H. E. (1938). *Am. Nat.* **62**, 228. (*188*)
HEARON, W. M. and MACGREGOR, W. S. (1955). *Chem. Rev.* **55**, 957. (*151*)
HEATHER, W. A. (1967). *Aust. J. biol. Sci.* **20**, 769, 1155. (*256, 265, 267*)
HECK, W. W. (1968). *A. Rev. Phytopath.* **6**, 165. (*253*)
HEDBERG, O. (1964). *Acta phytogeogr. suec.* **49**, 1. (*213*)
HEIDUSCHKA, A. and GARIES, M. (1919). *J. prakt. Chem.* **99**, 293. (*62*)
HEINEN, W. (1960). *Acta bot. neerl.* **9**, 167. (*286*)
— (1962). *Arch. Mikrobiol.* **41**, 268. (*176, 286*)
— (1963a). *Enzymologia* **25**, 281. (*286*)
— (1963b). *Acta bot. neerl.* **12**, 51. (*286*)
HEINEN, W. and VAN DEN BRAND, I. (1961). *Acta bot. neerl.* **10**, 171. (*286*)
— (1963). *Z. Naturf.* **18b**, 67. (*174*)
HEINEN, W. and LINSKENS, H. F. (1961). *Nature, Lond.* **191**, 1416. (*288*)
HEINEN, W. AND VRIES, H. DE (1966a). *Arch. Mikrobiol.* **54**, 339. (*286*)
— (1966b). *Arch. Mikrobiol.* **54**, 331. (*154, 286*)
HENSLOW, J. S. (1831). *Trans. Camb. phil. Soc.* **4**, 1. (*55*)
HERBIN, G. A. and ROBINS, P. A. (1968a). *Phytochemistry* **7**, 239. (*51, 141*)
— (1968b). *Phytochemistry* **7**, 257. (*142, 239*)
— (1968c). *Phytochemistry* **7**, 1325. (*142*)
HERBIN, G. A. and SHARMA, K. (1969). *Phytochemistry* **8**, 151. (*143*)
HERGERT, H. L. (1958). *Forest Prod. J.* **8**, 335. (*150*)
HERGERT, H. L. and KURTH, E. F. (1952). *TAPPI* **35**, 59. (*150*)
— (1953). *TAPPI* **36**, 137. (*150*)
HERING, T. F. (1965). *Trans. Br. mycol. Soc.* **48**, 391. (*286*)

HERRETT, R. A. and LINCK, A. J. (1961). *Weed Science* **9**, 224. *(241)*
HESLOP-HARRISON, J. (1963a). *Grana Palynol.* **4**, 7. *(119)*
— (1963b). *Symp. Soc. exp. Biol.* **17**, 315. *(119, 154)*
— (1968). *New Phytol.* **67**, 779. *(154, 155)*
HESLOP-HARRISON, J. and DICKINSON, H. G. (1969). *Planta* **84**, 199. *(155)*
HESSLER, W. and SAMMET, F. (1965). *Fette Seifen Anstr-Mittel* **67**, 552. *(49)*
HEYL, F. W. and LARSEN, D. (1933). *J. Am. pharm. Ass.* **22**, 510. *(63)*
HILDEBRAND, E. M. (1942). *J. agric. Res.* **65**, 45. *(270)*
— (1958). *Phytopathology* **48**, 262. *(268)*
HILDEBRAND, E. M. and MACDANIELS, L. H. (1935). *Phytopathology* **25**, 20. *(214, 269)*
HILL, A. S. and MATTICK, L. R. (1966). *Phytochemistry* **5**, 693. *(122, 143)*
HIRAI, T. and HIRAI, A. (1964). *Science, N.Y.* **145**, 589. *(268)*
HIRST, J. M., RICHE, H. H. LE and BASCOMB, C. L. (1961). *Pl. Path.* **10**, 105. *(290)*
HIRST, J. M. and STEDMAN, O. J. (1962). *Ann. appl. Biol.* **50**, 551. *(283, 290)*
HOLLOWAY, P. J. (1967). *Studies of the wettability of leaf surfaces.* Ph.D. thesis, University of London. *(32, 49, 126, 129, 130, 131, 132, 133, 134, 223, 225, 231, 232, 233)*
— (1969a). *Ann. appl. Biol.* **63**, 145. *(225, 231, 233)*
— (1969b). *J. Sci. Fd Agric.* **20**, 124. *(185, 232)*
HOLLOWAY, P. J. and BAKER, E. A. (1968a). Personal communication. *(53)*
— (1968b). *Pl. Physiol., Lancaster* **43**, 1878. *(47, 79)*
HOLLOWAY, P. J. and CHALLEN, S. B. (1966). *J. Chromat.* **25**, 336. *(41, 49)*
HOLLY, K. (1956). *Ann. appl. Biol.* **44**, 195. *(239)*
HOLMGREN, P., JARVIS, P. G. and JARVIS, M. S. (1965). *Physiologia Pl.* **18**, 557. *(216)*
HONKANEN, E. and VIRTANEN, A. I. (1960). *Acta chem. fenn.* **33 B**, 171. *(265)*
HORN, D. H. S., KRANZ, Z. H. and LAMBERTON, J. A. (1964). *Aust. J. Chem.* **17**, 464. *(46, 48, 50, 128, 129, 130, 132, 141, 163)*
HORN, D. H. S. and LAMBERTON, J. A. (1962). *Chemy Ind.* 2036. *(48, 127, 130, 132, 139)*
— (1963). *Chemy Ind.* 691. *(128)*
— (1964). *Aust. J. Chem.* **17**, 477. *(128, 130, 132, 140)*
HORN, D. H. S. and MATIC, M. (1957). *J. Sci. Fd Agric.* **8**, 571. *(126, 139)*
HORROCKS, R. L. (1964). *Nature, Lond.* **203**, 547. *(46, 202)*
HÖSTER-AUER, S. (1964). *Naturwissenschaften*, **51**, 267. *(148)*
HOUWINK, A. L. and ROELOFSEN, P. A. (1954). *Acta bot. neerl.* **3**, 385. *(76)*
HOWARD, A. J. and HAMER, D. (1962). *J. Am. Oil Chem. Soc.* **39**, 250. *(282)*
HOWES, F. N. (1936). *Bull. misc. Inf. R. bot. Gdns. Kew* No. 10, 503. *(292, 294)*
HUBER, B., KINDER, E., OBERMÜLLER, E. and ZIEGENSPECK, H. (1956). *Protoplasma* **46**, 380. *(96)*
HUELIN, F. E. (1959). *Aust. J. biol. Sci.* **12**, 175. *(47, 52, 144, 147, 153, 174)*
— (1964). *J. Sci. Fd Agric.* **15**, 227. *(220)*
HUELIN, F. E. and COGGIOLA, I. M. (1968). *J. Sci. Fd Agric.* **19**, 297. *(220)*
HUELIN, F. E. and GALLOP, R. A. (1951a). *Aust. J. scient. Res.* Ser. B **4**, 526. *(47, 52, 122, 144, 153, 174)*
— (1951b). *Aust. J. scient. Res.* Ser. B **4**, 533. *(122, 187)*
HUELIN, F. E. and MURRAY, K. E. (1966). *Nature, Lond.* **210**, 1260. *(124, 220)*
HUGHES, G. K. and RITCHIE, E. (1954). *Aust. J. Chem.* **7**, 104. *(151)*
HULL, H. M. (1958). *Weed Science* **6**, 133. *(100, 180)*
— (1964). *Proc. 7th Annu. Symp. S. Sect. Am. Soc. Pl. Physiol.* Emory Univ., p. 45. *(180, 238, 241)*
HULME, A. C. and EDNEY, K. L. (1960). In *Phenolics in plants in health and disease*, ed. J. B. Pridham. Pergamon Press, Oxford, p. 87. *(147, 264)*
HULME, A. C. and WOOLTORTON, L. S. C. (1958). *J. Sci. Fd Agric.* **9**, 150. *(147)*

HÜLSBRUCH, M. (1966). *Z. Pflanzenphysiol.* **55**, 181. (*81, 82, 101, 179*)
HUNSDIECKER, H. (1944). *Ber. dt. chem. Ges.* **77 B**, 185. (*70*)

I

ISOI, K. (1958). *J. pharm. Soc. Japan* **78**, 814. (*137*)
ISTRATI, C. I. and OSTROGOVICH, A. (1899). *C.r. hebd. Séanc. Acad. Sci., Paris* **128**, 1581. (*69*)
IVANOFF, S. S. (1963). *Bot. Rev.* **29**, 202. (*210, 212*)
IZMAILOV, N. A. and SHRAIBER, M. S. (1938). *Farmatsiya, Mosk.* No. 3, 1. (*37*)

J

JACKSON, R. T. and BROWN, H. D. (1963). *J. cell. comp. Physiol.* **61**, 215. (*240*)
JAMES, A. T. and MARTIN, A. J. P. (1952). *Biochem. J.* **50**, 679. (*42*)
JAROLIMEK, P., WOLLRAB, V. and STREIBL, M. (1964). *Colln Czech. chem. Commun. Engl. Edn.* **29**, 2528. (*51*)
JAROLIMEK, P., WOLLRAB, V., STREIBL, M. and ŠORM, F. (1964). *Chemy Ind.* 237. (*133, 134, 139*)
JARRETT, J. M. and WILLIAMS, A. H. (1967). *Phytochemistry* **6**, 1585. (*150*)
JARVIS, P. G. and JARVIS, M. S. (1963). *Physiologia Pl.* **16**, 501. (*216*)
JENNINGS, D. L. (1962). *Hort. Res.* **1**, 100. (*260*)
JENSEN, W. (1950a). *Paperi Puu* **32**, 293. (*151*)
— (1950b). *Paperi Puu* **32**, 261. (*152*)
— (1950c). *Paperi Puu* **32**, 291. (*152*)
JENSEN, W., FREMER, K. E., SIERILÄ, P. and WARTIOVAARA, V. (1963). In *The Chemistry of Wood*, ed. B. L. Browning. Interscience, New York and London, p. 587. (*151*)
JENSEN, W. and ÖSTMAN, R. (1954). *Paperi Puu* **36**, 427. (*152*)
JENSEN, W. and RINNE, P. (1954). *Paper Puu* **36**, 32. (*152*)
JENSEN, W. and TINNIS, W. (1957). *Paperi Puu* **39**, 261. (*152*)
JHOOTY, J. S. and McKEEN, W. E. (1965). *Phytopathology* **55**, 281. (*260*)
JOHNSON, G. and SCHAAL, L. A. (1952). *Science, N.Y.* **115**, 627. (*264*)
JOHNSTON, H. W. and SPROSTON, T. (1965). *Phytopathology* **55**, 225. (*47, 184, 266*)
JOHNSTON, R. D. (1964). *Aust. J. Bot.* **12**, 111. (*198*)
JOHNSTONE, K. H. (1931). *J. Pomol.* **9**, 30, 195. (*263*)
JORDAN, R. C. and CHIBNALL, A. C. (1933). *Ann. Bot.* **47**, 163. (*66*)
JUNIPER, B. E. (1959a). *Endeavour* **18**, 20. (*232, 245*)
— (1959b). *New Phytol.* **58**, 1. (*223, 232, 245*)
— (1960a). *J. Linn. Soc. (Bot.)* **56**, 413. (*181, 232*)
— (1960b). *Studies on structure in relation to phytotoxicity*. D.Phil. thesis, University of Oxford. (*181, 223, 225, 228, 232, 250*)
JUNIPER, B. E. and BRADLEY, D. E. (1958). *J. Ultrastruct. Res.* **2**, 16. (*23, 232*)
JUNIPER, B. E. and BURRAS, J. K. (1962). *New Scient.* **13**, 75. (*272, 273, 274*)
JUNIPER, B. E. and CLOWES, F. A. L. (1965). *Nature, Lond.* **208**, 864. (*107*)
JUNIPER, B. E. and ROBERTS, R. M. (1966). *Jl. R. microsc. Soc.* **85**, 63. (*107*)
JUNIPER, B. E., COX, G. C., GILCHRIST, A. and WILLIAMS, P. R. (1970). In *Techniques for Plant Electron Microscopy*. Blackwell Scientific Publications, Oxford. (*32*)
JYUNG, W. H. and WITTWER, S. H. (1964). *Am. J. Bot.* **51**, 437. (*240*)

K

KAGANOVICH, F. L., BEL'KEVICH, P. I. and RAKOVSKII, V. E. (1959). *Trudy Inst. Torfa Minsk.* **7**, 123, 131. (*282*)
KAMIMURA, S. and GOODMAN, R. N. (1964a). *Phytopath. Z.* **51**, 324. (*240, 242*)
— (1964b). *Physiologia Pl.* **17**, 805. (*239*)
— (1964c). *Phytopathology* **54**, 1467. (*242*)
KAMP, H. (1930). *Jb. wiss. Bot.* **72**, 403. (*98, 200*)
KANEDA, T. (1966). *Biochim. biophys. Acta* **125**, 43. (*167*)
— (1967). *Biochemistry* **6**, 2023. (*51, 167*)
— (1968). *Biochemistry* **7**, 1194. (*167*)
KAPIL, Y. P. and MUKHERJEE, S. (1954). *Proc. Sug. Technol. Ass. India* Pt. II, 220. (*126*)
— (1963). *Proc. annu. Conv. Sug. Technol. Ass. India* **31**, 199. (*126*)
KARIYONE, T. and AGETA H. (1959). *J. pharm. Soc. Japan* **79**, 47. (*132*)
KARIYONE, T., AGETA, H. AND ISOI, K. (1959). *J. pharm. Soc. Japan* **79**, 54 (*131, 139*)
KARIYONE, T., AGETA, H. and TANAKA, A. (1959). *J. pharm. Soc. Japan* **79**, 51. (*131*)
KARIYONE, T. and HASHIMOTO, Y. (1953). *Experientia* **9**, 136. (*136*)
KARIYONE, T. and ISOI, K. (1955). *J. pharm. Soc. Japan* **75**, 316. (*137*)
— (1956). *J. pharm. Soc. Japan* **76**, 473. (*133*)
KARIYONE, T., ISOI, K. and YOSHIKURA, M. (1959). *J. pharm. Soc. Japan* **79**, 61. (*133*)
KARIYONE, T., TAKAHASHI, M., ISOI, K. and YOSHIKURA, M. (1959). *J. pharm. Soc. Japan* **79**, 1340. (*138*)
KARRER, P. and SCHWAB, G. (1941). *Helv. chim. Acta* **24**, 297. (*139*)
KARSTEN, H. (1857). *Bot. Ztg* **15**, 313. (*56*)
— (1860). *Annln. Phys.* **109**, 640. (*56*)
KARTNIG, T. (1967). *Fette Seifen Anstr-Mittel* **69**, 401. (*51*)
KARTNIG, T. and SCHOLZ, G. H. (1965). *Fette Seifen Anstr-Mittel* **67**, 19. (*49*)
KAUFMANN, H. P. and DAS, B. (1963). *Fette Seifen Anstr-Mittel* **65**, 398. (*49*)
KAUSCH, W. and HAAS, W. (1965). *Naturwissenschaften* **52**, 214. (*52*)
KEPPEL, H. (1967). *Isotop. Pl. Nutr. Physiol.*, Proc. Symp., Vienna, 1966, p. 329. (*244*)
KERFOOT, O. (1968). *For. Abstr.* **29**, 8. (*197*)
KERR, A. and FLENTJE, N. T. (1957). *Nature, Lond.* **179**, 204. (*262*)
KETELLAPPER, H. J. (1963). *A. Rev. Pl. Physiol.* **14**, 249. (*9*)
KHASTGIR, H. N., PRADHAN, B. P., DUFFIELD, A. M. and DURHAM, L. J. (1967). *Chem. Commun.* 1217. (*149*)
KIMBER, G. and BOOTH, A. (1958). *Nature, Lond.* **181**, 1391. (*251*)
KLÖPPING, H. L. and VAN DER KERK, G. J. M. (1951). *Nature, Lond.* **167**, 996. (*266*)
KNECHT, E. and ALLAN, J. (1911). *J. Soc. Dyers Colour* **27**, 142. (*62*)
KNOLL, F. (1914). *Jb. wiss. Bot.* **54**, 448. (*277*)
KOLATTUKUDY, P. E. (1965). *Biochemistry* **4**, 1844. (*49, 163*)
— (1966a). *Biochemistry* **5**, 2265. (*163, 164, 166*)
— (1966b). *Front. Pl. Sci.* **18**, 6. (*163, 164*)
— (1967a). *Phytochemistry* **6**, 963. (*163, 166*)
— (1967b). *Biochemistry* **6**, 2705. (*163, 166*)
— (1968a). *Science, N.Y.* **159**, 498. (*163, 169, 170, 171*)
— (1968b). *Pl. Physiol., Lancaster* **43**, 375. (*163, 169, 171*)
— (1968c). *Pl. Physiol., Lancaster* **43**, 1466. (*163, 169*)
— (1968d). *Pl. Physiol., Lancaster* **43**, 1423. (*163, 170*)
KOLATTUKUDY, P. E., JAEGER, R. H. and ROBINSON, R. (1968). *Nature, Lond.* **219**, 1038. (*170*)
KOLJO, B. (1957). In *Die Chemie der Pflanzenzellwand*, ed. E. Treiber. Springer-Verlag, Berlin, p. 432. (*156*)

KOLJO, B. and SITTE, P. (1957). In *Die Chemie der Pflanzenzellwand*, ed. E. Treiber. Springer-Verlag, Berlin, p. 416. (*156*)
KONAR, R. N. and LINSKENS, H. F. (1966). *Planta* **71**, 356. (*215, 235*)
KÖNIG, J. (1906). *Z. Unters. Nahr.-u. Gennssmittel* **12**, 385. (*52*)
KÖNIG, J. and RUMP, E. (1914). *Z. Unters. Nahr.-u. Gennssmittel* **28**, 177. (*52*)
KONTAXIS, D. G. and SCHLEGEL, D. E. (1962). *Virology* **16**, 244. (*268*)
KOONCE, S. D. and BROWN, J. B. (1944). *Oil Soap* **21**, 231. (*125*)
— (1945). *Oil Soap* **22**, 217. (*125*)
KOSUGE, T. and HEWITT, W. B. (1964). *Phytopathology* **54**, 167. (*261*)
KOVÀCS, A. (1955). *Phytopath. Z.* **24**, 283. (*263*)
KOVÀCS, A. and CUCCHI, N. J. A. (1964). *Nature, Lond.* **204**, 1090. (*246*)
KOVÀCS, A. and SZEÖKE, E. (1956). *Phytopath. Z.* **27**, 335. (*263*)
KRANZ, Z. H., LAMBERTON, J. A., MURRAY, K. E. and REDCLIFFE, A. H. (1960). *Aust. J. Chem.* **13**, 498. (*50, 122, 126, 127, 139*)
— (1961). *Aust. J. Chem.* **14**, 264. (*133, 139*)
KREGER, D. R. (1948). *Rec. Trav. bot. neerl.* **41**, 606. (*34, 61, 126, 160, 164, 165, 169*)
— (1958). *J. Ultrastruct. Res.* **1**, 247. (*34*)
KREGER, D. R. and SCHAMHART, C. (1956). *Biochim. biophys. Acta* **19**, 22. (*34*)
KROH, M. (1963). *Proc. Symp. Pollen Physiol. Fertilization.* Nijmegen, 221. (*289*)
KÜGLER, K. (1884). *Arch. Pharm. Berlin* **222**, 217. (*69*)
KURTH, E. F. (1950). *J. Am. chem. Soc.* **72**, 1685. (*149*)
— (1967). *TAPPI* **50**, 253. (*151*)
KURTH, E. F. and HUBBARD, J. K. (1951). *Ind. Engng. Chem. ind. Edn.* **43**, 896. (*148, 150*)
KURTH, E. F. and KIEFER, H. J. (1950). *TAPPI* **33**, 183. (*149*)
KURTZ, E. B. (1950). *Pl. Physiol., Lancaster* **25**, 269. (*66, 178*)
— (1958). *J. Am. Oil Chem. Soc.* **35**, 465. (*6, 179*)
KUYKENDALL, J. R. and WALLACE, A. (1954). *Proc. Am. Soc. hort. Sci.* **64**, 117. (*243*)
KWIATKOWSKI, A. and LULINER-MIANOWSKA, K. (1957). *Acta Soc. Bot. Pol.* **26**, 501. (*154*)

L

LAMBERTON, J. A. (1961). *Aust. J. Chem.* **14**, 323. (*134*)
— (1964). *Aust. J. Chem.* **17**, 692. (*46, 128*)
— (1965). *Aust. J. Chem.* **18**, 911. (*127*)
LAMBERTON, J. A. and REDCLIFFE, A. H. (1959). *Chemy Ind.*, 1627. (*126, 139*)
— (1960). *Aust. J. Chem.* **13**, 261. (*51, 126*)
LAMBERTZ, P. (1954). *Planta* **44**, 147. (*84*)
LAMPORT, D. T. A. (1965). In *Advances in Botanical Research*, Vol. 2, ed. R. D. Preston. Academic Press, New York and London, p. 151. (*76*)
LASETER, J. L., WEBER, D. J. and ORÖ, J. (1968). *Phytochemistry* **7**, 1005. (*129*)
LASETER, J. L., WEETE, J. and WEBER, D. J. (1968). *Phytochemis ry* **7**, 1177. (*136*)
LAST, F. T. (1955). *Trans. Br. myc. Soc.* **38**, 221. (*254*)
LAST, F. T. and DEIGHTON, F. C. (1965). *Trans. Br. myc. Soc.* **48**, 83. (*255*)
LAW, J. A. (1965). *Chemy Ind.*, 171. (*292*)
LAWRIE, W., McLEAN, J. and PATON, A. C. (1964). *Phytochemistry* **3**, 267. (*149*)
LAWRIE, W., McLEAN, J. and YOUNES, M. EL-G. (1966). *Chemy Ind.*, 1720. (*125*)
— (1967). *J. chem. Soc.* C, 851. (*125*)
LEBEDEV, K. K. (1959a). *Trudy̆ Inst. Torfa, Minsk.* **7**, 19. (*14*)
— (1959b). *Trudy̆ Inst. goryuch. Iskop.*, 31. (*14*)

LEBEN, C. (1961). *Phytopathology* **37**, 553. (*255*)
— (1965). *A. Rev. Phytopath.* **3**, 209. (*256*)
— (1968). *Abstrs 1st Int. Congr. Pl. Path.*, London, p. 112. (*255*)
LEDBETTER, M. C. (1967). Personal communication. (*181*)
LEDERER, E. and LEDERER, M. (1954). *Chromatography*. Elsevier, London. (*44*)
LEE, B. (1925). *Ann. Bot.* **39**, 755. (*68, 70*)
LEE, B. and PRIESTLEY, J. H. (1924). *Ann. Bot.* **38**, 525.
(*2, 6, 59, 68, 95, 96, 100, 107, 144*)
LEGG, V. H. and WHEELER, R. V. (1925). *J. chem. Soc.* 1412. (*68, 70*)
— (1929a). *J. chem. Soc.* 2444. (*62, 68*)
— (1929b). *J. chem. Soc.* 2449. (*68*)
LEIGH, J. H. and MATTHEWS, J. W. (1963). *Aust. J. Bot.* **11**, 62. (*6*)
LEONARD, O. A. (1958). *Hilgardia* **28**, 115. (*238*)
LETHAM, D. S. (1958). *Nature, Lond.* **181**, 135. (*47*)
LEWIS, F. J. (1945). *Nature, Lond.* **156**, 407. (*228*)
LEYS, A. (1913). *J. Pharm. Chim., Paris* **5**, 577. (*62*)
LEYTON, L. and ARMITAGE, I. P. (1968). *New Phytol.* **67**, 31. (*197*)
LEYTON, L. and JUNIPER, B. E. (1963). *Nature, Lond.* **198**, 770. (*196, 205*)
LINDEMAN, L. P. and ANNIS, J. L. (1960). *Analyt. Chem.* **32**, 1742. (*44*)
LINSER, H. (1964). In *The Physiology and Biochemistry of Herbicides*, ed. L. J. Audus. Academic Press, New York and London, p. 483. (*240*)
LINSKENS, H. F. (1950). *Planta* **38**, 591. (*225, 231, 238*)
— (1952a). *Planta* **41**, 40. (*228, 231*)
— (1952b). *Naturwissenschaften* **39**, 65. (*223, 225, 231*)
— (1966). *Planta* **68**, 1. (*32*)
LINSKENS, H. F. and HAAGE, P. (1963). *Phytopath. Z.* **48**, 306. (*288*)
LINSKENS, H. F. and HEINEN, W. (1962). *Z. Bot.* **50**, 338. (*289*)
LINSKENS, H. F., HEINEN, W. and STOFFERS, A. L. (1965). In *Residue Reviews*, ed. Francis A. Gunther. Springer-Verlag, Berlin. Vol. 8, p. 136. (*223, 286*)
LINSKENS, H. F. and KRNER, H. (1966). *Nature, Lond.* **210**, 968. (*32*)
LITTLEWOOD, A. B. (1962). *Gas Chromatography. Principles, techniques and applications.* Academic Press, New York and London. (*44*)
LLOYD, F. E. (1942). *The Carnivorous Plants.* Chronica Botanica Co., Waltham, Mass., U.S.A. (*194, 272, 277*)
LOOMIS, W. E. and SCHIEFERSTEIN, R. H. (1959). *Proc. 9th Int. Bot. Congr.* Montreal, p. 235. (*100*)
LORD, K. A., MAY, M. A. and STEVENSON, J. H. (1968). *Ann. appl. Biol.* **61**, 19. (*214*)
LÜDTKE, M. (1961). *Melliand TextBer.* **42**, 667. (*52*)
LUDWIG, F. J. (1966). *Soap chem. Spec.* **43**, 70. (*50*)
LUNDQVIST, U. and VON WETTSTEIN, D. (1962). *Hereditas* **48**, 342. (*192*)
LUNDQVIST, U., VON WETTSTEIN-KNOWLES, P. and VON WETTSTEIN, D. (1968). *Hereditas* **59**, 473. (*192*)
LUPTON, F. G. H. (1967). *Wld Rev. Pest Control* **6**, 47. (*271*)
LÜTTGE, U. (1961). *Planta* **56**, 189. (*213*)
— (1962a). *Planta* **59**, 108. (*213*)
— (1962b). *Planta* **59**, 175. (*213*)

M

MACAULEY, B. J. and THROWER, L. B. (1966). *Trans. Br. mycol. Soc.* **49**, 509. (*286*)
MACEY, M. and BARBER, H. N. (1969). *Nature, Lond.* **222**, 789. (*170, 192*)

McCLURE, F. A. (1966). *The Bamboos—A Fresh Perspective.* Harvard University Press, Cambridge, Mass., U.S.A. (*104, 137*)
McCOLLOCH, R. J., HAMILTON, J. W. and BROWN, S. K. (1963). *Biochem. biophys. Res. Commun.* **11**, 7. (*138*)
McKAY, A. F. (1948). *J. org. Chem.* **13**, 86. (*138*)
McKENNA, E. J. and KALLIO, R. E. (1965). *A Rev. Microbiol.* **19**, 183. (*285*)
McLEAN, F. T. (1921). *Torrey bot. Club. Bull.* **48**, 101. (*270*)
McLEAN, J. and THOMSON, J. B. (1963). *Phytochemistry* **2**, 179. (*134*)
McNAIR, J. B. (1931). *Am. J. Bot.* **18**, 518. (*6*)
MACHADO, R. D. (1958). *Archos. Jard. bot., Rio de J.* **16**, 117. (*202*)
MACKIE, A. and MISRA, A. L. (1956). *J. Sci. Fd Agric.* **7**, 203. (*66*)
MAJOR, R. T. (1967). *Science, N.Y.* **157**, 1270. (*266*)
MAJOR, R. T., MARCHINI, P. and SPROSTON, T. (1960). *J. biol. Chem.* **235**, 3298. (*266*)
MAJUMDAR, G. D. and PRESTON, R. D. (1941). *Proc. R. Soc.* Ser. B **130**, 1201. (*81*)
MANGOLD, H. K. and MALINS, D. C. (1960). *J. Am. Oil Chem. Soc.* **37**, 383. (*49*)
MANN, C. E. T. and WALLACE, T. (1925). *J. Pomol.* **4**, 146. (*212*)
MANSFIELD, T. A. (1968). *Abstrs 1st Int. Congr. Pl. Path.*, London, p. 123. (*252*)
MARCELLIN, P. (1963). *Bull. Soc. fr. Physiol. vég.* **9**, 29. (*217*)
MAREKOV, N., STOÏANOVA-IVANOVA, B., MONDESHKY, L. and ZOLOTOVITCH, G. (1968). *Phytochemistry* **7**, 231. (*171*)
MARKLEY, K. S., HENDRICKS, S. B. and SANDO, C. E. (1932). *J. biol. Chem.* **98**, 103. (*63*)
— (1935). *J. biol. Chem.* **111**, 133. (*63*)
MARKLEY, K. S., NELSON, E. K. and SHERMAN, M. S. (1937). *J. biol. Chem.* **118**, 433. (*63*)
MARKLEY, K. S. and SANDO, C. E. (1931). *J. agric. Res.* **42**, 705. (*66, 187*)
— (1933). *J. agric. Res.* **46**, 403. (*52, 68*)
— (1934). *J. biol. Chem.* **105**, 643. (*63*)
— (1937). *J. biol. Chem.* **119**, 641. (*63*)
MARKLEY, K. S., SANDO, C. E. and HENDRICKS, S. B. (1938). *J. biol. Chem.* **123**, 641. (*63*)
MARKS, G. C., BERBEE, J. G. and RIKER, A. J. (1965). *Phytopathology* **55**, 408. (*260*)
MARTENS, P. (1933). *C.r. hebd. Séanc. Acad. Sci., Paris* **197**, 785. (*59*)
— (1934). *Cellule* **43**, 289. (*58, 100, 106*)
MARTIN, A. J. P. and SYNGE, R. L. M. (1941). *Biochem. J.* **35**, 1358. (*37*)
MARTIN, D. J. (1955). *Trans. Proc. bot. Soc. Edinb.* **36**, 278. (*290*)
MARTIN, H. (1964). *The Scientific Principles of Crop Protection.* Edward Arnold, London. (*234*)
MARTIN, J. T. (1957). *Proc. 4th Int. Congr. Crop Prot.*, Hamburg. (Braunschweig 1960) **2**, 1087. (*229, 246*)
— (1960). *J. Sci. Fd Agric.* **11**, 635. (*48*)
— (1961). *Rep. E. Malling Res. Stn. for 1960*, 40. (*223*)
— (1964). *A. Rev. Phytopath.* **2**, 81. (*13, 261, 287*)
— (1966). In *Radioisotopes in the detection of pesticide residues.* International Atomic Energy Agency. Vienna, p. 59. (*223*)
— (1967). *Proc. 4th Br. Insecticide and Fungicide Conference* **2**, 557. (*264*)
MARTIN, J. T., BAKER, E. A. and BYRDE, R. J. W. (1966a). *Ann. appl. Biol.* **57**, 491. (*265*)
— (1966b). *Ann. appl. Biol.* **57**, 501. (*265*)
MARTIN, J. T., BATT, R. F. and BURCHILL, R. T. (1957). *Nature, Lond.* **180**, 796. (*265*)
MARTIN, J. T. and FISHER, D. J. (1966). *Rep. agric. hort. Res. Stn Univ. Bristol for 1965*, 251. (*153*)
MARTIN, J. T. and SOMERS, E. (1957). *Nature, Lond.* **180**, 797. (*246*)
MARTIN, R. J. L. and STOTT, G. L. (1957). *Aust. J. agric. Res.* **8**, 444. (*203*)
MARTIN, R. L., WINTERS, J. C. and WILLIAMS, J. A. (1963). *Nature, Lond.* **199**, 110. (*281*)
MARTIN, R. O. and STUMPF, P. K. (1959). *J. biol. Chem.* **234**, 2548. (*163*)

MARTIN-SMITH, M., SUBRAMANIAN, G. and CONNOR, H. E. (1967). *Phytochemistry* **6**, 559. (*143*)
MATERNA, J. and RYŠKOVÁ, L. (1966). *Lesn. Čas.* **12** (**xxxix**), 453. (*45*)
MATIC, M. (1956). *Biochem. J.* **63**, 168. (*45, 52, 70, 144, 153*)
MATSUDA, K. (1962). *The biosynthesis of waxes in plants.* Ph.D. thesis, University of Michigan: *Diss. Abstr.* **23**, 810. (*161*)
MAZLIAK, P. (1958). *C.r. hebd. Séanc. Acad. Sci., Paris* **246**, 3368. (*47, 187*)
— (1960a). *C.r. hebd. Séanc. Acad. Sci., Paris* **251**, 2393. (*48, 123*)
— (1960b). *C.r. hebd. Séanc. Acad. Sci., Paris* **250**, 182, 2255. (*123*)
— (1961a). *C.r. hebd. Séanc. Acad. Sci., Paris* **252**, 1507. (*48, 123*)
— (1961b). *J. Agric. trop. Bot. appl.* **8**, 180. (*48, 50, 125*)
— (1962). *Phytochemistry* **1**, 79. (*47, 48, 50, 123, 125, 139*)
— (1963a). *Phytochemistry* **2**, 253. (*49, 123, 125*)
— (1963b). *Revue gén. Bot.* **70**, 437. (*124, 162*)
— (1963c). *Année biol.* **2**, 35. (*130, 180*)
— (1963d). *La cire cuticulaire des pommes* (*Pirus malus L.*), Ph.D. thesis, University of Paris. (*47, 48, 162, 186*)
— (1964). *C.r. hebd. Séanc. Acad. Agric. Fr.* **3**, 264. (*124*)
— (1965a). *Fruits* **20**, 49. (*133*)
— (1965b). *Fruits* **20**, 120. (*133*)
— (1965c). *Fruits* **20**, 605. (*162*)
— (1968). In *Progress in Phytochemistry*, ed. L. Reinhold and Y. Liwschitz. Interscience, New York and London. Vol. 1, p. 49. (*130, 148*)
MAZLIAK, P. and MARCELLIN, P. (1964). *Proc. 16th Int. hort. Congr.* **5**, 317. (*217*)
MAZLIAK, P. and POMMIER-MIARD, J. (1963). *Fruits* **18**, 177. (*123, 145, 158, 162, 187*)
MECKLENBURG, H. C. (1966). *Phytochemistry* **5**, 1201. (*48, 141*)
MECKLENBURG, R. A. and TUKEY, H. B. Jr. (1963). *Nature, Lond.* **198**, 562. (*10, 212*)
MECKLENBURG, R. A., TUKEY, H. B. Jr. and MORGAN, J. V. (1966). *Pl. Physiol., Lancaster* **41**, 610. (*244*)
MEIDNER, H. and MANSFIELD, T. A. (1968). *Physiology of Stomata.* McGraw-Hill, London, p. 66. (*209*)
MEIGH, D. F. (1964). *J. Sci. Fd Agric.* **15**, 436. (*124*)
— (1967). *J. Sci. Fd Agric.* **18**, 307. (*220*)
MEIGH, D. F. and FILMER, A. A. E. (1969). *J. Sci. Fd Agric.* **20**, 139. (*221*)
MEIGH, D. F. and HULME, A. C. (1965). *Phytochemistry* **4**, 863. (*188*)
MEINSCHEIN, W. G. and KENNY, G. S. (1957). *Analyt. Chem.* **29**, 1153. (*283*)
MENCE, M. J. and HILDEBRANDT, A. C. (1966). *Ann. appl. Biol.* **58**, 309. (*260*)
MERCER, F. V. and RATHGEBER, N. (1962). Int. Congr. Electron Microscopy, Philadelphia. Vol. 2, WW11. Academic Press, New York and London. (*210, 213, 214*)
METCALFE, C. R. and CHALK, L. (1950). *Anatomy of the Dicotyledons.* Clarendon Press, Oxford. (*82, 87, 90*)
MEYER, F. J. (1962). *Handbuch der Pflanzenanatomie.* Gebrüder Borntraeger, Berlin. (*73*)
MEYER, M. (1938). *Protoplasma* **29**, 552. (*60, 79, 80, 81, 235*)
MEYER, H. and SOYKA, W. (1913). *Mh. Chem.* **34**, 1159. (*62*)
MICHIE, M. J. and REID, W. W. (1968). *Nature, Lond.* **218**, 578. (*173*)
MIDDLETON, L. J. (1958). *Nature, Lond.* **181**, 1300. (*251*)
MIDDLETON, L. J. and SANDERSON, J. (1965). *J. exp. Bot.* **16**, 197. (*237, 239, 240, 244*)
MILLER, H. N. (1949). *Phytopathology* **39**, 403. (*287*)
MILLMAN, I. and YOTIS, W. (1958). *Proc. Soc. exp. Biol. Med.* **99**, 737. (*285*)
MITCHELL, J. W. (1936). *Bot. Gaz.* **98**, 87. (*216*)
MITSUI, T. and MATSUDA, J. (1942). *J. agric. Chem. Soc. Japan* **18**, 719. (*126*)

MIYOSHI, M. (1894). *Bot. Ztg* **52**, 1. (*261*)
— (1895). *Jb. wiss. Bot.* **28**, 269. (*261*)
MOLD, J. D., STEVENS, R. K., MEANS, R. E. and RUTH, J. M. (1963). *Biochemistry* **2**, 605. (*133, 141*)
MOLISCH, K. (1920). *Sber. Akad. Wiss. Wien* **129**, 261. (*114, 117*)
MONTEITH, J. L. (1963). In *The Water Relations of Plants*. British Ecological Society Symposium No. 3, ed. by A. J. Rutter and F. H. Whitehead. Blackwell Scientific Publications, Oxford, p. 37. (*199*)
MOOR, H., MÜHLETHALER, K., WALDNER, H. and FREY-WYSSLING, A. (1961). *J. Biophys. Biochem. Cytol.* **10**, 1. (*27*)
MOORBY, J. and SQUIRE, H. M. (1963). *Radiat. Bot.* **3**, 163. (*251*)
MOROZOVA, N. P., PLATONOVA, T. A. and SAL'KOVA, E. G. (1968). *Prikl. Biokhim. Mikrobiol.* **4**, 139. (*187*)
MOROZOVA, N. P. and SAL'KOVA, E. G. (1966). *Prikl. Biokhim. Mikrobiol.* **2**, 232. (*187*)
MORRIS, L. J. (1963). *J. Chromat.* **12**, 321. (*41*)
MORRISON, R. I. and BICK, W. (1966). *Chemy Ind.* 596. (*283*)
— (1967). *J. Sci. Fd Agric.* **18**, 351. (*283*)
MOSEMAN, J. G. and GREELEY, L. W. (1964). *Phytopathology* **54**, 618. (*255*)
MUDD, J. B. (1967). *A. Rev. Pl. Physiol.* **18**, 229. (*161*)
— (1968). *Abstrs 1st Int. Congr. Pl. Path.*, London, p. 134. (*253*)
MUELLER, L. E., CARR, P. H. and LOOMIS, W. E. (1954). *Am. J. Bot.* **41**, 593. (*23, 181, 232*)
MULLER, C. H. (1966). *Bull. Torrey bot. Club* **93**, 332. (*212*)
MÜLLER, H. (1915). *J. chem. Soc.* **107**, 872. (*139*)
MÜLLER, K. O. (1958a). *Aust. J. biol. Sci.* **11**, 275. (*264*)
— (1958b). *Nature, Lond.* **182**, 167. (*264*)
MUNDRY, K. W. (1963). *A. Rev. Phytopath.* **1**, 173. (*269*)
MURRAY, K. E. (1969). *Aust. J. Chem.* **22**, 197. (*221*)
MURRAY, K. E., HUELIN, F. E. and DAVENPORT, J. B. (1964). *Nature, Lond.* **204**, 80. (*124, 220*)
MURRAY, K. E. and SCHOENFELD, R. (1951). *J. Am. Oil Chem. Soc*, **28**, 461. (*62, 125*)
— (1953). *J. Am. Oil Chem. Soc.* **30**, 25. (*125*)
— (1955a). *Aust. J. Chem.* **8**, 432. (*124, 125*)
— (1955b). *Aust. J. Chem.* **8**, 437. (*125, 137*)

N

NAGY, B., MODZELESKI, V. and MURPHY, M. T. J. (1965). *Phytochemistry* **4**, 945. (*44, 51, 133, 139*)
NETOLITSKY, F. (1926). In *Handbuch der Pflanzen-Anatomie*, ed. K. Linsbauer. Gebrüder Borntraeger, Berlin. Vol. 10, Pt. 2. (*87*)
NEVENZEL, J. C. and RODEGKER, W. (1962). U.S. At. Ag. Contract A T (04–1) Gen–12. Univ. Los Angeles, California, p. 25. (*132*)
NEWCOMB, E. H. and BONNETT, H. T. (1965). *J. Cell Biol.* **27**, 575. (*107*)
NIENSTAEDT, H. (1953). *Phytopathology* **43**, 32. (*266*)
NOGARE, S. D. and JUVET, R. S. Jr. (1962). *Gas-liquid Chromatography. Theory and Practice.* Interscience, New York and London. (*44*)
NORMAN, A. G., MINARIK, C. E. and WEINTRAUB, R. L. (1950). *A. Rev. Pl. Physiol.* **1**, 141. (*237*)
NORRIS, R. F. and BUKOVAC, M. J. (1968). *Am. J. Bot.* **55**, 975. (*79, 239*)

O

O'BRIEN, T. P. (1967). *Protoplasma* **63**, 385. (*73, 77, 79, 81, 82, 181, 242*)
OHKI, K. (1932). *J. Tokyo Imp. Univ. Fac. Sci.* (*III Bot.*) **4**, 1 cited by McClure (1966). (*117*)
ORELLANA, R. G. and THOMAS, C. A. (1962). *Phytopathology* **52**, 533. (*261*)
ORGELL, W. H. (1955). *Pl. Physiol., Lancaster* **30**, 78. (*47, 79*)
ORGELL, W. H. and WEINTRAUB, R. L. (1957). *Bot. Gaz.* **119**, 88. (*238*)
ORÓ, J., LASETER, L. and WEBER, D. J. (1966). *Science, N.Y.* **154**, 399. (*136*)
ORÓ, J., NOONER, D. W. and WIKSTRÖM, S. A. (1965). *Science, N.Y.* **147**, 870. (*133, 289*)
OSMENT, D. J. (1963). In *Waterproofing and Water-Repellency*, ed. J. L. Moilliet. Elsevier, Amsterdam, London and New York, p. 384. (*199*)
ÔURA, H. (1953). In *Studies on Fogs*, ed. T. Hori. Tanne Trading Co. Ltd., Sapporo, Japan, p. 239. (*198*)

P

PALLAS, J. E., BERTRAND, A. R., HARRIS, D. G., ELKINS, C. B. and PARKS, G. L. (1962). U.S.D.A. Prod. Res. Report No. 87. (*209*)
PANT, D. D. and SRIVASTAVA, G. K. (1968). *J. Linn. Soc.* (*Bot.*) **61**, 201. (*281*)
PEARL, I. A. and DARLING, S. F. (1968). *Phytochemistry* **7**, 1851, 1855. (*150*)
PEARMAN, G. I. (1966). *Aust. J. biol. Sci.* **19**, 97. (*217*)
PEAT, J. E. (1928). *J. Genet.* **19**, 373. (*104, 189*)
PENMAN, H. L. (1963). *Vegetation and Hydrology*, in Technical Communication No. 53, Commonwealth Bureau of Soils, Harpenden. (*195, 198*)
PERIES, O. S. (1962). *Ann. appl. Biol.* **50**, 225. (*260*)
PFEIFFER, R. K., DEWEY, O. R. and BRUNSKILL, R. T. (1957). *Proc. 4th Int. Congr. Crop Prot.*, Hamburg **1**, 523. (*202, 245*)
PHILIPS, J. (1926). *Nature, Lond.* **118**, 837. (*194*)
— (1928). *Nature, Lond.* **121**, 354. (*194, 198*)
PHILLIPS, F. T. and GILLHAM, E. M. (1968). *Rep. Rothamsted exp. Stn for 1967*, 170. (*246*)
PICKETT-HEAPS, J. D. and NORTHCOTE, D. H. (1966). *J. Cell Sci.* **1**, 121. (*89, 90, 91*)
PIELOU, D. P., WILLIAMS, K. and BRINTON, F. E. (1962). *Nature, Lond.* **195**, 256. (*250*)
PIENIAZEK, S. A. (1944). *Pl. Physiol., Lancaster* **19**, 529. (*201*)
PIPER, S. H., CHIBNALL, A. C., HOPKINS, S. J., POLLARD, A., SMITH, J. A. B. and WILLIAMS, E. F. (1931). *Biochem. J.* **25**, 2072. (*63*)
PIPER, S. H., CHIBNALL, A. C. and WILLIAMS, E. F. (1934). *Biochem. J.* **28**, 2175. (*63*)
PISEK, A. and CARTELLIERI, E. (1932). *Jb. wiss. Bot.* **75**, 643. (*90*)
POHL, F. (1928). *Planta* **6**, 526. (*58*)
POLLARD, A., CHIBNALL, A. C. and PIPER, S. H. (1931). *Biochem. J.* **25**, 2111. (*64*)
— (1933). *Biochem. J.* **27**, 1889. (*65*)
PORTER, J. W. and ANDERSON, D. G. (1967). *A. Rev. Pl. Physiol.* **18**, 197. (*174*)
POSSINGHAM, J. V., CHAMBERS, T. C., RADLER, F. and GRNCAREVIC, M. (1967). *Aust. J. biol. Sci.* **20**, 1149. (*201*)
POWER, F. B. and CHESNUT, V. K. (1920). *J. Am. chem. Soc.* **42**, 1509. (*62*)
POWER, F. B. and MOORE, C. W. (1910). *J. chem. Soc.* **97**, 1099. (*62*)
PRASAD, R. and BLACKMAN, G. E. (1962). *Pl. Physiol., Lancaster* (Suppl.) **37**, xiii. (*240*)
PREECE, T. F. (1962). *Nature, Lond.* **193**, 902. (*47*)
PRESTON, R. D. (1952). *The Molecular Architecture of Plant Cell Walls.* Chapman and Hall, London. (*34*)

PRIESTLEY, J. H. (1921). *New Phytol.* **20**, 17. (*59, 69*)
— (1943). *Bot. Rev.* **9**, 593. (*6, 60, 100, 144, 174, 200*)
PRIESTLEY, J. H. and WOFFENDEN, L. M. (1922). *New Phytol.* **21**, 252. (*60*)
PROPHÈTE, H. (1926). *Bull. Soc. chim., Fr.* **39**, 1600. (*121*)
PURDY, S. J. and TRUTER, E. V. (1961). *Nature, Lond.* **190**, 554. (*45, 49, 114, 140, 141*)
— (1963a). *Proc. R. Soc.* Ser. B **158**, 536. (*45, 49, 129, 184*)
— (1963b). *Proc. R. Soc.* Ser. B **158**, 544. (*49, 129, 162*)
— (1963c). *Proc. R. Soc.* Ser. B **158**, 553. (*41, 49, 129, 143, 162*)

R

RADIN, N. S. (1965). *J. Am. Oil Chem. Soc.* **42**, 569. (*52*)
RADLER, F. (1965a). *Am. J. Enol. Vitic.* **16**, 159. (*48, 128*)
— (1965b). *Aust. J. biol. Sci.* **18**, 1045. (*128, 143*)
— (1965c). *Nature, Lond.* **207**, 1002. (*200, 204*)
— (1965d). *J. Sci. Fd Agric.* **16**, 638. (*48*)
RADLER, F. and HORN, D. H. S. (1965). *Aust. J. Chem.* **18**, 1059. (*46, 48, 49, 128*)
RAHMAN, W. and BHATNAGAR, S. P. (1968). *Aust. J. Chem.* **21**, 539. (*150*)
RALPH, C. S. and WHITE, D. E. (1949). *J. chem. Soc.* 3433. (*149*)
RAO, A. N. (1963). *Nature, Lond.* **197**, 1125. (*109*)
RAUHALA, V. T. (1961). *J. Am. Oil Chem. Soc.* **38**, 233. (*282*)
RAW, F. (1962). *Ann. appl. Biol.* **50**, 389. (*283, 290*)
RAZAFINDRAZAKA, J. and METZGER, J. (1963). *Bull. Soc. chim., Fr.* 1630, 1633. (*131*)
REICHENBERG, D. and SUTCLIFFE, J. F. (1954). *Nature, Lond.* **174**, 1047. (*242*)
RIBAS, I. (1951). *An. R. Soc. esp. Fis. Quim.* **47 B**, 61. (*152*)
RIBAS, I. and BLASCO, E. (1940). *An. R. Soc. esp. Fis. Quim.* **36**, 248. (*151*)
RIBAS, I. and GIL CURBERA, G. (1951). *An. R. Soc. esp. Fis. Quim.* **47 B**, 713. (*151*)
RIBAS, I. and SEOANE, E. (1954). *An. R. Soc. esp. Fis. Quim.* **50 B**, 963, 971. (*152*)
RICHMOND, D. V. and MARTIN, J. T. (1959). *Ann. appl. Biol.* **47**, 583. (*52, 138, 181, 263*)
RIVIÈRE, G. and PICHARD, G. (1924). *C.r. hebd. Séanc. Acad. Sci., Paris* **179**, 775 (*63*)
ROBERTS, M. F., BATT, R. F. and MARTIN, J. T. (1959). *Ann. appl. Biol.* **47**, 573. (*47, 52*)
ROBERTS, M. F. and MARTIN, J. T. (1963). *Ann. appl. Biol.* **51**, 411. (*183, 265*)
ROBERTS, M. F., MARTIN, J. T. and PERIES, O. S. (1961). *Rep. agric. hort. Res. Stn Univ. Bristol for 1960*, 102. (*45, 183, 261, 287*)
ROBERTS, E. A., SOUTHWICK, M. D. and PALMITER, D. H. (1948). *Pl. Physiol., Lancaster* **23**, 557. (*217, 234, 237*)
ROBINSON, SIR ROBERT (1963). *Nature, Lond.* **199**, 113. (*282*)
ROBINSON, R. K. (1967). *Ecology of the Fungi.* English Universities Press (Modern Biology Series), London. (*256, 261*)
ROELOFSEN, P. A. (1950). *Biochim. biophys. Acta* **6**, 340. (*96*)
— (1952). *Acta. bot. neerl.* **1**, 99. (*4, 47, 60*)
— (1959). *The Plant Cell-Wall.* Gebrüder Borntraeger, Berlin. (*77, 81, 82, 104*)
ROELOFSEN, P. A. and HOUWINK, A. L. (1951). *Protoplasma* **40**, 1. (*146*)
ROSEN, W. G., GAWLIK, S. R., DASHEK, W. V. and SIEGESMUND, K. A. (1964). *Am. J. Bot.* **51**, 61. (*107*)
ROWLEY, J. R. and SOUTHWORTH, D. (1967). *Nature, Lond.* **213**, 703. (*119*)
RUDOLF, K. (1925). *Bot. Arch.* **9**, 49. (*239*)
RUHLAND, W. (1915). *Jb. wiss. Bot.* **55**, 409. (*92*)
RUINEN, J. (1956). *Nature, Lond.* **177**, 220. (*255, 262*)

RUINEN, J. (1961). *Pl. Soil* **15**, 81. (*255*)
— (1963). *Antonie van Leeuwenhoek* **29**, 425. (*285*)
RUSSELL, G. E. and EVANS, G. M. (1968). *Ann. appl. Biol.* **62**, 465. (*262*)
RUZICKA, L., PLATTNER, P. A. and WIDMER, W. (1942). *Helv. chim. Acta* **25**, 1086. (*70*)

S

SADYKOV, A. S., ISAEV, K. I. and ISMAILOV, A. I. (1963). *Uzbek. khim. Zh.* **7**, 53. (*132*)
SAHAI, P. N. and CHIBNALL, A. C. (1932). *Biochem. J.* **26**, 403. (*64, 66, 159*)
SAKURAI, Z. (1933). *J. pharm. Soc. Japan* **53**, 579. (*137*)
SANDERS, E. H. (1955). *Cereal Chem.* **32**, 12. (*188*)
SANDERS, J. M. (1911). *Proc. chem. Soc.* **27**, 250. (*62*)
SANDO, C. E. (1923). *J. biol. Chem.* **56**, 457. (*47, 63*)
— (1924). *J. agric. Res.* **28**, 1243. (*147*)
— (1931). *J. biol. Chem.* **90**, 477. (*63*)
— (1937). *J. biol. Chem.* **117**, 45. (*147*)
SARGENT, J. A. (1965). *A. Rev. Pl. Physiol.* **16**, 1. (*241*)
— (1966). *Proc. 8th Br. Weed Control Conference* **3**, 804. (*226, 239, 240, 250*)
SARGENT, J. A. and BLACKMAN, G. E. (1962). *J. exp. Bot.* **13**, 348. (*236, 238, 240*)
— (1965). *J. exp. Bot.* **16**, 24. (*240, 241*)
SAUNDERS, P. J. W. (1968). *Abstrs 1st Int. Congr. Pl. Path.*, London, p. 170. (*251*)
SAVIDAN, L. (1956a). *Bull. Soc. chim., Fr.* 64, 67. (*48, 125*)
— (1956b). *Annls. Chim.* (13) **1**, 53. (*134*)
— (1959). *Bull. Soc. chim. Fr.* 1390. (*134*)
SAX, H. J. (1938). *J. Arnold Arbor.* **19**, 437. (*114, 115*)
SAX, K. and SAX, H. J. (1937). *J. Arnold Arbor.* **18**, 164. (*114, 115*)
SCHAAL, R. (1907). *Ber. dt. chem. Ges.*, 4784. (*62*)
SCHIEFERSTEIN, R. H. and LOOMIS, W. E. (1956). *Pl. Physiol., Lancaster* **31**, 240. (*201, 232*)
— (1959). *Am. J. Bot.* **46**, 625. (*6, 7, 45, 47, 178, 201*)
SCHMITZ, K. J. (1968). *Erwerbobstbau* **10**, 147. (*188*)
SCHNEIDER, W. (1960). *Beiträge zum chemischen Aufbau der Apfelschale.* Ph.D. thesis. Julius-Maximillian Univ. Würzburg. (*47, 145, 147*)
SCHNEPF, E. (1959). *Planta* **52**, 644. (*82*)
— (1963). *Flora, Jena* **153**, 1, 23. (*271*)
— (1964a). *Protoplasma* **58**, 193. (*87*)
— (1964b). *Planta* **60**, 473. (*213*)
— (1965). *Ztschr. Pflanzenphysiol.* **53**, 245. (*89, 209, 210, 271, 272*)
— (1966a). *Umschau* **66**, 522. (*213, 214, 271*)
— (1966b). *Ber. dt. bot. Ges.* **78**, 478. (*87, 213, 271*)
— (1968). *Planta* **79**, 22. (*139*)
SCHOPMEYER, C. S. (1961). *Forest Sci.* **7**, 330. (*196*)
SCHREINER, O. and SHOREY, E. C. (1910). *Bull. Bur. Soils U.S. Dep. Agric.* *74* (*283*)
SCHUETTE, H. A. and BALDINUS, J. G. (1949). *J. Am. Oil Chem. Soc.* **26**, 530, 651. (*132*)
SCHUETTE, H. A. and HANIF KHAN, M. (1953). *J. Am. Oil Chem. Soc.* **30**, 126. (*134*)
SCHUMACHER, W. and HALBSGUTH, W. (1939). *Jb. wiss. Bot.* **87**, 324. (*82*)
SCHUMACHER, W. and LAMBERTZ, P. (1956). *Planta* **47**, 47. (*84*)
SCOTT, D. H. and SARGANT, E. (1893). *Ann. Bot.* **7**, 243. (*194*)
SCOTT, F. M. (1963). *Nature, Lond.* **199**, 1009. (*107*)
— (1966). *Nature, Lond.* **210**, 1015. (*107*)

SCOTT, F. M., SCHROEDER, M. R. and TURRELL, F. M. (1948). *Bot. Gaz.* **109**, 381. (*81, 237*)
SCOTT, F. M., HAMNER, K. C., BAKER, E. and BOWLER, E. (1958). *Am. J. Bot.* **45**, 449. (*47, 107*)
SCOTT, R. W. and STROHL, M. J. (1962). *Phytochemistry* **1**, 189. (*136*)
SCURTI, F. and TOMMASI, G. (1913). *Annali Staz. chim.-agr. sper. Roma* **6**, 67. (*70*)
— (1916). *Gazz. chim. ital.* **46**, 159. (*70*)
SEN, J. (1954). *Bull. bot. Soc. Beng.* **8**, 149. (*14, 280*)
— (1956). *Bot. Rev.* **22**, 343. (*280*)
— (1961). *Riv. ital. Paleont. Stratigr.* **67**, 77. (*280*)
SEOANE, E. (1961). *Chemy Ind.*,1080. (*129*)
SEOANE, E., GIL CURBERA, G. and RIBAS, I. (1953). *An. R. Soc. esp. Fis. Quim.* **49 B**, 145. (*152*)
SEOANE, E. and RIBAS, I. (1951). *An. R. Soc. esp. Fis. Quim.* **47 B**, 61. (*152*)
SEOANE, E., RIBAS, I. and FANDINO, N. G. (1957). *Chemy Ind.*, 490. (*153*)
— (1959). *An. R. Soc. esp. Fis. Quim.* **55 B**, 839. (*153*)
SHARMA, G. K. (1967). *Cuticular variation in Kalanchoe and Datura*. Ph.D. thesis. University of Missouri. (*114*)
SHARP, E. L. (1965). *Phytopathology* **55**, 198. (*263*)
SHAW, G. and YEADON, A. (1966). *J. chem. Soc.* **C**, 16. (*154*)
SHELLHORN, S. J. and HULL, H. M. (1961). *Stain Technol.* **36**, 69. (*114*)
SHEPHERD, C. J. and MANDRYK, M. (1963). *Aust. J. biol. Sci.* **16**, 77. (*263*)
SHIINA, S. (1946). *J. Soc. chem. Ind. Japan* **49**, 18. (*134*)
SHIMADA, H. (1952). *J. pharm. Soc. Japan* **72**, 61, 63, 65, 67. (*151*)
SHUEL, R. W. (1961). *Pl. Physiol., Lancaster* **36**, 265. (*194, 236*)
SHULL, C. A. (1929). *Bot. Gaz.* **87**, 583. (*217*)
SHUTAK, V. G. and CHRISTOPHER, E. P. (1960). *Proc. Am. Soc. hort. Sci.* **76**, 106. (*220*)
SIDDIQI, A. M. and TAPPEL, A. L. (1956). *Archs Biochem. Biophys.* **60**, 91. (*174*)
SIDDIQUI, U. A. and AHSAN, A. M. (1967). *Pakist. J. scient. ind. Res.* **10**, 1. (*133*)
SIEGELMAN, H. W. (1955). *J. biol. Chem.* **213**, 647. (*147*)
SIFTON, H. B. (1963). *Can. J. Bot.* **41**, 199. (*179, 238*)
SILVA FERNANDES, A. M. S. (1964). *Chemical and physical studies on plant cuticles*, Ph.D. thesis, University of Bristol. (*185, 232, 234, 235*)
— (1965a). *Ann. appl. Biol.* **56**, 297. (*46, 232*)
— (1965b). *Ann. appl. Biol.* **56**, 305. (*235, 242, 244*)
SILVA FERNANDES, A. M. S., BAKER, E. A. and MARTIN, J. T. (1964). *Ann. appl. Biol.* **53**, 43. (*48*)
SILVA FERNANDES, A. M. S., BATT, R. F. and MARTIN, J. T. (1964). *Rep. agric. hort. Res. Stn Univ. Bristol for 1963*, 110. (*46, 47, 48, 184, 186*)
SIMIONESCU, C., DIACONESCU, E. and FELDMAN, D. (1960). *Rev. chim. Acad. rep. Pop. Roumaine* **5**, 57. (*133, 138*)
SINHA, S. (1965). *Indian Phytopath.* **18**, 1. (*256*)
— (1968). *Abstrs 1st Int. Congr. Pl. Path.*, London, p. 185. (*256, 262*)
SITTE, P. (1957). *Die Chemie der Pflanzenzellwand*, ed. E. Treiber. Springer-Verlag, Berlin, p. 439. (*156*)
SITTE, P. and RENNIER, R. (1963). *Planta* **60**, 19. (*4, 200*)
SIVADJIAN, J. (1956). *C.r. hebd. Séanc. Acad. Sci., Paris* **242**, 2478. (*206*)
SKENE, D. S. (1963). *Ann. Bot.* **27**, 581. (*188*)
SKOSS, J. D. (1955). *Bot. Gaz.* **117**, 55. (*45, 47, 52, 79, 81, 179, 205, 235, 236*)
SLATYER, R. O. (1960). *Bot. Rev.* **26**, 331. (*239*)
— (1965). Proc. Montpellier Symp. UNESCO Arid Zone Research No. 25. (*195*)
— (1967). *Plant-Water Relationships*. Academic Press, New York and London. (*200, 209*)

SMIRNOV, B. P. (1960). *Biokhimiya* **25**, 545. (*158*)
SMIT, J. and WIERINGA, K. T. (1953). *Nature, Lond.* **171**, 794. (*285*)
SMITH, H. H. (1946). *Bot. Gaz.* **107**, 544. (*229*)
SMITH, K. C. A. and OATLEY, C. W. (1955). *Br. J. appl. Phys.* **6**, 391. (*30*)
SOL, H. H. (1966). *Neth. J. Pl. Path.* **72**, 196. (*261*)
— (1968). *Abstrs 1st Int. Congr. Pl. Path.*, London, p. 187. (*261*)
SOMERS, T. C. and HARRISON, A. F. (1967). *Aust. J. biol. Sci.* **20**, 475. (*266*)
ŠORM, F. and BAZANT, V. (1950). *Colln Czech. chem. Commun. Engl. Edn.* **15**, 73. (*149*)
ŠORM, F., WOLLRAB, V., JAROLIMEK, P. and STREIBL, M. (1964). *Chemy Ind.* 1833. (*51, 127, 134, 139*)
SOSA, A. (1950). *Bull. Soc. Chim. biol.* **32**, 344. (*131*)
SPADA, A., COPPINI, D. and MONZANI, A. (1958). *Annali Chim.* **48**, 181. (*136*)
SPIERINGS, F. H. G. (1968). *Abstrs 1st Int. Congr. Pl. Path.*, London, p. 189. (*252*)
STACE, C. A. (1965). *Bull. Br. Mus. nat. Hist. Bot.* Vol. 4, No. 1. (*55, 98, 114, 115, 231*)
STADHOUDERS, A. M., HEINEN, W. and KRAAN, H. G. (1962). *Proc. K. ned. Akad. Wet.* **C 65**, 41. (*286*)
STAHL, E. (1953). *Z. Bot.* **41**, 123. (*214*)
— (1965). *Thin-layer chromatography.* Springer-Verlag, Berlin, Heidelburg and New York. (*44*)
STAHL, E., SCHRAETER, G., KRAFT, G. and RENZ, R. (1956). *Pharmazie* **11**, 633. (*37*)
STÅLFELT, M. G. (1956a). *Die cuticuläre Transpiration.* Handbuch der Pflanzenphysiologie. Springer-Verlag, Berlin. Vol. 3, p. 342. (*199, 200*)
— (1956b). *Die stomatäre Transpiration der Pflanze und die Physiologie der Spaltöffnungen.* Handbuch der Pflanzenphysiologie. Springer-Verlag, Berlin. Vol. 3, p. 351. (*200*)
— (1967). *Protoplasma* **64**, 452. (*201*)
STANLEY, J. S. (1958). *Mfg. Chem.* **29**, 385. (*234*)
STEBBINS, G. L. and KHUSH, G. S. (1961). *Am. J. Bot.* **48**, 51. (*90*)
STEVENS, N. E. (1932). *Am. J. Bot.* **19**, 432. (*179*)
STEVENSON, F. J. (1966). *Jour. Am. Oil Chem. Soc.* **43**, 203. (*15, 283*)
STEWART, D. R. M. (1965). *Bot. Jb.* **84**, 65. (*117*)
STEWART, J. M. and FOLLETT, E. A. C. (1966). *Can. J. Bot.* **44**, 421. (*280*)
STOCKING, C. R. (1956a). *Guttation and bleeding.* Encyclopaedia of Plant Physiology (Handbuch der Pflanzenphysiologie). Springer-Verlag, Berlin. Vol. 3, p. 489. (*94*)
— (1956b). *Excretion by glandular organs.* Encyclopaedia of Plant Physiology (Handbuch der Pflanzenphysiologie). Springer-Verlag, Berlin. Vol. 3, p. 503. (*94*)
STONE, E. C. (1958). *The Physiology of Forest Trees.* Ronald Press, New York, p. 125. (*196, 199*)
STRÁNSKÝ, K. and STREIBL, M. (1969). *Colln Czech. chem. Commun. Engl. Edn* **34**, 103. (*133, 143*)
STRÁNSKÝ, K., STREIBL, M. and HEROUT, V. (1967). *Colln Czech. chem. Commun. Engl. Edn* **32**, 3213. (*135, 143, 282*)
STRÁNSKÝ, K., STREIBL, M. and ŠORM, F. (1968). *Colln Czech. chem. Commun. Engl.* **33**, 416. (*282*)
STREIBL, M., JAROLIMEK, P. and WOLLRAB, V. (1964). *Colln Czech. chem. Commun. Engl. Edn* **29**, 2522. (*51*)
STREIBL, M. and KONEČNÝ, K. (1967). *Chemy Ind.* 546. (*51*)
STREIBL, M. and STRÁNSKÝ, K. (1968). *Fette Seifen Anstr-Mittel* **70**, 343. (*127, 171*)
STUMPF, P. K. (1965). In *Plant Biochemistry*, ed. J. Bonner and J. E. Varner. Academic Press, New York and London, p. 322. (*163*)

STUMPF, P. K. and JAMES, A. T. (1962). *Biochem. J.* **82**, 28P. (*158*)
— (1963). *Biochim. biophys. Acta* **70**, 20. (*158*)
STÜRCKE, H. (1884). *Justus Liebigs Annln. Chem.* **223**, 283, 312. (*61*)
SUCHORUKOV, K. T. (1960). *Proc. Symp. Scient. Probl. Pl. Prot.* Budapest, p. 71. (*262*)
SUCHORUKOV, K. T. and PLOTNIKOVA, Y. M. (1965). *Usp. sovrem. Biol.* **60**, 299. (*84*)
SWAN, E. P. (1968). *TAPPI* **51**, 301. (*151*)
SWIFT, M. J. (1965). *Nature, Lond.* **207**, 436. (*288*)

T

TASSILLY, E. (1911). *Bull. Soc. chim. Fr.* 608. (*62*)
TAYLOR, O. C. (1968). *Abstrs Ist. Int. Congr. Pl. Path.*, London, p. 199. (*252*)
TETLEY, U. (1931). *J. Pomol.* **9**, 278. (*118*)
TEUBNER, F. G., WITTWER, S. H., LONG, W. G. and TUKEY, H. B. (1957). *Q. Bull. Mich. St. Univ. agric. Exp. Stn* **39**, 398. (*241*)
THOMAS, B. A. (1967a). *J. nat. Hist.* **1**, 53. (*280*)
— (1967b). *Ann. Bot.* **31**, 775. (*280*)
THOMAS, D. A. (1961). *Aust. J. Sci.* **24**, 288. (*11*)
THOMAS, H. H. (1925). *Phil. Trans. R. Soc.* Ser. B **213**, 302. (*279*)
THOMAS, M. D. (1965). In *Plant Physiology*, ed. F. C. Steward. Academic Press, New York and London. Vol. 4A, p. 10. (*216*)
THOMPSON, K. F. (1963). *Nature, Lond.* **198**, 209. (*192, 270*)
— (1967). *Rep. Pl. Breed. Inst. 1965/6*, p. 7. (*270*)
THOMSON, W. W. and LIU, L. L. (1967). *Planta* **73**, 201. (*92, 215*)
THROWER, S. L., HALLAM, N. D. and THROWER, L. B. (1965). *Ann. appl. Biol.* **55**, 253. (*232*)
THURSTON, R., SMITH, W. T. and COOPER, B. P. (1966). *Entomologia exp. appl.* **9**, 428. (*271*)
THURSTON, R. and WEBSTER, J. A. (1962). *Entomologia exp. appl.* **5**, 233. (*271*)
TIETZ, H. (1954). *Höfchenbr. Bayer PflSchutz-Nachr.* **7**, 1. (*238*)
TINSLEY, T. W. (1953). *Ann. appl. Biol.* **40**, 750. (*269*)
TISCHER, J. (1960). *Pharmazie* **15**, 83. (*134*)
TOMANA, T. (1960). *J. hort. Ass. Japan* **29**, 273. (*220*)
TOPPS, J. H. and WAIN, R. L. (1957). *Nature, Lond.* **179**, 652. (*263*)
TRÉCUL, A. (1856). *C.r. hebd. Séanc Acad. Sci., Paris* **42**, 621. (*55*)
TRIBE, I. S. (1967). Personal communication. (*45, 50, 141, 180*)
TRIBE, I. S., GAUNT, J. K. and WYNN PARRY, D. (1968). *Biochem. J.* **109**, 8 P. (*7, 131, 132, 143*)
TROUGHTON, J. H. and HALL, D. M. (1967). *Aust. J. biol. Sci.* **20**, 509. (*180, 231, 232, 267*)
TRUTER, E. V. (1963). *Thin film chromatography.* Cleaver-Hume, London. (*44*)
TSUJIMOTO, M. (1935). *Bull. chem. Soc. Japan* **10**, 212. (*134, 138*)
TSWETT, M. (1906). *Ber. dt. bot. Ges.* **24**, 316, 384. (*36*)
— (1907). *Biochem. Z.* **5**, 6. (*36*)
TUKEY, H. B. Jr. and MECKLENBURG, R. A. (1964). *Am. J. Bot.* **51**, 737. (*213*)
TUKEY, H. B. Jr. and MORGAN, J. V. (1964). *Proc. 16th Int. hort. Congr.* **4**, 153. (*10, 138, 212*)
TUKEY, H. B. and TUKEY, H. B. Jr. (1959). *Proc. Am. Soc. hort. Sci.* **74**, 671. (*213*)
TUKEY, H. B. Jr., TUKEY, H. B. and WITTWER, S. H. (1958). *Proc. Am. Soc. hort. Sci.* **71**, 496. (*213*)

TULLOCH, A. P. and WEENINK, R. O. (1966). *Chem. Commun.* 225. (*48, 134, 139*)
TURRELL, F. M. (1947). *Bot. Gaz.* **108**, 476. (*92*)
TURNER, E. M. (1956). *Ann. appl. Biol.* **44**, 200. (*263*)

U

UENO, S. and TSUCHIKAWA, H. (1942). *J. Soc. chem. Ind. Japan* **45**, 203. (*134*)
ULOTH, W. (1867). *Flora, Jena* **50**, 385, 417. (*56*)
UPHOF, J. C. T. (1962). In *Handbuch der Pflanzen-Anatomie*, ed. K. Linsbauer. Gebrüder Borntraeger, Berlin. Vol. 4, Pt. 5. (*87, 92, 114, 117, 217, 272, 277*)
URSPRUNG, A. (1925). *Beih. bot. Zbl.* **41**, 15. (*228*)

V

VANDENBURG, L. E. and WILDER, E. A. (1967). *J. Am. Oil Chem. Soc.* **44**, 659. (*126*)
VANDENHEUVEL, W. J. A., GARDINER, W. L. and HORNING, E. C. (1965). *J. Chromat.* **19**, 263. (*50*)
VAN DER DRIFT, J. (1963). In *Soil Organisms*, ed. J. Doeksen and J. van der Drift. North-Holland Publ. Co., Amsterdam, p. 125. (*290*)
VAN DER HAAR, A. W. (1924). *Recl. Trav. chim. Pays-Bas Belg.* **43**, 543, 546, 548. (*63*)
VAN DIE, J. and SOEMARSONO, R. (1957). *Ann. bogor.* **2**, 225. (*136*)
VAN NOSTRAND, D. and GOODMAN, R. N. (1968). *The Biochemistry and Physiology of Infectious Plant Diseases.* Van Nostrand, Princeton, New Jersey, U.S.A. (*268, 269*)
VAN OVERBEEK, J. (1956). *A Rev. Pl. Physiol.* **7**, 355. (*223, 226, 228*)
VAN RHEE, J. A. (1963). In *Soil Organisms*, ed. J. Doeksen and J. van der Drift. North-Holland Publ. Co., Amsterdam, p. 55. (*290*)
VAN WISSELINGH, C. (1893). *Archs neerl. Sci.* **26**, 305. (*69*)
— (1895). *Archs neerl. Sci.* **28**, 373. (*58*)
VENKATA RAM, C. S. (1962). *Curr. Sci.* **31**, 428. (*265*)
VIGUERA LOBO, J. M. and SÁNCHEZ PARAREDA, J. (1963). *An. R. Soc. esp. Fis. Quim.* **59 B**, 123. (*134*)
VIGUERA LOBO, J. M., SÁNCHEZ PARAREDA, J. and SÁNCHEZ PARAREDA, I. (1964). *Grasas aceit.* **15**, 181. (*131*)
VLITOS, A. J. and CUTLER, H. G. (1960). *Pl. Physiol., Lancaster* **35**, Suppl. VI. (*262*)
VON HÖHNEL, F. (1877). *Sber. Akad. Wiss. Wien* **76**, 507. (*69*)
— (1878). *Öst. bot. Z.* **28**, 81, 115. (*67, 156*)
VON LANGE, O. L. and SCHULZE, E. D. (1966). *Forstwiss. ZentBl.* **85**, 27. (*75, 100, 101*)
VON MOHL, H. (1845). *Bot. Ztg* **3**, 1. (*55*)
VON RUDLOFF, E. (1959). *Can. J. Chem.* **37**, 1038. (*137, 139*)
VON SCHMIDT, M. (1904). *Mh. Chem.* **25**, 277, 302. (*69*)
— (1910). *Mh. Chem.* **31**, 347. (*69*)

W

WAIN, R. L. and WILKINSON, E. H. (1943). *Ann. appl. Biol.* **30**, 379. (*246*)
WAISEL, Y. and FRIEDMAN, J. (1965). *La-yaaran* **15**, 1. (*215*)
WALDRON, J. D., GOWERS, D. S., CHIBNALL, A. C. and PIPER, S. H. (1961). *Biochem. J.* **78**, 435. (*122*)

WALLACE, T. (1930). *J. Pomol.* **8**, 44. (*212*)
WALLIS, T. E. (1955). *Textbook of Pharmacognosy.* Churchill, London. (*295*)
WALTER, H. and STADELMANN, E. (1968). *Bioscience* **18**, 694. (*215*)
WALTON, J. (1953). *An Introduction to the study of Fossil Plants.* Adam & Charles Black, London. (*278*)
WANLESS, G. G., KING, W. H. and RITTER, J. J. (1955). *Biochem. J.* **59**, 684. (*48, 121, 122, 161*)
WARBURTON, F. L. (1963). *The effect of structure on waterproofing.* In *Waterproofing and Water-Repellency*, ed. J. L. Moilliet. Elsevier, Amsterdam, London and New York, p. 24. (*225*)
WARTH, A. H. (1956). *The Chemistry and Technology of Waxes.* 2nd ed. Reinhold, New York. (*126, 294*)
WASSERMANN, A. (1948). *Nature, Lond.* **161**, 562. (*243*)
WATSON, R. W. (1942). *New Phytol.* **41**, 223. (*179*)
WAY, M. J. and MURDIE, G. (1965). *Ann. appl. Biol.* **56**, 326. (*270, 271*)
WEAVER, R. J. and DE ROSE, H. R. (1946). *Bot. Gaz.* **107**, 509. (*241*)
WEBER, E. (1942). *Ber. schweiz. bot. Ges.* **52**, 111. (*58*)
WEINSTEIN, L. H. and McCUNE, D. C. (1968). *Abstrs 1st Int. Congr. Pl. Path.*, London, p. 214. (*253*)
WERNER, O. (1928). *Biologia gen.* **4**, 403. (*117*)
WHITE, D. E. and ZAMPATTI, L. S. (1952). *J. chem. Soc.*, 5040. (*149*)
WHITE, G. (1776). *The Natural History and Antiquities of Selborne.* Letter to the Hon. Daines Barrington. Selborne, Feb. 7. (*198*)
WIEDENHOF, N. (1959). *J. Am. Oil Chem. Soc.* **36**, 297. (*48, 126*)
WIESNER, J. (1871). *Bot. Ztg* **29**, 769. (*56*)
WILK, S., GITLOW, S. E. and CLARKE, D. D. (1967). Am. chem. Soc. 2nd Middle Atlantic Regional Meeting, New York. Abstracts of papers, p. 48. (*53*)
WILLARD, H. H., MERRITT, L. L. Jr. and DEAN, J. A. (1965). *Instrumental methods of analysis.* 4th ed. Van Nostrand, London. (*44*)
WILLIAMS, L. O. (1958). *Econ. Bot.* **12**, 103. (*293*)
WILLIAMS, P. R. and JUNIPER, B. E. (1968). *Proc. R. microsc. Soc.* **3**, 129. (*25*)
WILLIAMS, R. C. and WYCKOFF, R. W. G. (1946). *J. app. Phys.* **17**, 23. (*20*)
WILSON, A. R. and JARVIS, W. R. (1963). *Pl. Path.* **12**, 91. (*267*)
WILTSHIRE, S. P. (1915). *Ann. appl. Biol.* **1**, 335. (*263, 287*)
WITKAMP, M. and VAN DER DRIFT, J. (1961). *Pl. Soil* **15**, 295. (*290*)
WOLLRAB, V. (1967). *Colln Czech. chem. Commun. Engl. Edn* **32**, 1304. (*125*)
— (1968). *Colln Czech. chem. Commun. Engl. Edn* **33**, 1584. (*135*)
— (1969). *Phytochemistry* **8**, 623. (*135, 171*)
WOLLRAB, V., STREIBL, M. and ŠORM, F. (1965a). *Colln Czech. chem. Commun. Engl. Edn* **30**, 1654. (*134*)
— (1965b). *Colln Czech. chem. Commun. Engl. Edn* **30**, 1670. (*133*)
— (1967). *Chemy Ind.* 1872. (*135*)
WONG, C. L. and BLEVIN, W. R. (1967). *Aust. J. biol. Sci.* **20**, 501. (*217*)
WOOD, F. A. (1968). *Abstrs 1st Int. Congr. Pl. Path.*, London, p. 221. (*253*)
WOOD, R. K. S. (1960). In *Plant Pathology* ed. J. G. Horsfall and A. E. Dimond. Academic Press, New York and London. Vol. 2, p. 233. (*13*)
— (1967). *Physiological Plant Pathology.* Blackwell Scientific Publications, Oxford. (*264, 270*)
WOODING, F. B. P. and NORTHCOTE, D. H. (1964). *J. Cell. Biol.* **23**, 327. (*81*)
WORTMANN, G. B. (1965a). *Elektronmikroscopische Untersuchungen der Blattoberfläche und deren Veränderungen durch Pflanzenschutzmittel.* Diss. Landw. Hochschule, Hohenheim. (*232, 245*)

WORTMANN, G. B. (1965b). *Z. Pflkrankh. Pflpath. Pflschutz.* **72**, 641. (*245*)
WULFF, H. D. and STAHL, E. (1960). *Naturwissenschaften* **47**, 114. (*185*)

Y

YAMADA, Y., JYUNG, W. H., WITTWER, S. H. and BUKOVAC, M. J. (1965). *Proc. Am. Soc. hort. Sci.* **87**, 429. (*241*)
YAMADA, Y., RASMUSSEN, H. P., BUKOVAC, M. J. and WITTWER, S. H. (1966). *Am. J. Bot.* **53**, 170. (*241, 243*)
YAMADA, Y., WITTWER, S. H. and BUKOVAC, M. J. (1964). *Pl. Physiol., Lancaster* **39**, 28. (*241*)
— (1965). *Pl. Physiol., Lancaster* **40**, 170. (*241, 243*)
YARWOOD, C. E. (1956). *Phytopathology* **46**, 540. (*256*)
YOSHIDA, Z. (1953). In *Studies on Fogs*, ed. by T. Hori. Tanne Trading Co. Ltd, Sapporo, Japan, p. 1. (*197*)

Z

ZETZSCHE, F. (1932). *Handbuch der Planzenanalyse*, ed. G. Klein. **3**, Pt. 1, p. 205. (*70, 156*)
ZETZSCHE, F. and BÄHLER, M. (1931a). *Helv. chim. Acta* **14**, 642, 852. (*70*)
— (1931b). *Helv. chim. Acta* **14**, 846, 849. (*70*)
ZETZSCHE, F., CHOLATNIKOW, C. and SCHERZ, K. (1928). *Helv. chim. Acta* **11**, 272. (*70*)
ZETZSCHE, F., KALT, P., LIECHTI, J. and ZIEGLER, E. (1937). *J. prakt. Chem.* **148**, 267. (*154*)
ZETZSCHE, F. and LÜSCHER, E. (1937). *J. prakt. Chem.* **150**, 68. (*48*)
ZETZSCHE, F. and ROSENTHAL, G. (1927). *Helv. chim. Acta* **10**, 346. (*70*)
ZETZSCHE, F. and SCHERZ, N. N. (1932). *Handbuch der Pflanzenanalyse*, ed. G. Klein **3**, Pt. 2, p. 217. (*52*)
ZETZSCHE, F. and SONDEREGGER, G. (1931). *Helv. chim. Acta* **14**, 632. (*70, 149*)
ZETZSCHE, F. and VICARI, H. (1931). *Helv. chim. Acta* **14**, 58, 62. (*279*)
ZETZSCHE, F., VICARI, H. and SCHÄRER, G. (1931). *Helv. chim. Acta* **14**, 67. (*279*)
ZETZSCHE, F. and WEBER, K. (1938). *J. prakt. Chem.* **150**, 140. (*70*)
ZIEGENSPECK, H. (1928). *Bot. Arch.* **21**, 1. (*58*)
ZIEGLER, H. and LÜTTGE, U. (1966). *Planta* **70**, 193. (*92*)
— (1967). *Planta* **74**, 1. (*92*)
ZILL, L. P. and HARMON, E. A. (1962). *Biochim. biophys. Acta* **57**, 573. (*166*)

Biological Name Index

Figures in *italics* refer to illustrations

Subject Index

For information on the cuticle or bark of a particular species see entries under its name in the Biological Name Index (p. 329)

Figures in *italics* refer to illustrations

DATE DUE

DEC 04 1995	